Douglas Densmore, Abhijit Davare

A Platform-Based Design Methodology for the Electronic System Level

Douglas Densmore, Abhijit Davare

A Platform-Based Design Methodology for the Electronic System Level

Frameworks, Designs Flows, and Case Studies

VDM Verlag Dr. Müller

Imprint

Bibliographic information by the German National Library: The German National Library lists this publication at the German National Bibliography; detailed bibliographic information is available on the Internet at http://dnb.d-nb.de.

Cover image: www.purestockx.com

Publisher:
VDM Verlag Dr. Müller Aktiengesellschaft & Co. KG, Dudweiler Landstr. 125 a, 66123 Saarbrücken, Germany,
Phone +49 681 9100-698, Fax +49 681 9100-988,
Email: info@vdm-verlag.de

Produced in USA and UK by:
Lightning Source Inc., La Vergne, Tennessee, USA
Lightning Source UK Ltd., Milton Keynes, UK
BookSurge LLC, 5341 Dorchester Road, Suite 16, North Charleston, SC 29418, USA

ISBN: 978-3-8364-7314-9

For Diana

"And let us consider how we may spur one another on toward love and good deeds." - Hebrews 10:24

For my parents,

Jayant and Anjali Davare

Contents

List of Figures

List of Tables

Preface

"Before I refuse to take your questions, I have an opening statement." – Ronald Reagan,
40th President of the United States

Welcome to **A Platform-Based Design Methodology for the Electronic System Level**! This book represents the work and ideas of many professors, researchers, and students. Primarily it is the collection of work done at the University of California, Berkeley between the Fall of 2001 and the Spring of 2008. It is an elaboration of a collection of various research papers and doctoral dissertations. The goal of this book is to put in one place a collection of the most mature and interesting case studies done in both the METROPOLIS and METRO II design environments while at the same time discussing key methodological issues relating to Electronic System Level Design. This book is not designed to be a textbook and does not have exercises to be assigned to students. Rather it should serve as a window to the ideology of a group of researchers and illustrate their attempts to put that ideology into practice.

This book is of interest primarily to graduate students and researchers in the field of Electronic System Level Design, Computer Aided Design for VLSI, and the Design of Embedded Systems. The authors have made an attempt to make each chapter somewhat self sufficient so that it does not need to be read from cover to cover. The chapters are arranged however in such a way that they progress in a logical order. In the event that a concept is not clearly defined in a chapter, most likely it was first introduced and defined in an earlier chapter.

We hope that readers enjoy this book and that it inspires them to go off and do something wonderful!

Douglas Densmore and Abhijit Davare
1st June 2008

Author Information

"When you have a Ph.D., you call them hypotheses, not guesses." – Gregory Benford, Astro-
physicist and science fiction author

Douglas Densmore received his Bachelors of Science in Engineering (Computer Engineering) from the University of Michigan in April 2001. He received his Masters of Science in Electrical Engineering in May 2004 from the University of California at Berkeley. His masters thesis was entitled, "Platform Based Reconfigurable Architecture Exploration via Boolean Constraints" and demonstrated how Boolean Satisfiability could be used to produce configurations for programmable hardware. He received his PhD in Electrical Engineering from UC Berkeley as well in May 2007. His PhD thesis, entitled

"A Design Flow for the Development, Characterization, and Refinement of System Level Architecture Services", explored how electronic system level design methodologies can be abstract and modular while at the same time remaining accurate and efficient.

He is currently a UC Chancellor's post doctoral researcher at UC Berkeley studying under Prof. Alberto Sangiovanni-Vincentelli. His research area is in the development of System Level Design methodologies for electronic systems. Specifically, architecture modeling and refinement verification. His background and interests are in Computer Architecture, Logic Synthesis, Digital Logic Design and Synthetic Biology.

His industry experience includes four+ summers with Intel Corporation where he was involved in pre-silicon design efforts regarding chipset development, post-silicon validation of the Pentium 4 microprocessor, and chipset software validation. He has also worked as a researcher at Cypress Semiconductor and Xilinx Research Labs. He is currently a member of the Gigascale Systems Research Center (GSRC) and the Center for Hybrid and Embedded Software Systems (CHESS) at UC Berkeley. He has published work regarding a method of successive refinement verification of electronic systems, taxonomies of EDA design tools, and algebraic frameworks for the manipulation of functional design descriptions to expose computational parallelism. In addition he has a US patent pending regarding data characterization of programmable devices (such as field programmable gate arrays).

Abhijit Davare received his Bachelors of Science in Computer Engineering from the University of Pittsburgh in April 2002. He received his Masters of Science in Electrical Engineering and Computer Sciences from the University of California, Berkeley in December 2004. His Masters thesis entitled, "A Platform-based Design Flow for Kahn Process Networks", explored the use of Platform-based Design and the METROPOLIS environment for the design of multimedia embedded systems. Abhijit received a Ph.D. in Electrical Engineering and Computer Sciences from the University of California, Berkeley in December 2007. His dissertation entitled, "Automated Mapping for Heterogeneous Multiprocessor Embedded Systems", developed automated techniques for the mapping of embedded software onto heterogeneous multiprocessor hardware within the automotive and multimedia domains. The dissertation also provided a related design methodology which was used to drive the development of the METRO II design framework.

Abhijit is currently a Research Scientist in the Platform Validation group at Intel working on topics relating to validation tools, post-silicon coverage, and architecture for bug survivability. He is a recipient of the Best Paper Award at the 2007 ACM/IEEE Design Automation Conference, the Tong Leong

Lim Predoctoral Prize in Electrical Engineering and Computer Sciences at UC Berkeley in 2004, the California Microelectronics Fellowship in 2002-2003, and the Outstanding Senior Award in Computer Engineering at the University of Pittsburgh in 2002. He has authored over 15 refereed journal and conference publications in a diverse set of areas relating to multimedia systems design, automotive systems design, electronic system level design frameworks, desynchronization of synchronous circuits, low power circuit design, sensor networks, and mixed-signal simulation. His interests are in the area of electronic system-level design automation, embedded systems, sensor networks, and resilient design techniques.

Comments on the book can be sent to davare@cal.berkeley.edu.

About the cover

The cover image illustrates the concept of platform development very nicely (albeit in an unconventional manner). The individual food items represent the platform elements. These may include slices of bread, cheese, and condiments. The sandwich therefore is a platform instance as it is a legal assembly of the individual elements. This instance has a cost (e.g. calories, price) as well as performance metrics of interest (e.g. sweet, salty). It has the ability to support a number of applications including a lunch for the kids or a tasty midnight snack.

Acknowledgements

"Knowledge is in the end based on acknowledgement." – Ludwig Wittgenstein, Austrian philosopher

The authors would like acknowledge all those who have been involved in the work outlined in this book. First and foremost the authors are standing on the shoulders of Prof. Alberto Sangiovanni-Vincentelli. A large portion of this work is a result of his tireless efforts and outstanding vision of the future. Without his guidance and support this work would not have been possible. In addition, many of our colleagues at UC Berkeley such as Rong Chen, Jike Chong, Trevor Meyerowitz, Alessandro Pinto, Gerald Wang, Guang Yang, Haibo Zeng, Wei Zheng, Qi Zhu, Rhishi Limaye, Alex Elium, Jue Sun, and Rodny Rodriguez were involved in various aspects of the work presented here. Each of them contributed in their own way and helped make this possible. In particular Trevor, Alessandro, Guang, Haibo, and Qi helped to make the design frameworks outlined here a reality. In addition University of California, Irvine researcher Samar Abdi, University of California, Riverside student Xi Chen, Università Degli Studi di Trento student Alena Simalatsar, and Columbia University doctoral student Sampada Sonalkar were also key collaborators.

Industrial researchers such as Yoshi Watanabe, Shinjiro Kakita, Felice Balarin, Marly Roncken, John Moondanos, Adam Donlin, Patrick Lysaght, Marco Di Natale, Paolo Giusto, Sri Kanajan, Alex Kondratyev, Claudio Pinello, Sanjay Rekhi, Massimiliano D'Angelo, and Stavros Tripakis were invaluable in the development of many of the tools and case studies presented. They served as early mentors and were always quick to help out.

Professors such as Farinaz Koushanfar, Luciano Lavagno, Roberto Passerone, Jason Cong, Luca Carloni, Edward Lee, Jan Rabaey, John Wawrzynek, Alper Atamtürk, Lee Schruben, and Kurt Keutzer also provided valuable discussions and feedback on the work presented here.

This work was supported in part by the Center for Hybrid and Embedded Software Systems (CHESS) at UC Berkeley, which receives support from the National Science Foundation (NSF award #CCR-0225610), the State of California Micro Program, and the following companies: Agilent, DG-IST, General Motors, Hewlett Packard, Infineon, Microsoft, National Instruments, Intel, Xilinx, Cypress Semiconductor, and Toyota. This work was also supported in part by the MARCO-sponsored Gigascale Systems Research Center (GSRC) and a grant from the Consumer Electronics Group of Intel Corporation.

Doug would also like to thank Erika for her constant support and her amazing ability to make everything better. Eres un ángel en la tierra.

Chapter 1

Introduction

"It is far easier to start something than it is to finish." – Amelia Earhart, Aviation pioneer

Ever since the first electronic systems were created, their designers quickly saw the need to be able to model the devices whenever possible. Modeling was clearly seen as a means to be predictive and avoid the creation of devices which would not meet the required specifications. Early models were collections of mathematical descriptions which were solved by hand. As the field grew, so did the systems of equations required to understand the device behaviors. Later, as computers matured, these equations were transferred to computer programs solved automatically. Clever techniques had to be employed to solve these equations. They had to be solved in such a way that the space and time complexity required fit with the current computing paradigms (i.e. programming model) and did not violate current resource constraints (e.g. program memory). As individual devices became well understood, designers then moved onto collections of devices (function blocks, systems) and explored their interactions. It is in this tradition that Computer Aided Design (CAD) emerged as a required technological component of electronic system design. CAD tools eventually became part of a larger group of technologies known as Electronic Design Automation (EDA). EDA encompasses computer aided design, computer aided engineering (CAE), and computer aided manufacturing (CAM). EDA recognizes that electronic system design is a collection of specific design stages each with their own unique concerns. EDA is the supporting mechanism for the semiconductor industry which in turn fuels the general electronics industry. The earliest EDA efforts were a small collection of companies which replaced what had been traditionally performed as an in-house activity. Daisy Systems, Mentor Graphics, and Valid Logic Systems, all which began in the early 1980's, are often credited as being the founding companies of EDA. From these humble beginnings, a whole industry has been created worth over 1.6 billion dollars, incorporating over 80 EDA consortium member companies, and employing over 27,000 engineers and scientists (all as of Q4 2007 [Mut08]).

Despite the past success of this industry, new innovations are constantly required to keep pace with the new challenges and design requirements. EDA is at a critical point in its history. It is faced with a number of design decisions which are going to change the face of how electronic systems are specified, developed, and implemented. This chapter kicks off the discussion of EDA tool development needs and sets the tone for the solutions presented in future chapters.

This chapter will relate the current state of the EDA industry to a set of key challenges that are currently facing the industry. The emergence of programmable platforms will be introduced as one example of an attempt to address these challenges.

1.1   Chapter Organization

This chapter serves primarily to provide motivation for the rest of the work outlined in this book. Section 1.2 describes the current state of the EDA industry. This describes the current growth slow down and why this affects the larger embedded systems market. Section 1.3 describes how four separate concerns (heterogeneity, complexity, time-to-market, and the business climate) are creating a challenging era for embedded systems designers. Section 1.4 details how programmable platforms have a number of characteristics that make them uniquely positioned to address the challenges facing EDA. Finally, Section 1.5 sets up the organization for the rest of the book. It outlines not only the flow of the book but also the unique contributions each chapter provides.

1.2   The State of EDA

The Electronic Design Automation (EDA) industry provides the tools and techniques to enable the design of complex electronic systems. In previous years, tools were able to make incremental improvements to their approaches and designers were able to use existing design flows to produce products successfully (on time and at a profit). The success of these small improvements was able to sustain growth.

Currently, however, this industry is experiencing a slow down in growth. This slow down ranged from 1% [VH05] to -0.6% [SNBW05] growth in 2005 and only 3% [VH05] growth in 2006. This data is down from a growth spike of 7.6% in 2001 [Bal02]. Incremental approaches have not been able to keep up with not only the advanced capability offered to devices by Moore's law but also the increasing need to find performance gains from changes in progamming models as opposed to raw compute power.

In EDA there exists a constant tension between the designers of electronic systems and the designers of EDA tools. Designers by nature create their value by adding insight and design expertise into the design flow. Tools look to automate processes and to some extent remove designer influence. If tools are seen as either not more productive than human designers (i.e. do not decrease design time) or not as good (i.e. do not improve the quality of the design) then it is highly likely that they will not be successful. As designs move into the 21st century a number of trends and challenges are making it increasingly difficult for tools to keep pace with the needs of designers. As tools have increasing trouble keeping pace, mistrust between designers and tools can develop and hence, designers would rather stick with known methods and tools (hence the growth slow down). Tools must be created which add value above and beyond an ad-hoc collection of individual, older tools, flows, and designer expertise.

1.3   Trends and Challenges

To expand, it is our opinion that EDA must provide solutions to a set of new challenges, detailed in this section. These challenges deal with embedded systems which may contain software as well as hardware. Embedded systems are specialized to carry out specific tasks and are "embedded" in their environment. This is in contrast to personal computers or supercomputers which are general purpose and interact with users. Embedded systems are much more prevalent than their general-purpose counterparts; for instance, 98% of all microprocessors manufactured in a given year are used within embedded systems [MB06]. Embedded systems typically have strict performance requirements relating to issues such as latency, throughput, jitter, memory usage, and energy consumption. Due to their widespread usage and performance-critical nature, the design of embedded systems is both relevant and challenging. Deploying complex embedded applications on such platforms is especially challenging since these systems must meet strict performance constraints. We consider two broad classes of embedded systems in this work: multimedia and distributed systems. The challenges we describe arise from these systems.

1.3.1   Heterogeneity

The first factor that embedded systems have to contend with is *heterogeneity* in device types, systems fabrics, and technologies.

Heterogeneity is defined as "the quality or state of being heterogeneous" where *heterogeneous* is defined as "consisting of dissimilar or diverse ingredients or constituents" [Dic06]. In the case of embedded system design and electronic system design in general, there are primarily two broad classes

4

of heterogeneity. The first class deals with the various technologies integrated on a printed circuit board (PCB) or even the device die itself. Figure 1.1 shows "Existing and Predicted First Integration of SoC Technologies with Standard CMOS Processes". Notice by the release date of this work, all 11 of the presented technologies have been introduced. These technologies range all the way from basic CMOS logic to chemical sensors and electrobiological components. In order to make sure that these devices function properly, models must be created which can capture the complex interactions caused by such diverse combinations. As nano-technology continues to be developed [JO03] it is clear that integration heterogeneity issues will only continue to become more complex and critical.

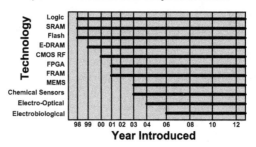

Figure 1.1: Technological SoC Heterogeneity [EKRZ04]

The second type of heterogeneity is inter-device heterogeneity. This description speaks to the many different types of individual components that are often assembled in a design (often on a single die). Figure 1.2 shows the Intel PXA270 System on a Chip (SoC). This integrated circuit is used in such devices as the Mypal A730 Personal Digital Assistant. This PDA has many state of the art features and is equipped with a digital camera and a VGA-TFT display. The primary issue in these types of systems is making sure that the communication between individual components can be sufficiently captured during simulation so that not only can system functionality can be verified during design but also that debugging is manageable. One must be able to isolate communication from computation, deal with different data types, and deal with different timing domains. Also it is important that each component be designed separately so that various product families can be developed with these components to service markets with different performance, power, and price requirements.

Heterogeneity is a factor that is not only difficult to manage but is increasingly becoming required. It is not practical or possible to have homogeneous systems for today's applications and in many cases the presence of heterogeneity may be seen as a design's strength.

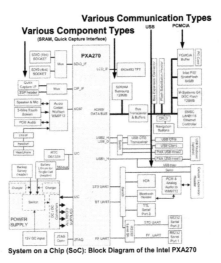

Figure 1.2: Device Component and Communication Heterogeneity [Int06b]

Heterogeneous multiprocessor architectural platforms are gaining prevalence for embedded systems. Greater parallelism is needed primarily due to increased energy efficiency requirements. Previously, greater performance for processing elements was achieved by increasing the clock frequency, exploiting instruction-level parallelism, introducing deeper pipelines, and carrying out speculative execution. For these techniques, the amount of performance gain achieved for a given increase in energy consumption has become smaller in recent years [OH05]. Since embedded devices are usually constrained by battery life and/or packaging cost, increased power consumption cannot be tolerated.

The alternative is to increase throughput by adding more parallelism to the system in the form of multiple cores. Each processor can run at a relatively low clock frequency and perform a portion of the specified application. Greater parallelism with relatively simple PEs running at lower frequencies provides increased computational capabilities with higher energy efficiency.

For embedded platforms, applications in a particular domain usually have a set of commonly used "kernels" or core computations that are carried out relatively frequently. In order to obtain high performance, PEs in the platform must have support for these kernels. Typically, parallel embedded platforms therefore contain a variety of PEs, customized for the common types of kernels found in applications for a domain [MP03].

Heterogeneity increases the programming burden as well. Code which is deployed on a PE cannot easily be migrated to another PE, especially at runtime. For embedded systems, this heterogeneous parallelism implies that parallel programming techniques from other communities may not be applicable. For instance, the supercomputing community generally deals with single-instruction-multiple-data (SIMD) types of programs on homogeneous parallel architectures [For93]. For embedded systems, the platforms are more irregular, and the focus is not on average-case performance.

The design challenges for heterogeneous parallel embedded systems become apparent by examining current design practice. The current flow for heterogeneous parallel platforms is typically an ad-hoc adaptation of the uniprocessor design flow. The uniprocessor flow typically involves manual implementation of code for the processor in a low-level language such as assembly code or C followed by extensive simulation to debug and meet performance constraints [Wol05].

The ad-hoc adaptation of this strategy for heterogeneous parallel platforms usually adds an initial partitioning step for the application that roughly divides functionality and assigns it to individual PEs. The manual implementation of code is carried out as before for each PE, and then simulation/testing is carried out on the entire system. However, due to the many possible interleaving patterns that may occur between the different PEs, simulation/testing is much less effective at finding problems in the implementation. If performance requirements are not met, then the code may have to be repartitioned, leading to another long design iteration. The main problems with this design practice are threefold: strong binding between the application and the architectural platform, lack of application verification, and inability to explore the design space efficiently.

The first problem occurs since design capture takes place only after rough partitioning has been carried out on the functionality. If portions of the application have to be migrated to other PEs, the existing design cannot be used. Similarly, a different architectural platform will require complete re-implementation. Therefore, early binding of this type between the application and the architectural platform limits the amount of design reuse [SM02].

The lack of application verification is partly due to the linkage between application and architecture but also because of the lack of a formal specification. Unless structured techniques are utilized, interactions between concurrently executing pieces of code cannot be analyzed [Lee06]. In the implementation, communication between PEs can cause serious problems in terms of synchronization issues, deadlock, and race conditions [HNO97]. The lack of application verification means that verification must be delayed until the implementation is complete, at which point it may not be known whether the error originated in the application specification or the implementation of this specification.

The inability to effectively explore the design space stems from the long design iterations and the reliance on simulation/testing to verify correctness of the system. Under these circumstances, each new point in the design space requires a significant amount of time and effort to produce. With this design practice, designer intuition is very important for producing a valid design. Since designer intuition cannot be relied upon, especially as new architectural platforms emerge, the lack of automated design space exploration techniques strongly limits design productivity [Gri04].

1.3.2 Complexity

The second factor facing embedded systems is *complexity* both in application (often manifested as software) and architecture (often manifested as hardware) designs.

Embedded applications are increasingly requiring more memory and compute power. Multimedia applications are an excellent example of this phenomenon. Computational requirements for embedded multimedia applications are increasing exponentially [NTL04] [HJKH03] [MBR02] [Mas01]. During the past 15 years, a variety of new protocols and standards have been introduced which feature rapidly increasing computational requirements. Figure 1.3 shows some of these trends for three classes of multimedia applications: video, cellular, and wireless LAN. Code size for these applications is also increasing, reflecting the trend that application complexity is increasing along with computational requirements.

In the automotive domain, the trend is similar. The value of software in a vehicle is expected to increase from 4% of the overall cost in 2000 to 13% of the overall cost in 2010 [HKK04]. Currently, a high-end automobile may contain over 270 different types of functionality that use 65 MB of binary code. This is expected to increase to 1 GB of code by 2010. This exponential trend is creating demand for the increasing number of transistors that can be integrated with scaling [Roh03]. Figure 1.4 illustrates a high level view of the sheer number of systems in a present day automobile.

IBM's Cell processor [Kah05] (an example of a cutting edge architecture design) is prominently featured in the Sony Playstation 3 and the most sophisticated devices in PCs today are related to graphics processing for videogames [GWH05]. Figure 1.5 provides a very clear illustration of the issues these complexity trends introduce. One of these is the increasing complexity of designs as measured by the number of transistors present in a device. This figure shows a 58% per year compounded complexity growth rate. However, the productivity rate (as measured in transistors per staff month) is only increasing at a 21% compounded growth rate. This growth rate mismatch leads to an increasing *productivity gap*. There is no inherent problem with the productivity gap. In theory this just means that all of the power of a device will not be realized. However in practice this gap leads to at least two side effects.

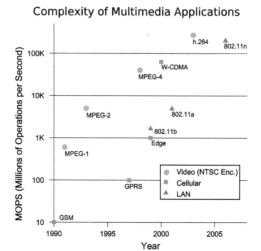

Figure 1.3: Exponentially Increasing Application Complexity

The first effect is that in the quest to utilize all that complexity, designs end up taking more time to develop. This is due to the fact that new architectural innovations must occur in order to take advantage of the added silicon. In the case of general purpose processors for example, companies like Intel are no longer pursuing advanced superscalar techniques but rather looking at multi- and many-core devices. These designs bring with them a whole set of verification, test, and design difficulties. In the event that productivity cannot keep pace it is very likely that design times will dramatically increase. This translates into lost revenue and lost opportunities for many companies.

In order to prevent this, the second effect is seen. Companies often respond by increasing the number of employees to tackle this problem. This leads to more development costs which end up raising the cost of the device. It is also not clear that this is simply a manpower issue. It is possible that more manpower will only exacerbate the complexity and management problems. In the event that the market will not bear this increased cost, either the employees cannot be hired or companies are not as profitable. Often what this means is that only the largest companies are able to compete in this space and as a result creativity and competition are not promoted. Innovation cannot occur and the small companies which may be ideally placed to look at new ideas are not viable.

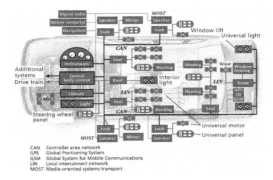

Figure 1.4: Complexity in the Automotive Domain [LH02]

1.3.3 Time to Market

The third factor facing embedded systems is *time to market pressures*. In semiconductor design for example, the design cycle times have decreased 33% since the early 1990's [Dat05]. A sampling of design cycle time decline is shown in Figure 1.6. This trend means that designs must avoid long development cycles and developer iterations in order to see profits necessary to justify new product development.

The first two factors, heterogeneity and complexity, were aspects of embedded system designs that were technology and application driven. This third factor, time to market pressure, is consumer driven and is in opposition to the other factors. Time to market is *why* a design method needs to be accurate and efficient. If it is not, there will be long iterations in the design and as a consequence, release dates will slip. Figure 1.7 shows three markets described by what the industry norm is regarding time between subsequent product releases (fast, medium, and slow). These markets could represent digital consumer devices (PDAs, cell phones), set-top equipment (televisions, DVD players), and automotive industries respectively. The Y-axis is what percent of revenue is lost if you are N months late (X-axis). While this is a fairly qualitative figure, the spirit of it remains. Essentially any longer than 12 months late is considered a product failure from a revenue standpoint. As little as 3 months late can be drastic as well (potentially losing as much as 15% of the expected product revenue). The lesson learned here is that time to market windows are small and the financial cost of missing them is extremely high.

Time to market issues cannot be ignored since they are why companies cannot take an arbitrary amount of time to produce designs. Granted it is not the only factor calling for an efficient design process (for example it would not be cost effective to manufacture an arbitrary number of devices at any design

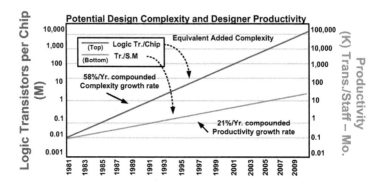

Figure 1.5: Growing Gap Between Device Capacity and Designer Productivity [fS99]

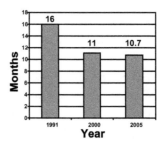

Figure 1.6: Semiconductor Design Cycle Time Decline [Dat05]

cycle speed in order to weed out process errors) but it is nonetheless a very powerful factor and the underlying influence behind almost all EDA efforts (tool design by nature looks to speed up the design process since time is often equated with designer effort).

1.3.4 Business Climate

In addition to the three trends outlined, it is important to note that there is a great deal of financial commitment and human resource effort involved in EDA. In 2005 the revenue in EDA was 3.9 billion dollars and was 4.3 billion dollars in 2006. It is projected as being as high as 7.4 billion dollars in 2009 [Goe05]. However, there are financial obstacles on the horizon which are exacerbated by the

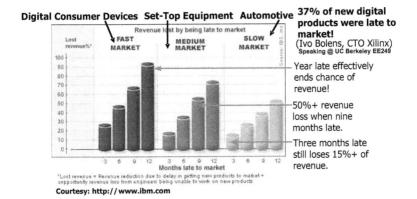

Figure 1.7: Time to Market Revenue Consequences [ES06]

challenges outlined.

Overall, a variety of technical and economic reasons have made application-specific utilization of transistors (e.g. ASICs) more difficult. First, time-to-market concerns and the need for flexibility may preclude the use of application-specific hardware. Second, the non-recurring engineering (NRE) costs associated with hardware fabrication often necessitate high product volumes to recoup the initial investment. NRE costs have continued to increase in recent years, with the mask costs alone for a single chip surpassing $1 million[Lam05]. Total design costs for designs implemented in 130 nm, 90 nm, and 65 nm technology are shown in Figure 1.8 [Kwo07]. Coupled with data indicating that the average selling price for ASICs is under $10 [Tur02] and not increasing, it becomes clear that implementing such an application-specific hardware device requires very high volumes to justify the initial expenditure.

The main strategy to overcome these trends is to manufacture programmable devices, which can be used with multiple applications, thereby increasing volumes and justifying the NRE costs. This trend is becoming clear when the number of total design starts is tallied, as shown in Figure 1.9 [Kwo07]. During the past few years, the number of unique designs implemented worldwide has slowly started to decrease.

The next section will introduce programmable platforms. These devices address the current business climate. In particular it will describe how they can be categorized and which types are of interest for the design flows and frameworks outlined in this work.

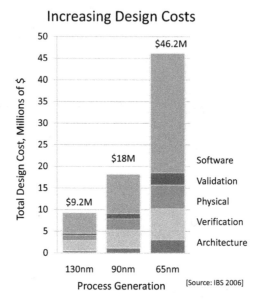

Figure 1.8: Design Costs Rapidly Increasing for Smaller Process Generations

1.4 Emergence of Programmable Devices

"Stay committed to your decisions, but stay flexible in your approach" – Tom Robbins, American author

EDA as a methodology in the past often was concerned with creating new electronic devices for each unique design requirement. Traditionally these have been ASICs or general purpose processing elements. In recent years however, programmable devices have been targeted by EDA. They are of particular interest to the work here because of the challenges outlined regarding time to market requirements and costs associated with new ASIC design starts. Programmable platforms have the ability to both bypass the expensive costs associated with ASIC design starts as well as offer flexibility.

When having a discussion about creating abstract, modular architecture service models which are still efficient and accurate one must quickly determine what types of implementation devices one is going to consider. One could consider static architecture service models. A static architecture service model for the purposes of this discussion is one which has its functionality bound during manufactur-

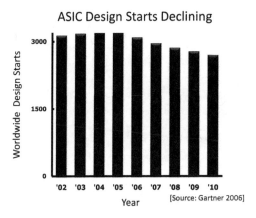

Figure 1.9: Number of Worldwide Design Starts Declining

ing. This is the case when speaking about General Purpose Processors (GPP) such as Intel's Pentium 4 [Int06a] or ARM style processor [ARM06]. ASIC designs could also be members of this group. These devices are perhaps programmable at the ISA level (GPPs) but one cannot change the computation fabric or interaction between computation or communication units after fabrication. They are usually either very special purpose (ASICs) or very generic (GPPs). Often they have a high design cost but are often cheaper to manufacture and recoup that design cost in sales volume. At the other end of the spectrum are programmable architectures or platforms (the term platform denoting a set of services which typically are not associated with traditional CPU architectures). A programmable platform is a system for implementing an electronic design. Examples of these are Platform FPGAs and ASIPs. These systems are distinguished by their ability to be programmed regarding their computation (functionality), communication (topology), or coordination (scheduling). Programmable platforms are increasing in use and popularity for several reasons: [Kue02], [DeH00]

- **Rapid Time-to-Market** - One can often eliminate fabrication time by using off the shelf parts. This also bypasses a large part of the verification time as well since parts are well understood and there is no lengthy post silicon verification phase.

- **Versatility, Flexibility** (increase product lifespan) - Design reuse within a programmable architecture family is often possible.

- **In-Field Upgradeability** - Many devices are reprogrammable using as little as a personal computer or a portable flash memory card.

- **Performance**: 2-100x compared to GPPs - Special purpose computation units can exploit spatial concurrency or dedicated hardware can be created.

Table 1.1 lists a set of characteristics that allow programmable platforms to achieve those advantages. However they naturally have some disadvantages as well:

- **Performance**: 2-6x slower than ASICs - Programmable architecture topology overhead related to programming the device may hurt performance. For example, FPGAs are unable to perform routing as efficiently as a custom ASIC due to their mesh-like structure.

- **Power**: 13x compared to ASICs - Programmable architecture fabric is not typically optimized for power although companies are starting to improve their power consumption dramatically.

Overall the strengths outweigh the weakness as both of the weaknesses are becoming less of an issue as technologies mature. Programmable Platforms often have a very regular device fabrics (FPGAs for example are famous for this). This regularity allows for advances in device technology (such as transistor scaling) to be taken advantage of with minimal design changes. An FPGA is able to double its computing capacity every 18 months with the same die size potentially. In fact, industry luminary Tsugio Makimoto of Sony Corporation has programmable platforms as a key extension of his now famous "Makimoto's Wave". Figure 1.10 illustrates this point. The wave demonstrates the observation that the electronics industry oscillates between standardization and customization. Standardization is used to proliferate designs and enable new companies and designers to enter into the marketplace. Customization occurs as a means for innovation and to enter new market areas where standards are not in place. Standardization is able to take advantage of factors such as regularity, automation, and predictability. All of those factors are reasons why this work explores programmable architectures services (a standardized approach). Tool development by its very nature is most productive during the standardization cycle of the wave.

Because of their increasing relevance and prevalence, programmable platforms are a natural target for EDA design flows. In addition, they directly target time to market issues. Also they often side step technology heterogeneity issues since they have regular design fabrics. Finally, they can be customized to directly address new complex applications. From a practical standpoint, if one were to create architectural models of a programmable device, these models by definition could be used to represent a very large set of individual architecture instances (i.e. each configuration). By modeling the primitives of

Makimoto's Wave

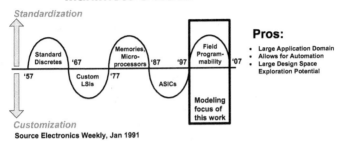

Standardization

| Standard Discretes | '67 | Memories, Micro-processors | '87 | '97 | Field Program-mability | '07 |

'57 Custom LSIs '77 ASICs

Modeling focus of this work

Pros:
- Large Application Domain
- Allows for Automation
- Large Design Space Exploration Potential

Customization

Source Electronics Weekly, Jan 1991

Figure 1.10: Makimoto's Wave and Programmable Devices [Mak00]

the programmable platform a very large design space can be easily created from a relatively small model set. However it is not enough to say that this work will focus on programmable platforms since this is still a broad classification. The discussion will now begin the process of narrowing down the focus within this space.

To begin, the characteristics of programmable platforms are shown in Table 1.1. These characteristics are intentionally vague and meant to contrast those not typically explored in static architectures. The architecture models to be described allow for all of these features as each is a very important aspect of a programmable platform. As mentioned, this table highlights the strengths of programmable platforms especially when dealing with concurrency and distributed control.

Characteristic	Description
Spatial Computation	Data processed by spatially distributing the computations
Configurable Datapath	Functionality and interconnection network of computational units is flexible
Distributed Control	Units process data based on local control
Distributed Resources	The required resources for computation are distributed throughout the device

Table 1.1: Characteristics of Programmable Platforms

Table 1.2 shows the wide range of programmable devices. As the table progresses, the level of abstraction increases as does the intended scope of the device (from component to whole system). For this work, *FPGAs, SoCs, and Hybrid Architectures* will be focused on. This work presented is purposefully device agnostic. However, the key issue here is abstraction (the granularity at which the device is modeled). This work is going to look at functional and transaction level models. Therefore it is inappropriate to talk about PLDs. In addition, analog issues will not be explicitly discussed in this work therefore Field Programmable Analog Arrays (FPAAs) will not be covered.

Device	Description
Programmable Logic Device (PLD)	PROMS, PLAs
	Examples: Flash Memory Devices from Intel [Int04]
Field Programmable Gate Array (FPGA)	Contains uncommitted configurable logic blocks (CLBs)
*FOCUS	*Examples*: Altera Cyclone FPGA [Alt04]
Field Programmable Analog Array (FPAA)	Contains uncommitted configurable analog blocks (CABs)
	Examples: Anadigm AN10E40 [Ana04]
System on a Chip (SOC)	Static and reconfigurable components at function unit level
*FOCUS	*Examples*: Cypress PSoC [Mic04]
Hybrid Architectures	Static and reconfigurable components at function and bit-level
*FOCUS	*Examples*: Xilinx Virtex II Pro [Cora]

Table 1.2: Programmable Platform Technology Classification

Table 1.3 illustrates the various aspects which need to be considered when creating a model of a programmable architecture. The left column indicates the various aspects of programmable platforms that are of interest in a modeling framework. A description and example of each is provided in the right column. This work will be dealing with *functional unit granularity and tight chip-level host coupling*. The other factors do not directly apply to this work. Reconfiguration methodologies are not directly discussed (but still can be modeled) and arbitrary memory organizations can be modeled.

Classification	Description
Granularity	Size of the smallest reconfigurable functional unit addressed by mapping tools
	Tradeoff between flexibility and performance overhead
	Examples: CLB, ADC, ISA (bit level, function unit, program control)
Host Coupling	Type of coupling to host processor
	Loose System Level/Loose Chip Level/Tight Chip Level
	Examples: Through I/O (SPLASH);
	Direct communication (PRISM); Same chip (GARP, Chameleon)
Reconfiguration Methodology	How the device is programmed
	Examples: bit stream (serial, parallel); dynamic; partial
Memory Organization	How computations access memory
	Examples: large blocks vs. distributed

Table 1.3: Example Programmable Platform Architecture Classifications

Finally, Table 1.4 shows the potential design levels (abstractions) upon which programmable devices can operate. There are two axes. The left column is the vertical axis which represents abstraction.

The other three right columns are the types of design element categories. This work will be concerned with both the *Microarchitecture level and the Process/Systems level*. System Level Design dictates that it only really makes sense to examine the levels above "Implementation". RTL based designs would be more concerned with "Implementation level" and their goal would be to integrate the ESL solution with a tool that could traverse this portion of the design flow.

Design Levels (Vertical Axis)	*Design Elements* (Horizontal Axis)		
	Communication	**Storage**	**Processing**
Implementation	Switches/Muxes	RAM Organization	CLB/IP Block
Microarchitecture *FOCUS	Crossbar/Bus	Register File Size Cache Architecture	Execution Unit Type Interpreter Levels
Instruction Set Architecture	Address Size	Register Set	Custom Instructions
Process Architecture *FOCUS **Systems Architecture *FOCUS**	Interconnection Network	Buffer Size	Number/Types of tasks

Table 1.4: Horizontal/Vertical Axis Classification Example [SVSK01]

In summary, this work will be concerned with modeling architecture services for FPGAs, SoCs, and Hybrid Architectures at the functional unit granularity with details present regarding the microarchitecure and system level. Specific examples will be discussed regarding the Xilinx Virtex II Platform FPGA [Cora] (hybrid architecture) in Chapter 5.

These sets of programmable platform categorizations were chosen since they are at the appropriate level of abstraction desired. Additionally, they are easily described as modular components. They are easy to characterize which will improve accuracy as well.

Programmable hardware devices can be termed as architectural platforms, since they can support a wide variety of applications, but usually within a specific domain [Hen03]. Design cost and effort now shifts to creating the software that is deployed on these platforms for a particular application [KMN+00]. Enabling increased design productivity for such platforms is therefore important.

1.5   Book Structure and Contribution

As has been discussed, EDA is facing some significant challenges. However we feel that this work will show that not only is there hope on the horizon but also introduce a number of innovative design flows and frameworks that can change the challenges presented into new and exciting opportunities. Explicitly this work illustrates:

- Functional modeling approaches which allow for design specifications to be captured using multiple models of computation at a variety of abstraction levels while maintaining a clear separation from any notion of the physical design implementation.

- Architecture service modeling approaches which allow for abstraction and modularity while maintaining accuracy and efficiency. These models can be augmented with rigorously characterized performance costs as well as tie to a number of different refinement verification methodologies.

- Mapping between functional models and architectural services can be formally defined and performed using a number of techniques including mathematical programming approaches. Mapping can be made extremely powerful and flexible by using an event based framework with declarative relationships defined between events.

To illustrate these contributions visually, the relationship between environmental factors, design solutions, techniques, and outcomes is outlined in Table 1.5. This table shows how the environmental and industrial factors discussed (heterogeneity, complexity, and time-to-market pressures) lead to the solutions (modularity and abstraction) that a EDA methodology should achieve. This work provides the techniques listed to achieve these goals and produce the stated outcomes (accuracy and efficiency). This is the central proposition of all of the work contained here.

Factors	Solutions	Supporting Techniques	Outcomes
Heterogeneity	Modularity	*Separation of Concerns* (Chapter 3)	Accuracy
Complexity	Abstraction	*Event Based Design Frameworks* (Chapter 4)	Efficiency
Time-to-Market		*Functional and Mapping Design Flow* (Chapter 5)	
		Architectural Service Design Flow (Chapter 5)	
		Architecture Service Refinement Verification (Chapter 6)	
		Architecture Service Characterization (Chapter 7)	

Table 1.5: Relationship Between Factors, Solutions, Supporting Techniques, and Outcomes

At this point the reader should now be familiar with the items necessary to understand the background, goals, and embedded system focus related to EDA. This final section will attempt to make very clear the contribution of this book. Thus far this chapter has established several things:

- Introduced heterogeneity, complexity, and time-to-market pressures as the motivating factors in this EDA research. These factors must be addressed in order for EDA to move forward to new growth areas and develop new methodologies for its continued success.

- Matched the design factors to the design solutions intended to resolve them. These are heterogeneity to modularity and complexity to abstraction.

- Described the business climate in which EDA finds itself and how the business climate has forced designers to think in new ways.

- Identified the outcomes that are desired. These are accuracy and efficiency and the ability to meet time-to-market demands. It is not enough to simply create abstract and modular designs without being accurate and efficient. It is clear that EDA growth is dependent on the ability to ensure these qualities.

- Identified that programmable platform architectures services are going to be the focus of the architecture service modeling in the methodology to be described. Not only do these devices look to address the new concerns of EDA, but they also possess key characteristics which make architecture modeling at the system level more accurate and efficient. Creating one set of programmable components takes the place of creating a very large set of static components.

Figure 1.11 shows how the this book is organized and how each aspect interacts with the others contained here. These topics are framed in a meet-in-the-middle approach to design which will be elaborated on in Chapter 3. The figured is labeled to indicate where in this book each individual concept is described in more detail.

1.5.1 Organization

The contributions to EDA as a result of this work are outlined concisely in Table 1.6 and summaries of aspects of the work found in this book can also be found in [Den07] and [Dav07].

The main contributions of this work are: a comprehensive examination of the ESL landscape, a distillation of requirements and descriptions of next-generation design frameworks, design flows for functional modeling and mapping as well as architecture service development, characterization and refinement techniques for architecture services, and explorations of these design flows in the multimedia and distributed system domains.

The discussion begins next in Chapter 2 where we introduce the notion of Electronic System Level (ESL) design. ESL with its notions of abstraction and modularity sets the tone for the rest of the chapters to come. We encourage the reader however to read the chapters in the order that they are of interest. Also we highly refer the reader to the references when possible for more details and definitions.

Chapter	Contribution/Impact
Electronic System Level Design **Chapter 2**	Definition of the new design paradigm Business and technological impact Five design taxonomies Six design scenarios
Platform-based Design **Chapter 3**	Separation of concerns PBD theory
Design Frameworks **Chapter 4**	Survey of related work METROPOLIS Design Environment METRO II Design Environment
Design Flows **Chapter 5**	Naïve design flow Functional and mapping design flow Architectural service design flow Design flow tradeoffs
System Level Service Refinement **Chapter 6**	Event based service refinement Interface based service refinement Compositional component based refinement
Architecture Service Characterization **Chapter 7**	Platform characterization Data extraction Data organization Data integration
Multimedia System Design **Chapter 8**	JPEG on Intel MXP5800 Motion-JPEG on Xilinx platforms H.264 deblocking filter UMTS METRO II design
Distributed System Design **Chapter 9**	FLEET architecture modeling SPI-5 packet processing Distributed supervisory control Experimental vehicle system

Table 1.6: Contributions of this Book

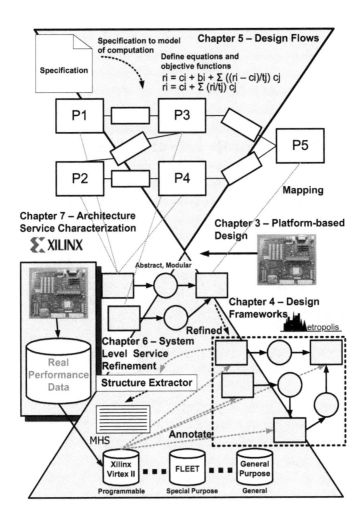

Figure 1.11: Overview of Concepts and Design Flows Covered in this Book

Chapter 2

Electronic System Level Design

"Science is the systematic classification of experience." – George H. Lewes, British Author

As any system that one designs gets larger and more complex, the natural response is to begin with more abstract representations of the design. Collections of these design representations are considered first often in a hierarchical fashion. Only when the larger pieces of the design are in place does one delve into the lower level details. This is seen in many design activities in the fields of engineering, business, and the physical sciences. This style of design is only useful however if the decisions made at the higher level of abstraction do not have to be redone when lower level decisions are made. The classification of these styles must be understood so that efficient design flows can be created. High level decisions must ultimately make the design process not only easier for the designer but result in high quality design decisions.

In Chapter 1 three key challenges were outlined for the design of embedded electronic systems (heterogeneity, complexity, and time to market pressures). To meet these three challenges, one of the key requirements is an increased level of abstraction when specifying and developing systems. Electronic system level (ESL) design embraces higher levels of abstraction. ESL designs, supporting environments, and toolsets have become increasingly visible in the EDA landscape. This chapter will examine ESL and its role in the future of embedded electronic system design.

This chapter will describe the ESL landscape and show a number of different perspectives on how ESL tools can be characterized and used in design flows.

2.1 Chapter Organization

This chapter's strength lies in its comprehensive set of taxonomies and discussions about ESL based design flows. These serve to give the reader a broad overview of the field in general. To this end, we begin with an introduction in Section 2.2. This discusses basic definitions as well as the business and technological impact of ESL. Next the related work on ESL classification and analysis is discussed in Section 2.3. The design aspects of abstraction and modularity which are key to ESL's success are discussed in Section 2.4. Five different taxonomies for ESL are presented in Section 2.5. This is followed by six related discussion of ESL in Section 2.6. These discussions include alternate design flows and tool "binning" approaches. Section 2.7 provides concluding thoughts and future work in the area of ESL classification and design.

2.2 Introduction

2.2.1 Definition

According to the International Technology Roadmap for Semiconductors (ITRS) in 2004 [fS04] ESL is defined as "a level above RTL including both HW and SW design". ESL is defined to "consist of a behavioral level (before HW/SW partitioning) and architectural level (after HW/SW partitioning)" and is claimed to increase productivity by roughly 200K gates/designer-year. The ITRS states that ESL will produce an estimated 60% productivity improvement over what they call "intelligent testbench" approaches (the previously proposed ITRS electronic system design improvement). While these claims cannot be verified as yet and do look quite aggressive, most agree that the overaching benefits of ESL are to:

- Raise the level of abstraction at which designers express systems;

- Enable new levels of design reuse;

- Provide for design chain integration across tool flows and abstraction levels.

ESL has come to replace the term *system level design* in many circles. As [BMP07] points out, Wikipedia has a definition for ESL but not one for system level design. ESL has outlived many other terms including Electronic System Level Design Automation (ESDA) and System Design Automation (SDA). ESL as a term is generally attributed to Gary Smith (until recently with Gartner/Dataquest). If one goes to any of the major EDA conferences, they will find that ESL is an umbrella term for any number of design solutions which claim to increase design productivity. If anything, one could make the case that the widespread adoption of ESL has actually created some confusion in the community and that efforts need to be made to standardize the discussion. This chapter hopes to make some inroads in the process.

2.2.2 Business Impact

The revenue from ESL tools is projected as being as high as 7.4 billion dollars in 2009 [Goe05]. According to [Kri05] the embedded hardware market (which uses EDA tools) will reach $78.7 billion in 2009 assuming an aggregate 14.2% growth rate. Figure 2.1 illustrates this tremendous growth of embedded integrated circuits, software, and printed circuit boards. This information clearly shows it is a very costly proposition to begin the process of shifting the entire industry to a new design methodology. It is not done on a whim or due to passing marketing pressures. However, the slow down in growth mentioned previously in Chapter 1's introduction has started the migration process to ESL and it appears that there is no turning back. The migration to a new methodology is very cognisant of and concerned primarily with four key factors. As mentioned these are heterogeneity, complexity, time to market pressures and deep submicron design effects.

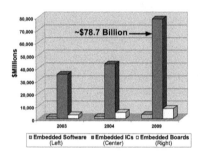

Figure 2.1: Global Embedded Systems Market [Kri05]

As a direct result of ESL increasing tool introduction, EDA growth is predicted to be 22% [Goe05] in 2007. Figure 2.2 shows not only the impact ESL will have on increasing EDA growth in the future (in terms of overall revenue projections), but it also shows how ESL tools are predicted to rival RTL tools (Register Transfer Level; usually specifying a relatively low abstraction level) in terms of revenue potential. This trend is very important as RTL is the current design benchmark in EDA.

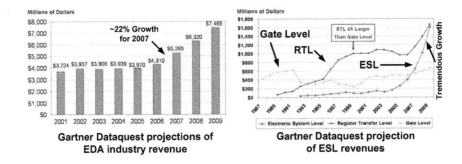

Figure 2.2: Overall EDA Revenue Growth and EDA Design Segment Growth [Goe05]

2.2.3 Technological Impact

ESL methodologies and tools are of increasing interest because they specifically look to exploit the "design gap" experienced by current design flows. More accurately this should be termed a "methodology gap" which exists between old design methodologies (i.e. RTL) and new design methodologies (i.e. ESL). Figure 2.3 presents a qualitative graph relating design complexity to designer productivity with both RTL and ESL design methods. Today, most designers work with RTL design tools and languages (VHDL and Verilog for example). They find themselves in the "methodology gap" where the system they are trying to create exceeds the capabilities of their design environment. This is not to say that the methodology gap cannot be crossed. On the contrary, the gap can be overcome with existing design methods but only at a significantly increased cost (both financially and in designer effort). Existing RTL design methods will continue to be employed until the additional cost of design overwhelms the commercial viability of the final design. This "maximum tolerable design gap" as shown in Figure 2.3 varies per technology, per market segment, or even per product and is always present at some level.

A transition from RTL to ESL is required to completely overcome the "methodology gap". A transition must occur since it is well accepted that design complexity will continue to increase (reflected in the continuance of Moore's Law). Today, the design community is approaching a point of inflection between the two methods - the rate of ASIC design starts in recent years is declining (see Figure 1.9) while implementation/programmable technology's growth (i.e. in the FPGA market) has continued to trend upwards [Dat08]. An important question, therefore, is "what limits the widespread adoption of ESL by the majority of designers?". A simplistic answer is that ESL design methods, tools, and lan-

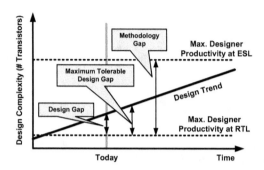

Figure 2.3: "Methodology Gap" Challenge in EDA

guages are simply not mature enough to convince designers to risk traversing the gap between the two methodologies. To further complicate the answer, we must respect that ESL methods must tackle multiple design problems. It is ultimately the design project itself which influences the relative importance of each problem. Therefore, a complex compromise must be struck between the ESL vendors who create a set of tools and the system designers who must work in ESL environments.

2.3   ESL Related Work

Since ESL burst onto the scene in the mid 1990's there has been no shortage of coverage on the subject. [BMP07] is an excellent reference (the best we know at the time of writing) on all things related to ESL. In this work the authors (Bailey, Martin, and Piziali) provide definitions of ESL, taxonomies, a history of ESL evolution, enablers of ESL, ESL flows, and a host of other discussions on ESL methodologies. The work here does not attempt to replace their work but rather to provide another view for the reader which is closely aligned with the other discussions in this book. For a more focused look at ESL exclusively we direct the reader to this source.

Other works which look at "What is ESL?" are [Man04] and [SNBW05]. These come from the analyst community and reflect a very commercial and financial view of ESL. They examine ESL's relationship with the business aspects of EDA growth and development.

ESL taxonomies are found in other locations including [BMA05]. This is a collection of taxonomies not only of ESL but of other design concepts across EDA. These include a model taxonomy, a functional verification taxonomy, a platform-based design taxonomy, and a hardware-dependent software taxonomy. This book provides an excellent classification of high-level design tools and whenever possible and appropriate we use its definitions. With respect to [BMA05], our approach places tools in a more general design context and is intended also to give guidelines on how to connect the available offers in a design flow.

In [SN05], two axes were used for ESL classifications. The first axis consists of three methodology components: an algorithmic methodology, a processor/memory methodology, and a control logic methodology. Each refers to the way in which a designer thinks about the design or its components. The second axis consists of the levels of abstraction at which the designs are expressed: the behavioral level, the architectural level, and the platform-based level. Approximately 50 approaches are examined. This creates the foundation for our second taxonomy presented in Section 2.5.2.

A similar approach to categorize industrial tools also based on two axes is presented in [Man04]. The first axis is the design style (embedded software, SoC (HW), behavioral, component) and the second axis is the language used to describe the design (e.g., C/C++, Verilog). Approximately 41 approaches are examined. This is the inspiration for our third taxonomy presented in Section 2.5.3.

In [Gri04], again two axes are used to classify ESL tools developed both in academia and in industry but now relate to levels of abstraction (e.g., system level and micro-architecture) and design stages (e.g., application, architecture, and exploration). Approximately 19 approaches are examined. Our fifth taxonomy in Section 2.5.5 is based on this work.

[DPSV06] is a subset of the taxonomies presented later in this chapter. Specifically it covers the first taxonomy (Section 2.5.1) and the first design discussion (Section 2.6.1). The reader can refer to that paper to get an abbreviated version of what is contained here.

2.4 ESL Design Components

Because of the potential to bridge the "methodology gap", ESL is being widely adopted and there have been a number of industrial and academic tools created to be ESL based solutions. Each approach attempts to solve a variety of design problems. However, there is by no means a unified view of how to best attack the forces driving ESL development. Fortunately there are a number of design scenarios (to be discussed in Section 2.6) which ultimately dictate which methodology is employed.

A major contribution of the first taxonomy presented in Section 2.5.1 is that it clearly demonstrates that all tools can be categorized around three orthogonal design aspects. These design aspects can be thought of as the *design components* which are useful in ESL.

Definition 2.4.1 Functionality - *this is "what" a system does. This can also be considered the application the design implements. Other common terms for this area are application domain or behavior.*

Definition 2.4.2 Architecture - *this is "how" a system carries out its operation. This can also be considered the services the system provides. Other terms for this area are platform components or services.*

Note that architectures can be traditional HW ASIC components, programmable processing engines, as well as general purpose processors (GPPs) capable of running software. All of this development is subject to abstraction in which case architecture services could be anything from logic gates to ISA instructions. The development of architecture service models a focus of this work particularly in Chapter 5, Section 5.7.

Definition 2.4.3 Mapping - *this is the process of assigning functionality to architecture (behavior to services). Often this is called binding as well and is traditionally seen as part of the synthesis process.*

Mapping is an assignment between behaviors in the functional model and services in the architectural model. Mapping can be "many-to-one". This allows "many" functional behaviors to be assigned to "one" architectural service. For example a DCT and FFT behavior can be mapped to a single abstract service dealing with signal processing.

There is a great deal of work related to each of these three areas as was shown in the taxonomy work [DPSV06]. Often ESL tools will fall into one of these categories only or perhaps combinations. The areas themselves will be touched on more specifically in Section 2.5 when each taxonomy approach is described in more depth. It should be pointed out that this work will focus individually on functional model creation and mapping in Chapter 5, Section 5.6. This will detail a number of ways in which to capture specifications and how to assign resources to functionality in an efficient way using a variety of techniques including mathematical programming. In addition, architecture service model development will be covered in Chapter 5, Section 5.7. This work will demonstrate how embedded system architecture service models can be created and how to formally verify properties of these models as it relates to refinement.

At this point is should be made very clear that this work is of interest since in order to legitimize ESL and to continue its adoption, functional models, architecture services, and mapping will need to be provided in such a way that various desired ESL characteristics attributed to abstraction can be maintained while achieving performance goals associated with RTL.

2.4.1 Abstraction

Abstraction allows the system to be described early and at a reasonable cost but it also casts a shadow of doubt over the accuracy of performance analysis data. Since the data gathered during simulation guide the selection of one system architecture over another, the veracity of data recovered from ESL performance analysis techniques with respect to the system feature being investigated must be considered carefully by the designer. Fear of inaccuracy in ESL performance analysis is a major impediment to the transition from RTL to ESL. Preventing this inaccuracy is paramount for ESL acceptance and legitimacy and is the major goal of this work. Abstraction specifically looks to deal with complexity while maintaining accuracy.

Definition 2.4.4 Abstraction - *the addition of system behaviors. A system is more abstract if it has more possible behaviors and less abstract if it has fewer possible behaviors.*

Abstraction does not have to do with code size, complexity of execution, or the number of "details" in this definition. Abstraction can be seen as a relaxation of constraints which expands the space of behaviors a system can exhibit. It is the process of obscuring aspects of the design in order increase the ability of the designer to only consider those which help to develop a design at that particular stage.

Abstraction could be a set of transistors being represented as logic gates, a set of bus transactions being reduced to a IP interface, or the operation of a processor being available as a set of abstract services (add, divide, etc). Abstraction will allow more device resources to be utilized more easily but it must be tempered by the level of controllability, observability, and accuracy. Higher levels of abstraction allow design changes to most dramatically effect the overall design but a designer also has the least insight into how precisely the changes brought about this change. The inverse is true for less abstraction. What is needed is something with the best of both techniques. This work will show how abstraction can be achieved while maintaining accuracy. Specifically maintaining relative accuracy or *fidelity*.

Definition 2.4.5 Fidelity - *requires that all pairs of corresponding measurements m_1, m_2 in a abstract model and p_1, p_2 on the actual implementation, hold $m_1 < m_2$ if and only if $p_1 < p_2$.*

2.4.2 Modularity

By keeping various aspects of the design separate (communication, computation, coordination), the now modular design allows for a smoother verification process, reuse, and abstraction. Modularity encourages reuse, localizes system functionality, provides more system observability, and helps to manage complex system development. However modularity can often be at odds with simulation efficiency. Overheads often associated with modularity may decrease simulation speed or enforce rigid syntactic or semantic requirements on the designer. If a design environment is to be widely accepted it must remain equally efficient (if not more so) as the current design environments it is replacing for the same amount of design productivity gains (both in terms of design time saved and design space explored). Preventing this inefficiency is also paramount for ESL acceptance and legitimacy and is partner to accuracy as a goal of this work. Modularity looks to deal with the design challenge of heterogeneity.

Definition 2.4.6 Modularity - *the ability to define clearly the boundary between interacting components both in terms of their communication, computation, and coordination. At these boundaries, components should be able to be tested and verified for correct functionality. In addition, there should be rules regarding how systems are composed of these components and how those boundaries can be changed during refinement.*

If a design is modular, one can test its components in isolation and will allow for reuse. Modularity allows communication issues to be isolated from computation issues as well. Throughout this work, modularity will be emphasized as it is a critical contribution in the design of system level architecture service simulation and verification techniques. Modularity will be constantly monitored in the context of maintaining an efficient simulation environment.

2.5   The ESL Taxonomies

This section will detail how one can classify ESL offerings. Specifically it will provide a set of taxonomies. A taxonomy is defined as "the science or technique of classification" [web08]. These taxonomies are useful for several reasons. The first reason is that a categorization will aid designers in choosing a tool before they begin the design process. If the wrong tool is chosen initially this could not only slow down the design process but also potentially lead to an incorrect design. The initial tool flow chosen often dictates the future validation abilities as well as the simulation power and portability of the design. Secondly, examining the number and types of tools allows one to see potential areas for improvement and research. If a particular area is deficient in some area, tool development in that area could be of use to the whole community. Finally, tool classifications give insight into the types of design practices currently employed in ESL. This can help one better understand why the EDA business model is constructed as it is as well as identify how practices can be improved.

What follows are five discussions of different taxonomies. These are not meant to be mutually exclusive but rather provide five different lenses by which to view the ESL space. The first presents a classic y-chart based approach. The second describes how designers envision the overall system. The third discusses entry languages and development focuses. The fourth places tools in the community in which they are developed. The final taxonomy discusses the overarching framework in which a tool is placed.

2.5.1   Taxonomy One: "Classic" Y-Chart

The design framework illustrated in Figure 2.4 is based on the platform-based design (PBD) paradigm as presented in [SVM01] and [SV02] and later in Chapter 3. The figure provides a description of each of the three areas.

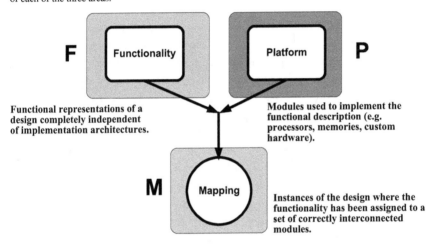

Figure 2.4: Platform-Based Design Classification Framework Elements

In this taxonomy, the design process is seen as a sequence of steps that repeat themselves as the design moves from higher levels of abstraction to implementation. The primary structure is shaped as a Y (in this sense it is similar to the famous Y-chart introduced by Gajski [GK83]) where the left branch expresses the functionality (the *what*) the designer wishes to implement, the right branch expresses the elements the designer is allowed to use to realize the functionality (the *how*), and the lower branch expresses the decision of which elements to use to implement the functionality (the *mapping*). In this context, the right branch of the framework is called a *platform*, and consists of:

1. A "library" of elements, including IPs and communication structures, and of composition rules that express which elements can be combined and how;

2. A method to assess the *quantities* associated with the use of each of the elements (e.g., power consumed and/or time needed to carry out a computation).

Each legal composition of elements from the platform is a *platform instance*. Mapping consists of the selection of the components for the design (choice of a platform instance) and of the assignment of parts of the functionality to each of these elements, so that the complete functionality (possibly with overlaps) is realized. This process is optimized with respect to a set of metrics and of constraints defined over the costs provided (quantities mentioned) and used to evaluate the feasibility and the quality of the design.

This view of the design process can be considered as an abstraction of a process that has actually been implicitly used for years at particular levels of abstraction. For example, we can interpret the logic synthesis process in this framework: RTL code (or Boolean functions) is used to represent the functionality of the design, the platform consists of a library of gates (or higher complexity logic blocks), and mapping is the actual logic synthesis step that implements the functionality as an interconnection of gates (platform instance) optimizing a set of metrics involving area, power, and timing. The mapped design (gate netlist) is then exported to the layout phase where this representation is mapped to a physical platform.

The PBD paradigm applies equally well to the application and algorithmic level where the functionality can be a "mathematical" description (for example, an MPEG encoding algorithm) and the platform may be a set of sub-algorithms that can be used to implement each of the functional blocks of the encoding method. The result of the mapping process is then fed to a lower level where the mapped platform instance becomes the left branch and the right branch is a new set of elements that can be used to implement the mapped platform instance. The process is iterated until the result of the mapping process is a fully implemented solution. Thus, the design process is partitioned into levels, where each level represents a particular abstraction, and where the corresponding platform and mapping process are designed to optimize some specific aspects of the design.

This framework "prescribes" a unified design methodology and as such, it can be used to identify where existing tools and flows fit and how they can be integrated in the overall system design process.

In this taxonomy we use the platform-based design paradigm to classify a large number of tools that are related to ESL. In doing so, we cast the present system level design efforts in a global framework that serves as the unifying element. We will see that the existing approaches may fall in more than one of the categories of the classification as they cover more than one "step" of platform-based design. This can be seen as a fault of the classification method since a classification may be considered effective if it partitions cleanly the various objects being classified. This taxonomy is based on the belief that partitioning the design steps *rather* than the tool coverage is more powerful as it identifies the role of the tools in the overall design paradigm. Indeed, the classification criteria can be seen as hints on how to connect different tools to yield an encompassing design flow. Note that, not surprisingly given the background of the authors, the METROPOLIS and METRO II environments reflect completely the design paradigm followed here and it can be used as the unifying framework for system design where various tools, libraries, and approaches could be embedded if the appropriate interfaces are built.

Bin Descriptions

The classification bins for this taxonomy reflect the Y-shaped diagram with an additional classification criterion related to the level of abstraction at which the tools work. In particular (see Figure 2.4):

- The F bin consists of functional representations of a design independent of implementation architectures and with no associated physical quantity (e.g., time or power). For example, a Simulink diagram expressing an algorithm for automotive engine control or a Ptolemy II description of an MPEG decoding algorithm belongs to the F bin[1]. Tools that manipulate, simulate, and (formally or not) analyze functional descriptions are assigned to this bin.

- The P bin represents the library of modules used to implement the functional description. The modules are architectural elements such as processors, memories, co-processors, FPGAs, custom hardware blocks, and interconnections such as busses and networks. The elements include also middleware such as operating systems for processors and arbitration protocols for busses since these software components "present" the architectural services offered by the hardware to the application software. Tools that are used to connect or otherwise manipulate the modules and analyze the property of the complete or partial platform instances so obtained will be classified in the P bin.

- The M bin represents mapped instances of the design where the functionality has been assigned to a set of correctly interconnected modules. The mapping process is represented by the connection among F, P and M. Any tool that assigns architectural elements to functionality and/or generates

---

[1] These diagrams may be refinements of a more abstract representation, for example, of a meta-model as in METROPOLIS.

the mapped view of the design is assigned to this bin. For example, a high-level synthesis tool is associated to M as the designer has assigned (perhaps manually) part of the functionality to a "virtual" hardware component in the platform and is asking the tool to generate the lower level view, in this case an RTL description of the design. By the same token, a code generation tool can be classified in M as the designer has assigned (perhaps manually) part of the functionality to a software programmable element of the library and is asking the tool to generate the lower level view. In this case, the view is a software program, be it assembly language, C, or higher level languages that are then compiled to move towards implementation. The compilation phase and the synthesis from RTL to gates for this paper is part of a "traditional design flow" and hence it is not part of our classification of ESL tools.

Some of the tools can handle two or even all three aspects of the platform-based design paradigm. To classify these tools, we introduce *meta-classes* (or meta-bins) indicated by combinations of F, P and M. For instance, a synthesis tool that handles functional components as well as their mapping to platform components is classified as meta-bin FM. Tools classified in meta-classes cover several parts of the PBD design flow. We believe that the designers using these tools can benefit from the view of design we propose by clearly decoupling function from architecture and mapping. In doing so, they can enhance reusability and correctness of their efforts.

To make the partitioning of the tools finer, we introduce another, orthogonal criterion for classification: the level of abstraction the tools operate at. While PBD does not limit the levels of abstraction the designers use *perse*, we find that most of the tools reviewed here work at three levels:

- The highest, or *system*, level S corresponds to heterogeneous designs where different models of computation are used to represent function, platforms, and mappings.

- The middle, or *component*, level C deals with subsystems composed of homogeneous components.

- The lowest, or *implementation*, level I deals with the final step of the design where the design team considers the job completed.

We present our classification starting from tools that fall in individual bins, which are meant to be part of a larger tool flow or work in a very specific application domain. Then, we continue with tools that cover larger portions of the design flow space.

Bin F

Tools in this bin are often intended to capture designs and their specifications quickly without making any assumptions about the underlying implementation details. At this level, behavioral issues such as concurrency and communication concepts (for example, communication protocols) may be included in the descriptions. Some tools are limited to handle one model of computation (for example, Finite State Machines) while other are more general and can handle a set of MoCs or have no restrictions. For example, the Simulink representation language handles discrete data-flow and continuous time. Hence it is a limited *heterogeneous* modeling and analysis tool. Ptolemy II with its actor oriented abstract semantics can instead handle all MoCs. Depending on the supported model of computation, design entry for each tool may start at a higher or lower level of abstraction.

Table 2.1: Meta-Bin F classification

Provider	Tool Name(s)	Focus	Abstraction	Website
Mathworks	Matlab	High-level technical computing language and interactive environment for algorithm development, data visualization, analysis, and numeric computation	S- Matlab language, vector and matrix operations	www.mathworks.com/products/matlab
Scilab	Scicos	Graphically model, compile, and simulate dynamical systems	S- Hybrid systems	www.scilab.org
Novas Software	Verdi	Debug for SystemVerilog	I - Discrete event	www.novas.com
Mentor Graphics	System Vision	Mixed-signal and high level simulation	S - VHDL-AMS, SPICE, C	www.mentor.com/products/sm/systemvision
EDAptive Computing	EDAStar	Military and aerospace system level design	S - Performance models	www.edaptive.com
Time Rover	DBRover, TemporalRover, StateRover	Temporal rules checking, pattern recognition, knowledge reasoning	C - Statechart assertions	www.time-rover.com

Continued on Next Page...

Table 2.1 – Continued

Provider	Tool Name(s)	Focus	Abstraction	Website
Maplesoft	Maple	Mathematical problem development and solver	S - Mathematical equations	www.maplesoft.com
Wolfram Research	Mathematica	Graphical mathematical development and problem solving with support for Java, C or .Net	S - Mathematical equations	www.wolfram.com
Mesquite Software	CSIM19	A process-oriented, general purpose simulation toolkit for C/C++	S - C/C++	www.mesquite.com
Agilent	Agilent Ptolemy	Functional verification	C - Timed synchronous data flow	www.agilent.com
National Instruments	LabView	Test, measurement, and control application development	S - LabView programming language	www.ni.com/labview
Academic				
UC Berkeley	Ptolemy II	Modeling, simulation, and design of concurrent, real-time, embedded systems	S - All models of computation	ptolemy.eecs.berkeley.edu
Royal Institute of Technology, Sweden	ForSyDe	System design starts with a synchronous computational model that captures the functionality of the system	C - Synchronous model of computation	www.imit.kth.se
Mozart Board	Mozart	Advanced development platform for intelligent, distributed application	S - Object oriented GUI using Oz	www.mozart-oz.org
Languages				
Celoxica	HandelC	Designed for compiling programs into hardware images of FPGAs or ASICs	C - Communicating sequential processes	N/A
UC Irvine	SpecC	ANSI-C with explicit support for behavioral and structural hierarchy, concurrency, state transitions, timing, and exception handling	C - C language based	www.ics.uci.edu/~specc

Continued on Next Page...

Table 2.1 – Continued

Provider	Tool Name(s)	Focus	Abstraction	Website
INRIA	Esterel	Synchronous reactive programming language	C - Synchronous reactive	`www-sop.inria.fr/meije/esterel/esterel-eng.html`
University of Kansas	Rosetta	Compose heterogeneous specifications in a single declarative semantic environment	S - All models of computation	`www.sldl.org`
Mozart Board	Oz	Advanced, concurrent, networked, soft real-time, and reactive applications	C - Dataflow synchronization	`www.mozart-oz.org`
Various	ROOM	Real time Object Oriented Modeling	S - Object oriented	N/A

Bin P

In this category fall the providers of platforms and/or of platform components, as well as tools and languages that can be used to describe, manipulate, and analyze unmapped platforms. Similarly to the F bin, tools that fall in the P bin may span several layers of abstraction, and support different kinds of architectural components. For instance, Xilinx and Altera are mainly concerned with programmable hardware devices, while Tensilica focuses on configurable processors. Others, like Sonics, Arteris, and Beach Solutions, focus instead on the integration and communication components. This category is characterized by configurability, which ensures the applicability of a platform or of components to a wide variety of applications and design styles.

Table 2.2: Meta-Bin P classification

Provider	Tool Name(s)	Focus	Abstraction	Website

Continued on Next Page. . .

Table 2.2 – Continued

Provider	Tool Name(s)	Focus	Abstraction	Website
Prosilog	Nepsys	Standards-based IP libraries and support tools (System C)	C - RTL and transaction level SystemC; VHDL for SoCs	N/A
Beach Solutions	EASI-Studio	Solutions to package and deploy IP in a repeatable and reliable manner	C - Interconnection	www.beachsolutions.com
Altera	Quartus II	FPGA's, CPLD's, Structured ASIC's	C - IP blocks, C, or RTL; FPGA	www.altera.com
Xilinx	Platform Studio	IP integration framework	C - IP blocks, FPGA	www.xilinx.com
Mentor Graphics	Nucleus	Family of real-time OS's and development tools	S - Software	www.mentor.com/products/embedded_software/nucleus_rtos
Sonics	Sonics Studio	On-chip interconnection infrastructure	I - Bus functional models	www.sonicsinc.com
Xilinx	ISE, EDK, XtremeDSP	FPGA's, CPLD's, Structured ASIC's	I - IP blocks, C, or RTL; FPGA	www.xilinx.com
Design & Reuse	Hosted Extranet Services	IP delivery systems	S - All types of IP	www.design-reuse.com
Stretch Inc.	Software Configurable Processor compiler	Compile a subset of C into hardware to use as instruction extensions	C - Software configurable processors	www.stretchinc.com
ProDesign	CHIPit	A transaction-based verification platform	C - FPGA based rapid prototyping	www.prodesign-usa.com
Languages				

Continued on Next Page...

Table 2.2 – Continued

Provider	Tool Name(s)	Focus	Abstraction	Website
SPIRIT Consortium	SPIRIT	An IP exchange and integration standard written in XML	S - Various IP levels	www.spiritconsortium.com

Bin M

This bin is populated by tools dedicated to the refinement of a functional description into a mapped platform instance, including its performance evaluation and possibly the synthesis steps required to proceed to a more detailed level of abstraction. The tools in this category vary widely with respect to the particular design style, model of computation, and application area that is supported, and are typically very specific to provide the necessary quality of results.

Table 2.3: Meta-Bin M classification

Provider	Tool Name(s)	Focus	Abstraction	Website
Mathworks	Real Time Workshop	Code generation and embedded software design	S - Simulink level models	www.mathworks.com
dSpace	TargetLink	Optimized code generation and software development	S - Simulink models	www.dspace.com
ETAS	ASCET	Modeling, algorithm design, code generation and software development with particular emphasis on the automotive market	S - Ascet models	www.etas.com/en/products/ascet_software_products.php
Y Explorations	eXCite	Takes virtually unrestricted ISO/ANSI C with channel input/output behavior and generates Verilog or VHDL RTL output to be passed to logic synthesis	S - C Language input	www.yxi.com
AccelChip	Accel Chip and Accel Ware	DSP synthesis; Matlab to RTL	C - Matlab	www.accelchip.com

Continued on Next Page...

Table 2.3 – Continued

Provider	Tool Name(s)	Focus	Abstraction	Website
Forte Design Systems	Cynthesizer	Behavioral synthesis	C - SystemC to RTL	www.forteds.com
Future Design Automation	System Center Co-development Suite	ASCI-C to RTL synthesis toolset	C - C to RTL	N/A
Agility (was Catalytic)	MCS, RMS	Synthesis of DSP algorithms on processors or ASICs	I - Matlab algorithms	www.agilityds.com
ACE Associate Compiler Experts	CoSy	Automatic generation of compilers for DSPs	I - DSP-C/Embedded C language extensions	www.ace.nl
ARC (was Tenison)	VTOC	RTL to C++/SystemC	I - RTL, Transactional	www.arc.com/software/simulation/vtoc.html
Sequence Design	ESL Power Technology, Power Theater, CoolTime, Coolpower	Power analysis and optimization	I - SystemC level	www.sequencedesign.com
PowerEscape (with CoWare)	Powerescape Architect, Powerescape Synergy, Powerescape Insight	Memory hierarchy design, code performance analysis, complete profiling	C - C code	N/A
CriticalBlue	Cascade	Design flow for application specific HW acceleration co-processors for ARM processors	I - C code to Verilog/VHDL	www.criticalblue.com
Synfora	PICO Express	C to RTL or C to SystemC (TLM)	I - Pipeline processor arrays	www.synfora.com
Actis	AccurateC	Static code analysis for SystemC	C - C Syntax and semantic checking	www.actisdesign.com

Continued on Next Page...

Table 2.3 – Continued

Provider	Tool Name(s)	Focus	Abstraction	Website
Impulse Acceleration Technologies	CoDeveloper	C-to-FPGA	C - C code	www.impulsec.com
Poseidon Design Systems	Triton Tuner and Triton Builder	Design flow for Application Specific HW acceleration co-processors	C and SystemC	www.poseidon-systems.com
SynaptiCAD	SynaptiCAD line of products	Testbench generators and simulators	C - RTL and SystemC	www.syncad.com
Avery Design Systems	TestWizard	Verilog HDL, VHDL, and C-based testbench automation	I - RTL and C	www.avery-design.com
Emulation and Verification Engine	ZeBu	Functional verification	I - HW emulation	www.eve-team.com
Academic				
University of Illinois Urbana Champaign	IMPACT Complier	Compilation development for instruction level parallelism	S - C code for high performance processors	www.crhc.uiuc.edu/Impact/

Meta-Bin FP

This category consists of languages that can be used to express both functionality and architecture. Typically, these languages are capable of expressing algorithms, as well as different styles of communication and structure, for different models of computation. Assertions, or constraints, come as a complement to the platform description. In the case of UML, the semantics are often left unspecified.

Table 2.4: Meta-Bin FP classification

Provider	Tool Name(s)	Focus	Abstraction	Website
Mathworks	Simulink, State Flow	Modeling, algorithm design, and software development	S - Timed dataflow, FSM	www.mathworks.com

Continued on Next Page...

Table 2.4 – Continued

Provider	Tool Name(s)	Focus	Abstraction	Website
Languages				
Open SystemC Initiative	SystemC	Provides hardware-oriented constructs within the context of C++	S - Transaction-level to RTL	www.systemc.org
Object Management Group	UML	Specify, visualize, and document models of software systems	S- Object oriented, diagrams	www.uml.org
Accellera	SystemVerilog	Hardware description and verification language extension of Verilog	S - Transaction level, RTL, assertions	www.systemverilog.org

Meta-Bin FM

This meta-bin reflects tools that provide some combination of functional description and analysis as well as mapping and synthesis capabilities. In this case, the platform architecture is typically fixed. This lack of flexibility is offset by the often superior quality of the implementation results that can be obtained.

Table 2.5: Meta-Bin FM classification

Provider	Tool Name(s)	Focus	Abstraction	Website
Agility (was Celoxica)	DK Design Suite	Algorithmic design entry, behavior design, simulation, and synthesis	C - HandelC based	www.agilityds.com
Bluespec	BlueSpec Compiler, BlueSpec Simulator	BlueSpec SystemVerilog rules and libraries	S - SystemVerilog and Term Rewriting Synthesis	www.bluespec.com
Telelogic (was I-Logix)	Rhapsody and Statemate	Real-time UML embedded applications	S - UML based	modeling.telelogic.com

Continued on Next Page...

Table 2.5 – Continued

Provider	Tool Name(s)	Focus	Abstraction	Website
Mentor Graphics	Catapult C	C++ to RTL synthesis	C - Untimed C++	www. mentor.com
Esterel Technologies	SCADE Esterel Studio	Code generation for safety critical applications like avionics and automotive	I - Synchronous	www. esterel-technologies. com
Calypto	SLEC System	Functional verification between System-level and RTL-level	C - SystemC/RTL	www. calypto. com

Meta-Bin PM

This meta-bin reflects tools that provide some combination of architectural services and mapping. These tools have a tight coupling between the services they provide and how functionality can be mapped to these services. They require the use of other tools for some aspect of system design (often in how the design functionality is specified).

Table 2.6: Meta-Bin PM classification

Provider	Tool Name(s)	Focus	Abstraction	Website
ARM	RealView MaxSim	Embedded microprocessors and development tools. System-level development tools	C - C++ Arm processor development	www.arm. com
Tensilica	Xtensa, XPRES	Programmable solutions with specialized Xtensa processor description from native C/C++ code	C - Custom ISA processor, C/C++ code	www. tensilica. com
Summit (acquired by Forte Design Systems)	System Architect, Visual Elite	Efficiently design and analyze the architecture and implementation of multi-core SoCs and large-scale systems	C - SystemC	N/A
VaST Systems Technology	Comet, Meteor	Very high performance processor and architecture models	S - Virtual processor, bus, and peripheral devices	www. vastsystems. com

Continued on Next Page...

Table 2.6 – Continued

Provider	Tool Name(s)	Focus	Abstraction	Website
Virtio	Virtio Virtual Platform	High-performance software model of a complete system	I - Virtual platform models at SystemC level	www.virtio.com
Cadence	Incisive	Integrated tool platform for verification including simulation, formal methods and emulation	S - RTL and SystemC, assertions	www.cadence.com
Mentor Graphics	Platform Express	XML-based integration environment for existing IPs	C - XML-based structure	www.mentor.com
Spiratech (acquired by Mentor)	Cohesive	Protocol Abstraction Transformers	C - Transaction Level, IP Blocks	N/A
ARC International	ARC	Embedded microprocessors and development tools	I - ISA extensions, microarchitecure Level	www.arc.com
Arithmatica	CellMath Tool Suite	Proprietary improvements to implementing silicon computational units	I - Microarchitecure datapath computation elements and design	www.arithmatica.com
Target Compiler Technologies	Chess (compiler) / Checkers (ISS)	Retargetable tool-suite, intended for the development, programming and verification of embedded IP cores	I - Mapping of C code to processors written in nML	www.retarget.com
Arteris	Danube, NoCexplorer	Synthesis of NoC	C - NoC dataflow	www.arteris.net
ChipVision Design Systems	Orinoco	Pre-RTL power prediction for behavioral synthesis	C - System C algorithm input	www.chipvision.com
Wind River Systems	Various Platform Solutions	Provides various platforms targeted at different design segments (auto, consumer)	I - Software API	www.windriver.com

Continued on Next Page...

Table 2.6 – Continued

Provider	Tool Name(s)	Focus	Abstraction	Website
CoWare	ConvergenSC	Capture, design, and verification for SystemC	S - SystemC functionality input; SystemC, HDL services	www.coware.com
Carbon Design Systems	VSP	Presilicon validation flow	C - Verilog/VHDL, bus protocols	www.carbondesignsystems.com
GigaScale IC	InCyte	Chip estimation and architecture analysis	S - High level chip info (i.e. gate count, I/O, IP blocks)	www.gigaic.com
Virtutech	Virtutech Simics	Building, modifying, and programming new virtual systems	I - C Language and ISA	www.virtutech.com
National Instruments	LabView 8 FPGA	Create custom I/O and control hardware for FPGAs	C - LabView graphical programming	www.ni.com/fpga
CoWare	LisaTek	Embedded processor design tool suite	C - Lisa architecture description language	www.coware.com
Academic				
Virginia Tech and University of Wisconsin	SOAR (was MESH)	Enable heterogeneous micro design through new simulation, modeling and design strategies	C - C input; Programmable heterogeneous multiprocessors	www.ece.wisc.edu/soar
UCLA	xPilot	Automatically synthesize high-level behavioral descriptions to silicon platforms	C - C, SystemC	cadlab.cs.ucla.edu/soc

Meta-Bin FPM

Entries in this category are the frameworks that fully support the PBD paradigm. In particular, METROPOLIS and METRO II embody fully the paradigm and hence they covers all bins and all layers of abstraction. In this category we include design space exploration tools and languages that are capable of separately describing the functionality and a set of possible architectures for the implementation. These tools are also capable of supporting the mapping of functionality onto the platform instances to obtain metrics regarding the performance of the implementation.

Table 2.7: Meta-Bin FPM classification

Provider	Tool Name(s)	Focus	Abstraction	Website
CoFluent Design	CoFluent Studio	Design space exploration via Y chart modeling of functional and architecture models	S - Transaction level SystemC	www.cofluentdesign.com
MLDesign Technologies	MLDesigner	Integrated platform for modeling and analyzing the architecture, function, and performance of high level system designs	S - Discrete event, dynamic data flow, and synchronous data flow	www.mldesigner.com
Mirabilis Design	VisualSim product family	Multi-domain simulation kernel and extensive modeling library	S - Discrete-event, synchronous data flow, continuous time and finite state machine	www.mirabilisdesign.com
Synopsys	System Studio	Algorithm and Architecture capture, performance evaluation	S - SystemC	www.synopsys.com
Academic				
UC Berkeley	Metropolis	Operational and denotational functionality and architecture capture, mapping, refinement and verification	S - All models of computation	www.gigascale.org/metropolis

Continued on Next Page...

Table 2.7 – Continued

Provider	Tool Name(s)	Focus	Abstraction	Website
UC Berkeley	Metro II	Three phase execution semantics, event based mapping, and support for heterogeneous IP import	S - All models of computation	www.gigascale.org/metropolis
Seoul National University	PeaCE	Codesign environment for rapid development of heterogeneous digital systems	S - Objected-oriented C++ kernel (Ptolemy Based)	peace.snu.ac.kr
Vanderbilt University	GME, GREAT, DESERT	Meta-programmable tool for navigation and pruning of large design spaces	S - Graph transformation, UML/XML based, and external component support	repo.isis.vanderbilt.edu
Delft University of Technology	Artemis, Compaan/Laura, Sesame, Spade	Workbench enabling methods and tools to model applications and SoC-based architectures	C - Kahn process networks	ce.et.tudelft.nl/artemis
UC Berkeley	MESCAL	Programming of application-specific programmable platforms	S - Extended Ptolemy II, network processors	www.gigascale.org/mescal

Concluding Thoughts

This first taxonomy presented over 90 different tools. While the lifetime of the tools on this list will vary, it is interesting to note the current landscape. While naturally we have neglected some offerings (apologies in advance) currently there are 72 industrial offerings (77%), 12 academic offerings (13%), and 10 languages (10%). This distribution is expected since industry has a much larger financial and employee base than academia. The largest bin is PM with 22 offerings (23%) followed by M with 21 (22%), F with 20 (21%), P with 11 (12%), FPM with 10 (11%), FM with 6 (7%), and finally FP with 4 (4%). It is not surprising that PM is the largest since it is most similar in ideology to the RTL flows which ESL is trying to leverage at the lower levels of abstraction. FP tools are uncommon since tools with functional and architectural capabilities often have a mapping strategy as well. FPM (complete solutions) are not as well represented as other solutions since the EDA industry traditionally has relied of multiple tool providers to solve various design problems. In fact 6 of the 10 FPM offerings are academic.

2.5.2   Taxonomy Two: Ideology vs. Programming Model

The second taxonomy presented attempts to expose the fact that designers often begin with very specific desires regarding what aspect of a design they want to specify. This can be viewed as their ideology and then they go about implementing that ideology This can be viewed as their methodology.

Figure 2.5 is the first of several perspectives on how to categorize ESL tools outside of the traditional y-chart approach. Along the "x-axis" there are three distinctions along *ideological* lines. The first are those tools which provide *behavioral* design environments. These are designs which capture what a system should do but do not provide notions of cost or implementation constraints. Often these are more mathematical based systems and focus on ensuring system description mechanisms rather than paths to implementations. To relate this concept to the first taxonomy these are tools in the F bin.

The second category are *architectural* design environments. These focus more on the specifics of how designs are implemented and the costs involved in implementation. Often these provide libraries of devices with already defined paths to implementation. The creation of designs in these environments are inherently constrained by the notion of a physical implementation. To relate this concept to the first taxonomy these are tools in the P bin.

The third category are *platform-based* design environments. These are a combination of the previous two design environments. To be a true member of this category, the design must allow that the functional aspect of the design be described separately from the architectural aspect. These are then joined together by a separate mapping process. To relate this concept to the first taxonomy these are tools in the FPM bin.

These three areas could be viewed as aspects of the y-chart approach. The first category are functional, the second category are architectural, and the third category encompasses all three. One can imagine that the y-chart has been extended with an additional axis.

This y-axis is divided by the programming model. This is the methodology portion of the design. In this case it is classified as *algorithmic*, *processor/memory*, and *control logic* methodologies. Algorithmic approaches are those in which designs are specified procedurally. Often these are sequential descriptions but concurrent descriptions are allowed as well. There is not a notion of objects or individual components but rather just a system of equations and relations. Processor/memory approaches are those which examine processing engines which execute sequential code. These individuals PEs then communicate through memory. Control logic simply describes how the design should operate but abstracts away data and computation into very basic operations. Control logic focuses on the coordination of the system. These systems are often used for describing industrial or automotive systems.

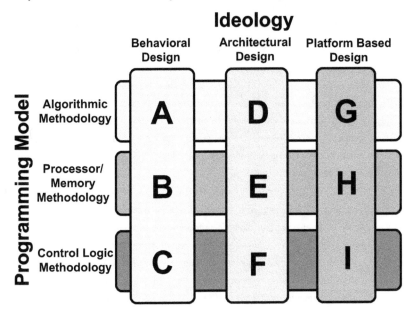

Figure 2.5: Ideology vs. Programming Model

In Figure 2.5 the uppercase letters A-I are used to designate the bin a particular tool belongs to. Table 2.8 provides an example of tools that belong to each of the nine bins of this taxonomy.

2.5.3 Taxonomy Three: Entry Language vs. Development Focus

The third taxonomy exposes the fact that the actual language used to specify the design may span a number of development focuses. That is to say that one language may be useful for a variety of design styles and at variety of design abstraction levels.

Figure 2.6 both shows how designs can be specified by a design entry language and how they

	Provider	Tool		Provider	Tool
A	Mesquite Software	CSIM19	G	Stretch Inc.	Software Configurable Processor compiler
B	Mathworks	Real Time Workshop	H	Emulation and Verification Engine	ZeBu
C	Telelogic	Rhapsody and Statemate	I	VaST Systems Technology	Comet, Meteor
D	AccelChip	Accel Chip and Accel Ware			
E	Arteris	Danube, NoCexplorer			
F	Bluespec	BlueSpec Compiler			

Table 2.8: Example Binning for Taxonomy Two

are composed semantically. Along the x-axis is C and higher, RTL, and other. The *C and higher* designation includes not only C, C++, C#, and SystemC but also languages such as Java. These languages are typically object oriented and allow sequential code to be encapsulated by individual threads for the illusion of concurrency. *RTL* or register transfer languages are those which are typically concerned with lower level details and a path to implementation via synthesis. Some higher level languages such a SystemVerilog are included here along with the standard RTL languages Verilog and VHDL. Finally the *other* designation allows for environments such as Matlab, Simulink, Modelica, etc.

Along the y-axis are component, embedded, SOC, HW/SW Co-design designations. These indicate the scope of the environment. Component focuses on developing individual pieces of a design intended to work with other components in either their native or other design environments. Embedded indicates that designs are collection of components with design metrics which reflect embedded design concerns (size, power). SOC reflects heterogeneous designs which a variety of components, timing domains, and models of computation. Finally HW/SW codesign reflects methodologies which partition a design into both hardware and software pieces and looks at how to relate the two aspects.

The designations along the y-axis are not mutually exclusive and are provided to indicate which community the tool are targeted toward.

In Figure 2.6 the Roman numerals I-XII indicate a particular bin a tool can belong to. Table 2.9 provides an example of tools that belong to each of the 12 bins of this taxonomy.

2.5.4 Taxonomy Four: Community vs. Design Flow

The fourth taxonomy simply serves to illustrate the focus of the academic community vs. the industrial community across specific design flows.

Figure 2.7 shows academic and industrial tools on the "x-axis" along with design and simulation, test and verification, and behavioral synthesis along the "y-axis". Academic tools are simply those

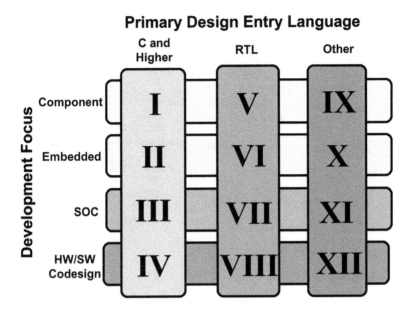

Figure 2.6: Entry Language vs. Development Focus

which are created in an academic environment while industrial tools are commercial tools created by an independent company (both free and otherwise). This distinction often indicates the goal of the project (commercial and proprietary vs. research based and open source). The y-axis categories indicates the intended aspect of the design flow in which the tool should be used. Design and simulation approaches are usually used for early design space exploration when the design is still very undefined and a designer is looking to make key decisions regarding the design. Test and verification look to prove properties of a design or to verify the correctness of a design. These can be formal, semi-formal, or ad-hoc in nature. Behavioral synthesis tools look to transform purely behavior specifications into hardware.

The individual bins for this taxonomy in Figure 2.7 are labeled with the acronym for the y-axis category and a "!" for academic and a "$" for industrial. Table 2.10 provides an example of tools that belong to each of the 6 bins of this taxonomy.

	Provider	Tool		Provider	Tool
I	PowerEscape (with CoWare)	Powerescape Architect	IX	AWR	Visual System Simulator
II	Target Compiler Technologies	Chess	X	Wind River Systems	Various Platform Solutions
III	Synfora	PICO Express	XI	Esterel Technologies	SCADE Esterel Studio
IV	Virtutech	Virtutech Simics	XII	Prosilog	Magillem
V	Arithmatica	CellMath Tool Suite			
VI	ARC	VTOC			
VII	Novas Software	Verdi			
VIII	Aptix	Zaiq			

Table 2.9: Example Binning for Taxonomy Three

	Provider	Tool		Provider	Tool
!DS	Royal Institute of Technology, Sweden	ForSyDe	$DS	Tensilica	Xtensa, XPRES
!TV	UCB, U of Penn, SUNY Stony Brook	Mocha	$TV	Avery Design Systems	TestWizard
!B	UCLA	xPilot	$B	MLDesign Technologies	MLDesigner

Table 2.10: Example Binning for Taxonomy Four

2.5.5 Taxonomy Five: DSE Approach vs. Framework

The fifth and final taxonomy shows that frameworks at different levels of abstraction may use unique or shared methods to do design space exploration (DSE).

In Figure 2.8 the "y-axis" indicates system level frameworks, microarchitectual frameworks, and related frameworks (those which do not fit in either category). Also shown are application, architecture, and exploration mode approaches to design space exploration along the x-axis. Shown in the intersection of these categories are examples of specification mechanisms for the given classification.

If we examine the vertical *application model* designation the first two levels (system and microarchitectual) can be thought of as changes in abstraction. System level application models are Kahn Process Networks (KPN), Dynamic Data Flow (DDF), Petri Nets, C language (and its variants), Matlab, and Discrete Event systems for example. At a more detailed level often the C language is used or assembly languages. Related frameworks also use C or more abstract generic structures such as Directed Acyclic Graphs (DAGs).

The *architectural model* design space exploration model often includes hardware description languages (HDLs), architecture description languages (ADLs), or highly specific performance models.

Exploration mode refers to how simulation or validation is performed in order to reach a desired goal or optimization. This process can be done using scripting languages and environments (Perl, Tcl/Tk), manual scripts (collections of various tools) and heuristics for optimization.

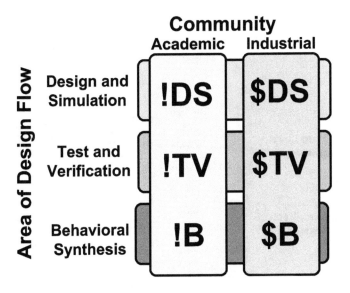

Figure 2.7: Community vs. Design Flow

In each of the classifications presented here, the techniques provided in Figure 2.8 are not exhaustive and only meant to give a couple select examples.

2.6 Design Discussions

In this section there are six discussions related to ESL. Four of these discussions illustrate the ways in which the taxonomies can be useful in helping a designer assemble a complete design flow for the development of an embedded system. These flows are much more on the qualitative side but should give the reader some insights into the usefulness of such an approach. The other two discussions illustrate interesting binning approaches not covered in the previous taxonomies.

2.6.1 Design Discussion One: Three Design Scenarios

In this section, we use the PBD framework of Figure 2.4 to map three different design flow scenarios on the ESL tool landscape. We accomplish this by indicating the meta-bins from the first

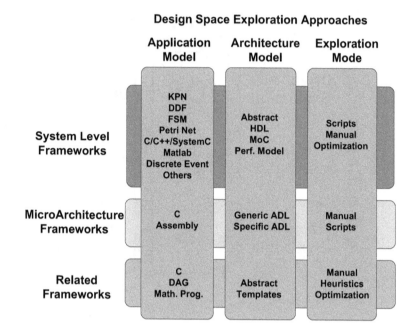

Figure 2.8: DSE Approach vs. Framework

taxonomy and the levels of the hierarchy where activities take place, as shown in Figure 2.9.

Scenario 1 - New Application Design from Specification

The important characteristics of this scenario are:

- The need to start from a high-level specification;

- The desire to capture and modify the initial specification quickly;

- The ability to express concurrency, constraints, and other behavior-specific characteristics efficiently;

- The capture of abstract services that can be of use in "implementing" the high-level specifications into a more detailed functional view.

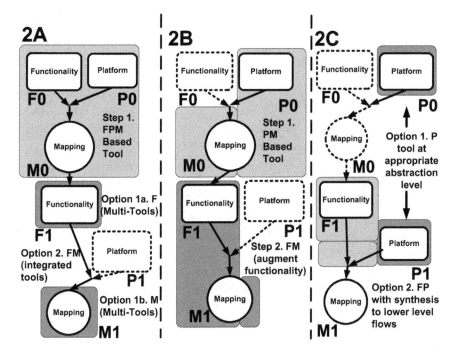

Figure 2.9: Design Scenarios for Discussion 1

The flow thus starts at the higher levels of abstraction in bin F of our first taxonomy that can be expanded into a "Y" diagram of the same structure as the one described in Figure 2.4, that offers:

- Flexible specification capture;
    - No ties to particular implementation style or platform.
- Provide services that aid in taking the abstract design toward a more constrained version (e.g., algorithms that can implement the functionality);
- Independent mapping of the functionality onto algorithmic structures which allow for the reuse of the functional specification.

    To exemplify this scenario, let us examine an example in the multimedia domain: the implementation of a JPEG encoder on a heterogeneous multi-processor architecture such as the Intel MXP5800 architecture that consists of eight Image Signal Processors (ISP_1 to ISP_8) connected with programmable Quad Ports (8 per ISP)[2].

The encoder compresses raw image data and emits a compressed bitstream. The first step in the scenario is to choose a particular model of computation to describe the functionality of the design. To be more efficient in applying the design paradigm proposed here, the designer should use a model of computation that could also be used to describe the capabilities of the architecture. In so doing, the designer eases the mapping task and the analysis of the properties of the mapped design. In addition, an automatic synthesis step could be developed to carry out the mapping process automatically.

Since the application is data-streaming in nature and will be mapped onto a highly concurrent architecture, a Kahn Process Networks (KPN) representation is natural. In KPN, a set of processes communicate via one-way FIFO channels. Reads from channels block when no tokens are present – processes cannot query the channel status. However, this model is Turing-complete, so the scheduling and buffer sizing problems are still undecidable. The KPN model of the JPEG encoder algorithm is completely independent of the target architecture satisfying the requirements set above. Ptolemy II could be used to capture this model and to simulate the behavior of the selected algorithm. To allow a better analysis and to refine the model towards implementation, this model can be mapped into another dataflow model, similar to cyclo-static dataflow [BELP95], where only one writer is permitted per channel, but multiple reader processes are allowed. For all channels, each reader process can read each data token exactly once. Also, limited forms of data-dependent communication are allowed. To enable support for executing multiple processes on a single processing element, this MoC has support for multitasking. In particular, a process may only be suspended between firings. Scheduling, buffer sizing, and mapping are decidable problems for this MoC. This model can be easily expressed in Ptolemy II and can also be described in Simulink and SPW. This first step of mapping a more flexible model for the functionality into a more restricted one that is easier to implement and to analyze, is a crucial step in any system level design.

Subsequently, the mapped specification becomes the functional representation for the diagram of Figure 2.4 so that the flow may continue at lower levels of abstraction with tools in meta-bin FM for an integrated solution, or bin F followed by M for a multi-tool solution. Here, since the architecture is mostly fixed, a specialized and efficient approach is more appropriate. Figure 2.9a shows a potential traversal of the framework. For our JPEG case, we can use the METROPOLIS environment to map the functionality onto the MXP5800 to analyze potential problem with the architecture or to optimize the

---

[2]This test case was presented in [DZMSV05].

coding of the application for the chosen platform instance.

Scenario 2 - New Integration Platform Development

This scenario describes the development of a new integration platform, i.e., a "hardware architecture, embedded software architecture, design methodologies (authoring and integration), design guidelines and modeling standards, virtual components characterization and support, and design verification (hardware/software, hardware prototype), focusing on a particular target application" [CCH+99]. As opposed to the first scenario, this is not driven by the design of a particular application but rather by the development of a "substrate" for the realization of several applications. Characteristic of this scenario is the service/mapping-centric requirements that point to tools in meta-bin PM for the development and analysis at the desired level of abstraction. The *user* of the platform would instead proceed in meta-bin FM to map the the desired functionality to the platform instance of choice. Figure 2.9b illustrates the meta-bin flows that support these development requirements.

Consider as test case the development of a new ECU platform for an automotive engine controller. The application code has already been developed for a given platform but a Tier 1 supplier wishes to improve the cost and performance of its product not to lose an important OEM customer. If the PBD paradigm used in this work is employed, the application itself has been made as independent on the ECU platform as possible. Then, in collaboration with a Tier 2 supplier (a chip maker) the Tier 1 supplier determines qualitatively that a dual core architecture should offer a better performance at a lower manufacturing cost. The dual core architecture is captured in a tool for platform development such as LisaTek. If the dual core is based on ARM processing elements, the ARM models and tool chains can also be used. An appropriate novel RTOS may be developed to exploit the multi-core nature of the implementation. At this point, the designers map the application onto one of the possible dual core architectures exploring the number of bits supported by the CPU as well as the set of peripherals to integrate and the interconnect structure. For each choice, the mapped design is simulated with the engine control software or a subset of it that has been selected for its capability of stressing the architecture. These simulations can be done with the ARM tools or with VAST offerings to obtain rapidly the important statistics such as interconnect latency and bandwidth, overall system performance, and power consumed. At the end of this exercise the Tier 2 supplier is fairly confident that its architecture is capable of supporting well a full fledged engine control algorithm. This product may be used by any other Tier 1 supplier for its engine control offering.

Scenario 3 - Legacy Design Integration

The final scenario represents a common situation for many companies that wish to take their existing designs and integrate them into new ESL flows. In this case, it is difficult to separate functionality and architecture as in most embedded systems, the documentation refers to the final implementation and not to its original specifications and the relative implementation choices. If the design has to be modified to implement additional features, it is very difficult to determine the effect of the presence of the new functionality on the existing design. We need then to establish a "reverse" engineering process where functionality is extracted from the final implementation. The most effective way to do so may be to start from scratch with the description of the functionality using tools in the F bin. An alternative may be an effective encapsulation of the legacy part of the design so that the new part interacts cleanly with the legacy part. Existing components can then be considered as architectural elements that must be described using tools included in the P bin. This, in turn, can be done at different levels of abstraction. Because legacy components typically support a specific application, mapping is often not needed, while functional/architectural co-simulation is used to validate a new design. Meta-bin FP at the system level is therefore the appropriate flow model. Figure 2.9c illustrates this scenario.

2.6.2 Design Discussion Two: Alternate Binning Approach

In Figure 2.10 an additional binning scheme is shown. This builds upon the y-chart classification of the first taxonomy. However, in this case there are explicitly three levels of abstraction. The first level (A) is the most abstract. Functional tools in this category will allow for tools to express multiple models of computation (MoCs). These tools would fall into bin 1A. Service tools also support multiple MoCs and support packages (validation for example). These tools are in bin 2A. Finally mapping tools in this area (3A) must support approaches from the 1A and 1B bins. At this level of abstraction designs have the largest set of behaviors. They may exhibit non-determinism as well.

In the second level (B), the MoC expressibility is much more restricted. In each case the tools at this level operate using a single model of computation. A benefit at this level of abstraction is the fact that analysis tools may be more rigorous since the designers can make assumptions about the operational semantics the design will be specified with.

At the third level even more flexibility (and therefore behavior) is removed. At this level the designs begin to target specific devices (i.e. specific processor or FPGA). Additionally the mapping step here often transforms the model into an more traditional representation (RTL) for use with well established, automatic flows. Below the third level, the abstraction options presented would be so limited that we would not consider the tools ESL level.

2.6.3   Design Discussion Three: Meta Binning

In Figure 2.11 7 meta-bins are shown. This is yet another potential binning approach but unlike the others presented in this chapter thus far, this allows for a more coarse binning that not only allows for the grouping of various areas of the "y-chart" at a given abstraction level, but also grouping across levels. In practice this is a much more practical approach since it is rare to find tools that fit neatly into more granular bins. The bins are labeled as "complete" if they encompass functional, service, and mapping individual bins. They are labeled as "composite" if they encompass two bins and "individual" if they only use one bin.

The first meta-bin contains tools which can purely be put in the highest level of the abstraction hierarchy (A). These can cover all three areas of the y-chart (complete), collections (composite), or only individual areas (individual). These tools are often concerned more with specification and early design space exploration as opposed to synthesis.

The second meta-bin are those tools which cover all three areas of the second level (B) of abstraction. These are interesting tools since they offer a complete solution. This meta-bin is more specific as compared to the first meta-bin since tools at this time are more likely to be able to offer complete solutions at this level of abstraction as opposed to higher levels. These are typically based on only one model of computation.

The third meta-bin are those tools which only belong to one of the y-chart areas of the second level (B). These are highly specific tools with an emphasis of filling a niche in the ESL market.

The fourth meta-bin are those tools which cover all three areas of the third level (C). These tools are able to cover all three areas since they are typically only extensions of RTL level tools and flows. Often they provide simply a more abstract syntax by which to express the design.

The fifth meta-bin are those tools which combine two C level approaches. Often these are functionality and mapping or architecture services and mapping. The mapping process at this level often can be closely coupled with the specification of other parts of the design flow since the final implementation target is often known.

The sixth meta-bin contain only one C level approach. These are typically verification tools that are not part of the core of the design flow but rather additional analysis or optimization pieces. These are highly related to the final implementation of the design and to be used only once design space exploration has been done to the designer's satisfaction.

The seventh and final meta-bin are those which are in the same type of bin but span multiple levels of abstraction. These tools are very interesting and extremely powerful. By focusing only in one area they are able to identify the key design aspects which must be extracted at each level in order to perform their operations.

Naturally this is not an exhaustive combination of how meta-bins could be formed but it represents the most likely scenarios in which tools are created that the authors have seen. These meta-bins will be used in Sections 2.6.4, 2.6.5, and 2.6.6.

2.6.4   Design Discussion Four: Connection to Pre-Existing ESL Flow

This discussion is a design scenario which desires to connect a tool/design or a set of tools/designs to an existing ESL flow. This often occurs if a company transitions to a new set of tools for increased features and performance or if a tool becomes obsolete. In Figure 2.12 two potential options are presented.

The first option starts with the tools at the highest level of abstraction. These are tools from meta-bin 1 of Section 2.6.3. These will be used to capture multiple models of computation quickly. This can be done to re-specify the original design in the new design environment. This is also an opportunity to make the design more reusable for the future. The outcome of a mapped solution at this level can then be the functional description at the next level. The next level will use a meta-bin 3 approach. This individually binned tool should be capable of taking the design to an existing flow.

Similarly using the secondary option, one can start with the second level with a more restricted MoC and proceed in a similar fashion. This is useful if the design fits very naturally within one MoC. This will then transition to a meta-bin 6 tool (a single binned, functional tool perhaps) with the capabilities to connect to a traditional flow.

2.6.5   Design Discussion Five: Functional Entry Points into ESL Design

In many cases the design entry point will either be a functional specification or a model which assumes very little implementation details. This occurs since in many scenarios the design is simply a high level industry standard specification (i.e. JPEG) or a set of requirements (i.e. real-time processing). In this case it is important that the designer be able to proceed from this functional description all the way to a final design. What must be decided upon is at which level of abstraction should this initial specification be performed.

In Figure 2.13 three entry points are shown. The first entry point is at the highest level of abstraction. Here multiple MoC are accepted. This allows a wide variety of services to be used in the mapping. The new design entry will start with meta-bin 1A. It will then use a subset of tools from 1A to do mapping and service development. A composite tool from 1A may be a good choice since it will be an integrated solution. It is at this point that the hardware platform begins to take shape.

The second potential entry point is at the second level of abstraction. This often occurs with designs targeted toward a specific architecture implementation. Typically now APIs are the service level upon which the design is mapped. What is being developed at this level are a set of APIs for an existing embedded architecture platform.

Finally at the lowest level services are virtual components and mapping creates a specific HW/SW instance. This is done if a specific microarchitecture is already known.

2.6.6   Design Discussion Six: Connection to a Legacy Flow

In Figure 2.14 a connection to a legacy flow is shown. Legacy flows are flows which are well established but perhaps are not being currently developed or supported. There exists scenarios when one of these flows may have to be interfaced with. This discussion proposes a way in which this may be done.

To begin, the level of abstraction A (the highest) is not discussed since frequently this will be too high a level of abstraction to connect with pre-ESL tools. It will be the job of the design team to move to the B level of abstraction using the techniques described in other sections.

Often the design will start with meta-bin 3. These individual bins can be used if a manual flow is required at the next step. The goal is to quickly get to the lowest level of abstraction which is "above" the legacy flow. This may require manual steps to take the results of the tool and transform them to input for the legacy flow.

Once the design is at the "lowest" level of abstraction (C) various meta-bins can be used. The complete set of bins at level C can be used (meta-bin 4) as a complete solution or a subset (meta-bins 5 and 6) as well. The legacy flow may be responsible at some point in the design to provide information about its design aspects (MoC required, costs, constraints, etc) to the tools. This may have to be extracted manually by the designer.

2.7   Conclusion

This chapter has shown that ESL is an important design paradigm for embedded systems. ESL not only has had a tremendous financial impact on EDA but will continue to do so in the foreseeable future. ESL is both financially relevant to EDA as well as technically vital to the growth of electronic systems. Many companies have emerged which explicitly only offer ESL solutions and other, more established companies are creating their own offerings to compete in this space.

ESL itself as a term however is very generic. We have illustrated via a collection of taxonomies how one can classify tools which claim to fall into the ESL space. These taxonomies are helpful for both adding clarity as well as illuminating potential research and commercial opportunities. These taxonomies were accompanied by a collection of discussions which illustrated how ESL can be useful as a design methodology. The next chapter will discuss a particular ESL methodology termed Platform-based Design.

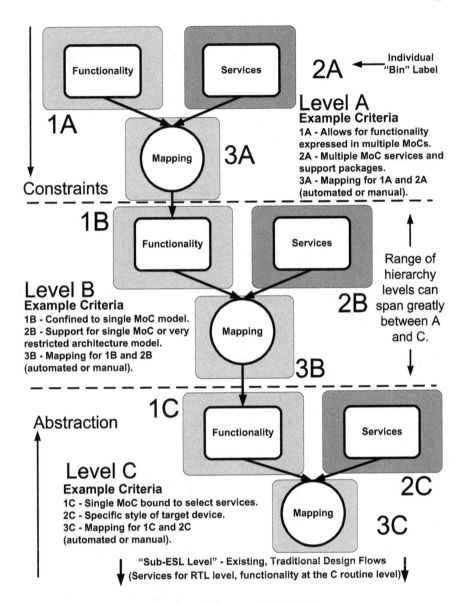

Figure 2.10: Design Discussion 2 Binning Scheme

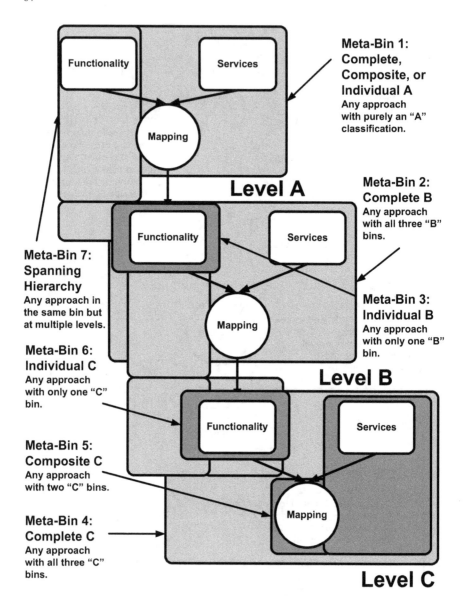

Figure 2.11: Design Discussion 3 Meta-Bins

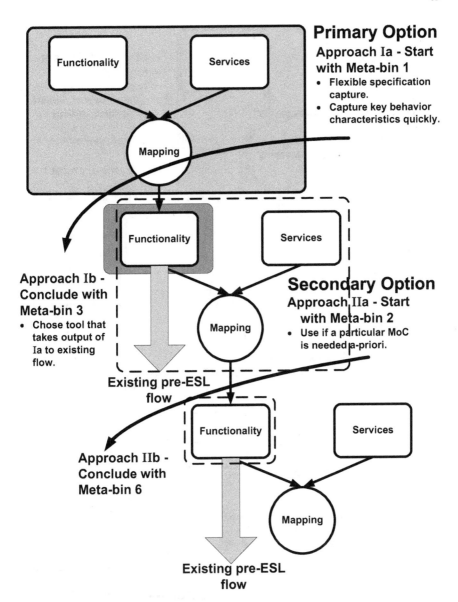

Figure 2.12: Design Discussion 4 Flow

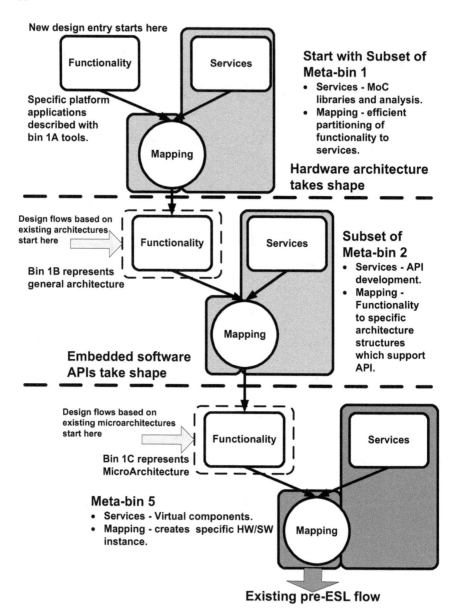

Figure 2.13: Design Discussion 5 Flow

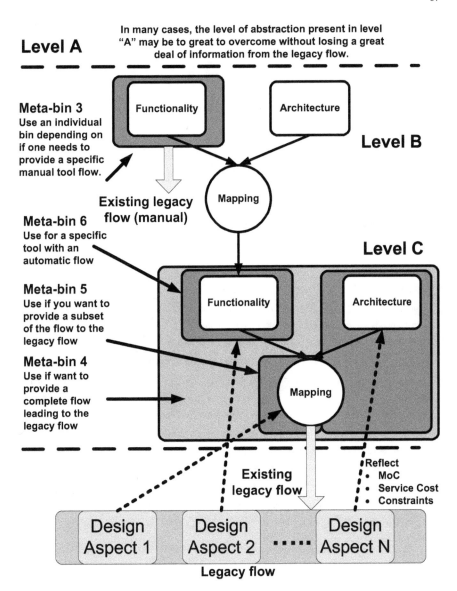

Level A

In many cases, the level of abstraction present in level "A" may be to great to overcome without losing a great deal of information from the legacy flow.

Meta-bin 3
Use an individual bin depending on if one needs to provide a specific manual tool flow.

Functionality

Architecture

Level B

Mapping

Existing legacy flow (manual)

Meta-bin 6
Use for a specific tool with an automatic flow

Level C

Meta-bin 5
Use if you want to provide a subset of the flow to the legacy flow

Functionality

Architecture

Meta-bin 4
Use if want to provide a complete flow leading to the legacy flow

Mapping

Existing legacy flow

Reflect
• MoC
• Service Cost
• Constraints

Design Aspect 1

Design Aspect 2

Design Aspect N

Legacy flow

Figure 2.14: Design Discussion 6 Flow

Chapter 3

Platform-Based Design

"The future lies in designing and selling computers that people don't realize are computers at all" – Adam Osborne, Personal Computing Pioneer

Separating independent aspects of a design problem is a natural means of decomposition. Separation not only allows independent tasks to be done simultaneously, but also allows modularity since the design does not have to be redone in case one of the sub-problems is modified. Separation allows for many people to work on design problems concurrently and also allows tools to analyze aspects of the design while others are being created. Separation as a methodology is present in many computer science disciplines such as object oriented software development and computer aided design for VLSI.

However, when high performance is required, designers often resist separation. Often optimization approaches require information from a variety of sources in a design, thereby encouraging consolidation. Combining aspects often makes local and global optimizations easier and the work of the designer (in the short term) is made more straightforward. This chapter will argue that the benefits of separation far outweigh the benefits of consolidation. We argue that not only does separation give the designer all the benefits of modularity, but it also can lead to higher performance designs by enabling a more rigorous design space exploration process.

The rest of this chapter will discuss key elements of a methodology called Platform-based Design and how they can avoid the traditional challenges associated with design separation and performance loss.

This chapter will describe the key aspects of the platform-based design methodology and describe why this approach is focused on throughout this work.

3.1 Chapter Organization

This chapter begins by laying out the fundamentals of PBD in Section 3.2. Next, the key concept of "separation of concerns" is outline in Section 3.3. Section 3.4 provides a more theoretical exploration of PBD. This is followed by related work in Section 3.5. This primarily details other prominent descriptions of PBD and tools which employ this methodology. Finally Section 3.6 provides conclusions and future directions.

3.2 Overview

In the past, HW/SW co-design approaches have treated hardware and software in a very segregated manner. Software approaches are not timed and often have sequential specification. This makes them very difficult to use to design hardware which by its nature is timed and concurrent. Hardware approaches are often too rigid in their semantics to be useful for software designs. Also the level of abstraction in these approaches is not high enough for a wide variety of design scenarios. Ideally a methodology would:

- Treat hardware and software equally. Both should be first class citizens and techniques should allow each to be specified in a natural way for each domain.

- Favor high levels of abstraction for the initial design specification. This will aid in analysis and reuse.

- Offer effective architecture design exploration. Architecture models should be parameterizable and composable.

- Achieve detailed implementation by synthesis or manual implementation. There should be some path to a final, implemented design.

Chapter 2 discussed Electronic System Level (ESL) design and its increasingly important role in EDA. It is important to understand that ESL is a large design umbrella defined by a generic set of goals with a number of possible approaches enclosed within it. In fact, a very large number of industrial and academic offerings fall within the category of ESL. In [DPSV06] (the taxonomy covered in Chapter 2), over 90 tools and environments were categorized. The approaches differ by their ability to support (F)unctional modeling, (P)latform services, or (M)apping capabilities. Approaches could be combinations of these distinctions. According to the terminology used in that source, we are examining an FPM approach here. FPM approaches are attractive since this investigation can be carried out in a single unified environment. In particular this work will be focusing on a particular FPM style within ESL called *Platform-Based Design* [SV02].

The Platform-based design [KMN+00] [KMN+00] paradigm has been proposed to cope with constantly increasing complexity, safety and security requirements, and time-to-market pressures (the challenges from Chapter 1). Increasingly, embedded system designers are turning to more rigorous design methods. These favor the adoption of higher levels of abstraction in system specification, correct-by-construction deployment, and reusability. PBD targets all of these needs.

In this paradigm, a platform is designed with sufficient flexibility to support the implementation of an entire set of products. The product design problem then involves configuring the platform, and deciding which parts of the product's functionality are to be implemented by which platform resources. Typically, designers evaluate several configurations before selecting one that meets design goals. This process is known as *design space exploration*. It requires building a series of models, one for each combination of configurations to be evaluated. Developing these models is traditionally time consuming and error prone. Therefore, it is natural to re-use them as much as possible. However, it is often hard to do so because modifying a configuration or a part of the description of a model usually requires extensive changes to models in the rest of the design.

Definition 3.2.1 Platform - *A platform is defined to be a library of components that can be assembled to generate a design at a particular level of abstraction.*

This "library" should have both computation elements as well as communication elements. A platform instance is a set of components that is selected from the library (the platform) and whose parameters are set.

PBD allows for a designer to think of traditional software and hardware aspects of the design separately. Algorithms are decoupled from the elements which implement them. For the purposes of this work, "system level" will primarily refer to the level of abstraction employed. Computation will typically take place at the granularity of function calls. Communication operations will be considered as transactions (as opposed to bit-level or register interactions).

Definition 3.2.2 Platform Based Design - *a design methodology which is not a fully top-down nor a fully bottom-up approach in the traditional sense; rather, it is a meet-in-the-middle process as it can be seen as the combination of two efforts. The first effort maps an instance of the functionality of the design into an instance of the platform and propagates constraints. The second effort builds a platform by* *choosing the components of the library that characterizes it and an associated performance abstraction.*

3.3 Separation of Concerns

"One must separate from anything that forces one to repeat No again and again" – Friedrich Nietzsche, German Philosopher

A key tenet of Platform-Based Design (PBD) is what is termed the *orthogonalization or separation of concerns*. A solution to the re-use problem is to orthogonalize concerns and keep various aspects of a design separate. There are several concerns in embedded system design that can be orthogonalized. The three main concerns are the following:

- *Functionality versus Architecture*: Functionality represents the application that the designers want the system to carry out (what something does), while the architecture represents a configuration of resources that can implement this functionality (how it does it). For example, multiplication itself is very well defined functionally. However, the architecture which implements it may be a series of adders or a dedicated multiplier. The functional portion of the design exercises *services*, which can be provided by different architectural models – or platforms – with different costs. A particular mapping of a functional model with an architectural model corresponds to a system model. The architecture determines the performance in terms of the quantities of interest (e.g. energy, time) while the functionality determines which services are used and in what way. By allowing an architectural model to be reconfigurable or instantiated in different ways, we can easily represent a family of parameterizable architectural platforms. Then, the mapping must also choose an appropriate platform instance from the choices available.

  Since the only interaction between the architectural and functional models takes place due to the mapping of services together, once these are agreed upon, separate groups of developers can code, debug, and maintain the functional and architectural models. The separation between functionality and architecture is also captured in the Y-chart approach [GK83] [KDvdWV02].

- *Behavior versus Cost*: Behavior reflects the services offered by the component, while cost represents the expense of providing these services. A bus protocol is an example of a behavior. Performance is a cost of that behavior. Bus transaction latency times (performance) are a function of many things not specified by the behavior (for example clock speed is not a behavior). Cost can be defined in terms of time, power, chip area, or any other quantity of interest. This orthogonalization allows the framework to easily support the usage of "virtual" components and facilitates back-annotation to accurately model cost-metrics. Virtual components are architectural resources that do not reflect existing physical designs (hardware/software). A designer can configure and utilize virtual components in a system, and dictate the final parameters as constraints for implementation once she is assured that the component can be successfully used. Even if an architectural component is available and its behavior known, its performance can be obtained at various levels of accuracy. A separation between behavior and cost allows this component to be used even if accurate numbers are not available. For instance, a synchronous bus component can be used without knowing the exact number of cycles taken for a transfer. An estimate can be used and system evaluation can proceed. Once cycle-accurate numbers become available, they can be substituted without requiring additional changes to the system.

- *Computation vs. Coordination vs. Communication*: The behavior of a design is often specified as a set of concurrent processes, where each object executes a sequential program and communicates with other processes. A process may share resources with other processes, this often requires coordination among multiple processes. Coordination can be described separately from the sequential programs for the individual processes. Computational activities are usually highly design-specific while coordination schemes are usually standardized.

Specification of computation and coordination may be carried out in different ways. It is often convenient to model coordination using declarative constraints, rather than imperative programs. For instance, it is simpler to declare that two actions should be mutually exclusive, rather than write a program for a protocol that realizes the exclusion. Such declarative statements are very useful during the initial stages of the design flow, when conciseness of specification is more important than performance estimation.

3.4 Theory

The platform-based design process often begins with a functional specification. In Figure 3.1, the left hand side of the figure illustrates the functional space. The "dot" indicates the selection of a particular function (i.e. MJPEG decoding). This selection of a desired functionality then manifests itself as a constrained version of the functionality (as shown by the network of connected components). These constrains often impose various communication semantics (blocking read and write for example) or computation restrictions (arithmetic precision). This is seen as a refinement process which makes the functionality semantically compatible with the platform for mapping purposes.

The platform space is shown on the right hand side of Figure 3.1. There are a library of elements (bottom section of the figure) which, when composed, create the individual instances (individual squares in the top portion of the figure). The complete set of all possible platform instances make up the entire platform (enclosed top portion of figure encompassing each individual instance). This can be expressed more formally as the algebraic closure of the platform.

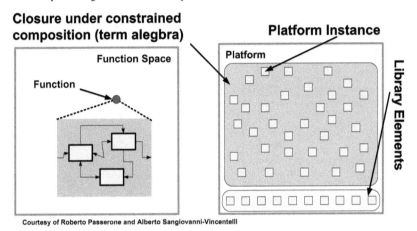

Courtesy of Roberto Passerone and Alberto Sangiovanni-Vincentelli

Figure 3.1: PBD Theory: Platform Instance

Once a platform instance has been selected, the particular behaviors which the instance exhibits will be known. In Figure 3.2, the individual instance selected on the right exhibits a subset of the behaviors of the entire platform (shown on the left). Some of the platform behaviors are non-deterministic as well.

Once a platform instance has been selected and the functional description is known, the mapping process can proceed. In Figure 3.3, the right hand side shows the platform instance behaviors. It also shows all possible functional behaviors (projected from the design on the left hand side). The intersection is the permissible behaviors of a mapped system. In order to exhibit the largest set of mapped behaviors, the mapped instance should be selected at the "highest" point in the intersection.

Platform-based design can be visually depicted as in Figure 3.4. This illustration captures both the top-down and bottom-up approaches which make up the "meet-in-the-middle" methodology.

74

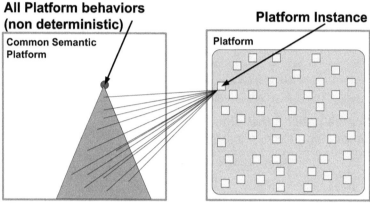

Figure 3.2: PBD Theory: Platform Behavior

The top-down approach starts with what can be called the *application space*. In this space an individual *instance* is selected. This describes the functionality of the design. The top-down refinement process restricts the functionality as necessary to make designs less abstract so it can be realized by the platform components.

The bottom-up approach starts with what can be called the *architecture space*. An individual architecture instance will export up its performance costs. Instances are individual targets which represent the selection of various platform components during the mapping process.

This view has been termed "fractal" as it repeats at various abstraction levels. Figure 3.5 illustrates this at three levels by turning these triangles on their side. It should be assumed that the triangles higher in the design are at a higher level of abstraction as compared to the lower triangles. During a refinement process, the mapped functionality (the meeting point of the triangles) of the abstract design becomes the functionality at the lower level of refinement. At this stage another platform is then considered and the process is iterated until all the components are implemented in their final form.

In this example, the figure is labeled with three networking abstraction levels (Network, Data Link, and Physical Layer). However a more concrete example could be as follows. Imagine at the top level of abstraction, the functional space is a particular algorithm and the platform is a collection of generic APIs. The mapped instance (a collection of generic APIs which implements the algorithm) will

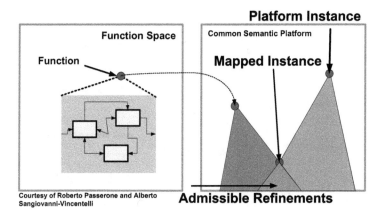

Figure 3.3: PBD Theory: Platform Mapping

now become the functional space at the next level of abstraction. The platform at this level may be functional digital logic blocks which implement the APIs. Once a collection of functional blocks are assembled to cover the required APIs, the next level may be collections of logic gates which must be mapped to the functional blocks. This continues until the entire design is specified in a representation which is appropriate for manufacturing/implementation.

3.5 PBD Related Work

3.5.1 Academic

Platform-based design as described in this work can be directly attributed to Prof. Alberto Sangiovanni-Vincentelli. In this vein there are several papers which outline his vision. In [SV02], he defines the methodology in general and describes a number of various platform examples such as system platform, network platforms, and silicon implementation platforms. In [SVM01] another overview is provided. This was one of the earlier works (2001) which discussed the issues here (separation of concerns and design space exploration). [SV07] discusses system level design and provides example applications of the framework. This paper has a slightly different approach as it talks very much about the state of EDA and what must be done for it to say competitive.

[CBP+05] is a collaboration between Sangiovanni, Luca Carloni, Fernando De Bernardinis,

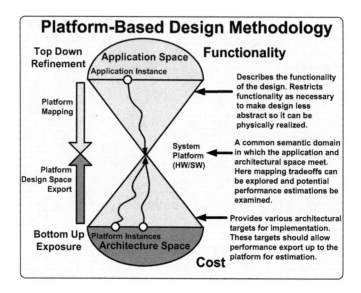

Figure 3.4: Platform-Based Design Methodology [SV02]

Claudio Pinello, and Marco Sgroi which covers not only the basics of PBD but also discusses network platforms, fault tolerant platforms, and analog platforms.

[KMN+00] is a more general work in which various approaches are described. This is a paper in which K. Keutzer, S. Malik, R. Newton, J. Rabaey and A. Sangiovanni-Vincentelli each use the idea of *orthogonalization of concerns* to frame their research agendas. Amongst those are wireless networks at the Berkeley Wireless Research Center (BWRC), automotive designs at Magneti-Marelli, and the MESCAL approach (work developed by K. Keutzer).

Model driven development (MDD) [KSLB03] is the methodology most similar to PBD. MDD is a methodology which advocates the creation of a platform-independent application model and then refinements which are more platform-specific. Vanderbilt University has created a tool flow based on MDD which makes use of formally analyzable models during the design process.

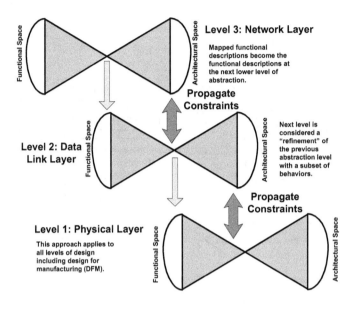

Figure 3.5: Fractal Nature of PBD

3.5.2   Industrial

In industry, the related work is less theoretical and methodological and more focused on domain specific families of solutions. Two excellent examples in multimedia domains are provided by Philips [Yos02] and ST Microelectronics [RB03].

The Philips Digital Video Platform (DVP) was one of the earliest examples of a platform. The platform is based on Nexperia Home Entertainment Engines, which are a family of compatible system-on-a-chip integrated circuits. These comply with the DVP architecture. DVP defines a dual processor, multi-bus architecture, with common bus interfaces for on-chip co-processors and peripherals. Programmable CPU cores allow implementation of new capabilities and standards. Nexperia DVP supports software upgrades to update applications and add new services. In Figure 3.6 the general structure of DVP is shown. On the right are the software components. These include application development which is supported by real-time operating systems like VxWorks and WinCE as well as Philips' own Nexperia streaming software components. These are developed to efficiently map to the hardware platform. The hardware (on the left) is a MIPS and TriMedia based CPU system with programmable device IPs connected to memory and communication buses.

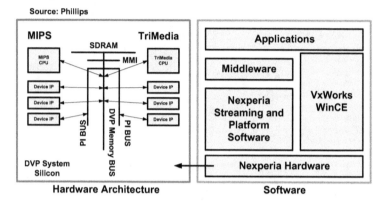

Figure 3.6: Philips Nexperia and PBD

Another example of a platform is ST Microelectronics' OMEGA family of processors for set-top-box (STB) decoding. Here the development process is seen as a HW/SW stack. ST Microelectronics customers add value by creating reference designs, applications, and middleware. ST in turn provides a complete "Software Above Chip" platform which provides not only the silicon but also the RTOS, firmware, and drivers (though STAPI). This is an excellent example of the phenomenon of hardware manufacturers being required to provide more and more software infrastructure to entice customers to use their products.

3.6 Conclusion

This chapter discussed how Platform-Based Design can be used in the process of designing an embedded electronic system. Key concepts were the separation of concerns, application and architecture mapping, and the fractal nature of PBD. In addition we have provided examples of related work in both academically and industrially. This chapter will be important as we discuss two frameworks which explicitly use PBD as a design flow. These frameworks are METROPOLIS and METRO II. Both of these are deeply rooted in PBD and were created to be key enablers of this methodology.

Future directions for Platform-based design depend heavily on the success of the current frame-

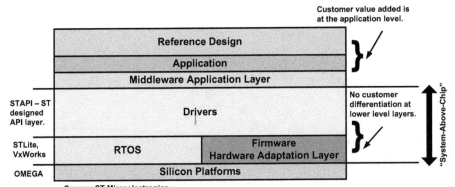

Figure 3.7: ST Microelectronics and PBD

works that implement it. One of the current concerns is the reliability of simulation data measured at high levels of abstraction. It will be required that designers of PBD tools take careful measurements to understand the implications of the operations they abstract and make sure to have the appropriately calculated performance costs. In addition, in moving from one level of abstraction to another will require sets of formal methods to ensure that design properties are held from one level to the next. Each of these topics is covered later in Chapters 6 and 7.

Chapter 4

Design Frameworks

"It is the framework which changes with each new technology and not just the picture within the frame." – Marshall McLuhan, Educator

After any methodology or ideology is developed it quickly becomes necessary that tools and people be organized to carry out the many associated tasks. In the case of Platform-based Design, these tools need to be constructed in such a way that the core concepts are maintained.

In order to achieve these goals, PBD is a three stage process: top down application development, bottom up performance export, and defining a common semantic meeting point to explore functionality and architecture mappings. Figure 3.4 illustrates this methodology and provides the needed description.

An ESL tool using a Platform-Based Design methodology is METROPOLIS [BHL+03]. As mentioned, METROPOLIS is an FPM (Functionality, Platform, and Mapping) ESL solution. It was developed at UC Berkeley from 2000 to 2006 and is available through the Gigascale Systems Research Center (GSRC). The design environment is shown in Figure 4.1 along with its organizational structure in terms of primary and support activities. The beginning of a METROPOLIS design flow starts by describing either a functional model or architectural model in the METROPOLIS Meta-Model (MMM) language. Working at the MMM level to develop functional models and architecture service models will be the scope of the discussion in Chapters 5, 8, and 9. From this user input, the Meta-Model compiler decomposes the description into an abstract syntax tree (AST). This AST can be fed into any number of backends in order to simulate the design, perform synthesis, or for verification tasks. The majority of this work is interested in using METROPOLIS for *design space exploration* (DSE). Chapter 6 will include a discussion of how backends can be used to verify refinements of architecture services.

Another ESL tool using Platform-Based Design is METRO II. Also an FPM based approach, it is the successor to METROPOLIS. Primarily it looks to streamline METROPOLIS and provide better support for heterogeneous IP import, cleaner separation of annotation and scheduling activities, and a three phase simulation engine. Chapter 5 will provide a proposal on how architecture service models may be modeled using this tool.

Definition 4.0.1 Design Space Exploration - *the process of looking at a variety of designs and using the results of simulation, verification, or other analysis methods to make decisions regarding which design should be selected, which modifications can be made to existing designs to increase performance, and observe potential design issues that may have been overlooked during specification. This process is done prior to committing to a particular design with the intention of physically creating it or its prototypes.*

It should be noted that the work covered in this book can be made independent of METROPOLIS or METRO II. It is true that there are aspects of these environments exploited to achieve the goals outlined previously. However these can be applied to other tools as well. More specifically, a design environment with the following characteristics would also be able to take advantage of the techniques outlined by this work:

1. *Support for multiple models of computation* - this work requires both *tagged signal modeling* semantics as well as *data flow modeling*.

2. *Explicitly separate an architecture model's behavior and how its operation is scheduled* - this work requires this separation to meet its performance and reuse goals. The refinement formulation is also highly dependent on this distinction.

3. *Event based synchronization* - this work requires that elements which form architecture services be coordinated with events.

More specifics about METROPOLIS and METRO II execution and modeling will be covered in this chapter and can be found in [BHL$^+$03] and [DDM$^+$07].

This chapter will describe the METROPOLIS and METRO II design environments. All the needed background will be provided so that case studies using these tools can be described in future chapters.

4.1 Chapter Organization

This chapter begins by describing related work in Section 4.2. This describes a number of frameworks which have similar goals and approaches to METROPOLIS and METRO II. In Section 4.3, the METROPOLIS design environment is described. This includes the goals of the framework (4.3.1), design activities (4.3.2), and the limitations of the framework (4.3.3). In Section 4.4 the METRO II design framework is described. The goals of the framework (4.4.1), its building blocks (4.4.2), execution semantics (4.4.3), mapping semantics(4.4.4), and implementation details (4.4.5) are all described. Finally in Section 4.5, conclusions and future work are provided.

4.2 Related Work

In this section, an overview of related system-level design frameworks is provided. When discussing related work in system level architecture service modeling there are many comparisons and there are many approaches that can be examined. These approaches can be divided into those which are industrially developed and those which are academic based. Additionally, these approaches can be placed in a platform-based design flow. This placement allows them to be divided into those which just allow a functional description (F), platform description (P), or a platform description plus mapping capabilities (PM). This is exactly how the taxonomy in [DPSV06] is constructed. This section will present a brief overview of that work along with providing other insights into how this work fits into the existing system level design landscape.

SystemC [Ini07] is by far the most recognized system level architecture development language. SystemC is a set of libraries built on top of C++ which allows for concurrent module simulation, event synchronization, and a variety of elements which facilitate architecture descriptions. The core is an event driven simulator using events and processes. The extensions include providing for concurrent behavior, a notion of time sequenced operations, data types for describing hardware, structure hierarchy, and simulation support. Accellera's SystemVerilog [Acc07] is an extension of Verilog which adds system level features. For example it can co-simulate with C/C++/SystemC code, includes support for assertion based verification (ABV), and provides extended data types and eases restrictions on type usage. Unified Modeling Language (UML) [Lan07] is another well known language which is in this space. UML allows for the abstract specification of a system using different graphical views of the system. It is used to illustrate the system topology and the relationship between components.

In the industrial domain, tools which focus on platform descriptions (P) include such tools as Prosilog's Nepsys [Pro07]. This tool relies on IP libraries based on SystemC. It works at the component transaction level. Beach Solutions' EASI-Studio [Sol07] focuses on interconnection issues at the component level and provides solutions to package IP in a repeatable, reliable manner. The suite provides a collection of tools which help to manage the design. These tools include data import features, graphical interface capture, and IP watermarking. Of particular interest are its Specification Rule Checks (SRC) which ensure adherence to naming conventions, name uniqueness, address space uniqueness, and that parameter values are resolvable. Sonics' Sonics Studio [Son07] works at the implementation level by using bus functional models (BFM). This tool includes a graphical, drag-and-drop environment for configuring SoC designs. This environment also provides monitor functions and simulation support for IP blocks.

Industrial domain tools for creating platform descriptions with mapping capabilities (PM) include VaST Systems Technology's Comet/Meteor [Sys07]. The Comet tool focuses on high performance processor and architecture models at the system level. This tool uses virtual processors, buses, and peripheral devices. Meteor is an embedded software development environment. It also accepts virtual system prototypes for cycle accurate simulation and parameter driven configuration. Finally, Summit's System Architect [Des07c] looks at multi-core SoCs and large scale systems and is a SystemC component based system. Summit was acquired by Mentor Graphics.

Finally, industrial tools with functional, platform, and mapping capabilities (FPM) include MLDesign's MLDesigner [Tec07]. This tool allows for discrete event, dynamic dataflow, and synchronous dataflow model of computation to be described. It is intended to be used for a "top-down" design flow starting from initial specification to final implementation. It includes an integrated development environment (IDE) to integrate all aspects in one package. Mirabilis Design's Visual Sim [Des07b] product family adds continuous time and finite state machine (FSM) models of computation natively to this list of supported MoCs (they are also available in the experimental library of MLDesigner). The design process in Visual Sim begins by constructing a model of the system using the parameterizable library provided. This model can be augmented as well with C, C++, Java, SystemC, Verilog, or VHDL blocks. The library blocks operate semantically using a wide variety of models of computation as listed. The design is then partitioning into software, middleware, or hardware. Finally the design is optimized by running simulations and adjusting parameters of the library elements. The last industrial FPM tool is Cofluent's Systems Studio [Des07a]. It provides transaction level SystemC models which perform design space exploration in the Y-chart modeling methodology. The functional description is a set of communicating processes executing concurrently. The platform model is a set of communicating processes and shared memories linked by shared communication nodes. The platform model has performance attributes associated with it as well. This approach is very similar to METROPOLIS but does not support as wide a variety of models of computation or as rich a constraint verification infrastructure.

The academic domain has many offerings as well. An academic tool which captures the functionality of a design (F) is ForSyDe [SJ04] [RSSJ03]. This tool was developed at the Royal Institute of Technology, Sweden. It performs modeling, simulation, and design of concurrent, real-time, embedded systems. It has support for a wide variety of synchronous MoCs. It uses transformation rules to proceed from a functional specification to collections of process networks.

ForSyDe focuses on formal design transformations that enable design refinement. This allows the designer to start with an abstract definition of the design and proceed toward implementation. At each step, there are two types of transformations that can be made. Semantic preserving transformations do not change the behavior of the model, while design decision transformations are unrestricted. The focus of ForSyDe is on the verification aspects of design, and not the automation.

A tool which allows platform descriptions as well as mapping (PM) is Carnegie Mellon's MESH [PT02]. MESH [CPT03] is a design framework that separates the design into three parts: the application layer, the physical layer and the scheduling layer. These three layers are roughly equivalent to the functional model, the architectural model, and mapping within the Platform-based design methodology. MESH stands for Modeling Environment for Software and Hardware. This approach examines heterogeneous system design at the component level through C input. In MESH, threads are modeled as an ordered sets of N events. MESH is interested as well in the development of benchmarks, called scenarios, which evaluate collections of heterogeneous programs. Stanford's Rapide [LV95] [LKA+95] is an Executable Architecture Definition Languages (EADL). It utilizes an event based execution model for distributed, time sensitive systems. Rapide is a PM approach as well.

Ptolemy II [LXL01] is a meta-modeling framework which focuses on simulation and the interaction between different models of computation. While the focus is not on function-architecture separation and mapping, several hardware targets have been explored in the context of the precursor Ptolemy framework [Mur].

Tools which add the ability to specify functional descriptions as well (FPM) include Seoul National University's PEACE [HLY+06]. This codesign environment is Ptolemy based [BHLM02]. It touts an open-source framework, a reconfigurable framework (design steps are decoupled so that users can introduce their own steps), a separate Java based GUI (named Hae) from the kernel, an objected-oriented C++ kernel, support for multilingual system design (dataflow graphs for functional representations and FSMs for control), and automatic hardware/software synthesis as its strengths.

UC Berkeley's MESCAL [MKM+02] is an approach for the programming of application specific programmable platforms. The main domain of concentration is network processors, which can be considered a specialized type of multimedia applications. Past work has been carried out on customizing instruction sets of processors according to the application. Recently, the investigation of FPGAs as an implementation fabric and automated allocation techniques have also been explored [JSRK05]. Mescal has extended Ptolemy II [IHK+01] for its implementation. Vanderbilt's GME/GREAT/DESERT [LMB+01] are a set of tools for pruning the design space. Aspects of these tools are UML and XML based and they are focused on domain specific modeling and program synthesis.

Polis [BCG+97] was a design environment which was one of the first to allow for function-architecture separation. Designs in this framework are based on the communicating finite-state machines (CFSM) model of computation [BHJ+96]. Architectural components can only be chosen from a set of predefined components, limiting the expressiveness. METROPOLIS is the direct successor of the Polis effort.

The Compaan/Laura [SZT+04] approach uses Matlab specifications to synthesize Kahn Process Network (KPN) [Kah74] models, which are then implemented on a specific architectural platform as hardware and software. The architecture platform consists of a general purpose processor along with an FPGA, which communicate via a set of memory banks. Software runs on the general purpose processor, while the hardware is synthesized into VHDL blocks which are realized in the FPGA. The partition between hardware and software occurs relatively early in the design flow and is based on workload analysis. The types of optimizations that are carried out automatically relate to loop analysis. The software implementation makes use of the YAPI [KES+00] library and does not consider deadlock caused by insufficient buffer space.

The Spade [LSvdWD01] and Sesame [PEP06] [vHPPH01] approaches within the Artemis [PHL+01] project from the Delft University of Technology [LvdWD01] focus on synthesizing Kahn Process Networks (KPN) specifications in hardware/software. Spade also employs a Y-chart based approach to design with functionality and architecture separated. In this case they are termed *workload* and *resources* respectively. It employs trace driven simulation where time can be accounted for and performance data collected. The most relevant optimization approach from their work utilizes an evolutionary algorithm to minimize a multi-objective non-convex cost function. This cost function takes into account power and latency metrics from the architectural model. The optimization problem is solved using a randomized approach based on evolutionary algorithms.

All of these academic and industrial tools work at the system level in terms of the level of abstraction employed. As mentioned, each of these approaches are placed in a taxonomy in [DPSV06]. Without going into all the details contained in that work, one can say that the following issues are investigated: model of computation supported, support for quantity annotation, mapping support, specific device support (ASIC, FPGA, etc), level of abstraction supported, and underlying semantics. The reader would be well served to look at that work as it covers 90+ tools.

Table 4.1 has a small sample of the comparisons that can be made between METROPOLIS & METRO II and the other academic (top half) and industrial (bottom half) approaches. The issues outlined,

Event Based, Mapping, Independent Schedulers, and Pure Architecture model, were chosen since each one will be integral to providing the desired outcomes outlined in Chapter 1. "Event based" refers to the fact that synchronization between concurrently-executing portions of the design is done via notification and wait statements using a unified concept such as events. "Mapping" allows for functionality to be assigned to services. "Independent Schedulers" support indicates that scheduling is explicitly separate. Finally a "pure architecture model" indicates that there are two models (functional and architectural) for each system kept explicitly separate. A "+" indicates that the tool supports this concept while a "-" indicates that it does not explicitly support this. Naturally if two tools share the same markings, it does not mean that they are equivalent in their other features.

	Event Based	Mapping	Ind. Sched.	Pure Arch. Model
METROPOLIS	+	+	+	+
METRO II	+	+	+	+
ForSyDe [SJ04]	+	+	-	-
Rapide [LV95] [LKA+95]	+	+	-	+
Spade [LvdWD01]	-	+	-	+
Nepsys [Pro07]	-	-	-	-
Comet/Meteor [Sys07]	-	+	-	+
Systems Studio [Des07a]	-	+	-	+

Table 4.1: Comparison of Architecture Service Modeling Approaches

These criteria were selected since mapping is going to be important to the work presented here. It will allow the architecture model to be completely separate from the functional model. Independent Schedulers and Event Based Semantics are crucial to the characterization and annotation methods to be presented. Knowing which approaches support which aspects illustrates which other design environments may be amenable to the work presented here.

Compared to these other design environments, the METROPOLIS framework is unique in the respect that it provides meta-modeling capabilities and is geared toward function-architecture mapping. Meta-modeling capabilities allow reasoning about designs which are expressed using different models of computation. The function-architecture mapping is a natural method by which these diverse models come together. Thus, METROPOLIS is a singular framework which will allow the implementation of mapping approaches for systems from different embedded systems domains. METRO II is a refinement of METROPOLIS that eases specification while relaxing analysis capabilities.

4.3 The METROPOLIS Design Framework

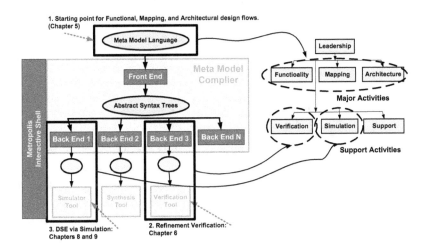

Figure 4.1: METROPOLIS Design Environment and Organization

The METROPOLIS Design Framework is an embodiment of the Platform-based design methodology. The methodology is flexible enough to be applied to many different types of design problems, but the METROPOLIS framework is focused on electronic system-level design (ESL) [BMP07]. The METROPOLIS framework consists of a specification language – the Metamodel [Tea04] – as well as a compiler and a set of plugins that can interface with external tools. Figure 4.1 shows the architecture of the METROPOLIS design framework. Functional, architectural, and mapping specification for the system is first captured within the Metamodel language. The front end allows this specification to be compiled into a set of abstract syntax trees. Various backends, including simulation, synthesis, and verification tools can then access this information. The backends are invoked by the user using the interactive shell.

In this section, we will cover the goals of the METROPOLIS framework and how they manifest themselves in the infrastructure and the METROPOLIS Metamodel Specification language.

4.3.1 Goals of the Framework

The METROPOLIS framework is geared toward attacking common electronic design problems that occur at the system level. With that aim in mind, the major goals of the framework are three-fold: Facilitating design reuse to enable the design of large systems, preserving analysis capabilities by capturing the design specification with formal semantics, and enabling declarative statements in the specification to formally capture capture constraints, assertions, and performance annotations. Design reuse follows from the platform-based design methodology discussion in Chapter 3. In the following sections, we will examine the other goals in further detail. An overview of the platform-based design methodology implemented in METROPOLIS is given in [Pin04].

Preserving Analysis Capabilities

A major complication with large, heterogeneous systems lies in the task of verification. Ensuring that the system performs according to the specification can easily take a majority of the total design time. In an attempt to remedy these problems, the METROPOLIS design framework stresses the usage of formally defined models of computation for modeling. The specification language used in METROPOLIS, the METROPOLIS Metamodel [Tea04] [BLea02], has formally defined semantics that allow the expression of many different models of computation. Each statement in this language has a formal representation in the form of action automata. This formal underpinning can be used for different types of analysis, including refinement verification [Den04].

Event interactions and annotations

One of the unique aspects of the METROPOLIS framework is the support for both operational code and declarative statements in the specification. Most other system-level design environments only allow operational code as the specification mechanism. Supporting declarative statements allows the designer to succinctly specify behavior or assertions in the design. This is especially important in the initial phases of the design process, when the designer may be more interested in specifying *what* properties the components of the design need to have, rather than *how* those properties will be manifested in an implementation.

Currently in the METROPOLIS framework, support is provided for declarative statements in two different logics, Linear Temporal Logic (LTL) [SC82] and the Logic of Constraints (LOC) [BBL+01]. Both of these logics allow for statements to be made about event instances in the design. Event instances are generated whenever a thread of control in the design executes any action. Event instances may be annotated with quantities, which may represent a diverse set of indices, from access to a shared resource to energy or time. LTL statements can be used to specify mutual exclusion constraints and synchronization between events. LOC may be used to make statements about quantity annotations on events.

4.3.2 Design activities within the METROPOLIS framework

After having described the goals of the METROPOLIS framework, we turn to the design activities that are important for the task of implementing functional applications on architectural platforms. Specifically, we look at three major issues: modeling the behavior – including the computation and communication – in the functional and architectural models, annotating costs to operations in the architectural model, and the mechanism by which the functional and architectural models are associated together. An overview of these design activities within the context of a simple case study is provided in [DDSZ04].

Describing the Functional and Architectural Models with Processes and Media

Processes are objects that possess their own threads of control. Media are passive objects that provide services to processes and other media. A process cannot connect to other processes directly. Instead, an intermediate medium must be used to manage the interaction between multiple processes. When an object wishes to utilize the services provided in another medium, it must communicate with that medium by using ports which have associated interfaces. The concept of ports and interfaces is widely used for design specification in frameworks like SystemC [Ini07]. Media implement certain interfaces which then become services provided to processes or other media. For instance, a media may implement read and write services which can be used by processes.

Processes in the architectural model may represent tasks which are executing on architectural media such as processors. By themselves, these processes do not carry out any useful work, they just execute a nondeterministic sequence of operations. In this sense, the set of possible behaviors in the architectural model encompasses all legal traces of operations.

Definition 4.3.1 Process - *A process represents computation. This is an active object (thread) and groups of processes run concurrently. Processes cannot communicate to each other directly. Processes can be synchronized by constraints or a special "await" statement.*

In Figure 4.2, the process declaration (Cpu) is shown. Two ports are available (port0 and port1) with access to CpuAPI and CpuAccess interfaces. A parameter is available to customize this process as well as a constructor. The "meat" of this process would be provided in the section contained in the

```
process Cpu {
   port CpuAPI port0;
   port CpuAccess port1;
   parameter int MODE;
   Cpu(int mode) {MODE = mode;} // constructor
   void thread() useport port0, port1 {...} // defines execution
}
```

Figure 4.2: METROPOLIS Process Example Code

"thread()" section. Processes have access to "await statements". Their syntax is, *await(guard, test_list, set_list)* . In order for an await statement to be evoked, the "guard" must be true and no interface in the "test_list" can currently be in use. Once these conditions are true, the code within the await statement can execute and no other process can use the interfaces in the "set_list" while the await code is executed.

The threads contained in METROPOLIS processes are scheduled to run by a *manager* with controls the simulation flow. There are two phases in METROPOLIS. In the first phase, the threads run until each is blocked. In the second phase the manager must decide which of these processes should be selected to resume running. This process is described in much more detail in [BLP+02a].

Definition 4.3.2 Medium - *Media are the manner in which processes communicate to one another. Media may also be connected to other media. Media are passive objects in that they do not have their own threads of execution. They implement interfaces which are extended through the use of ports.*

```
medium Bus implements CpuAPI {
   parameter int BITWIDTH;
   Bus(int size) {BITWIDTH = size;}
   public eval void busRequest(int processID) {...}
   public update void driveData(int addr, int data) {...}
}
```

Figure 4.3: METROPOLIS Medium Example Code

Figure 4.3 illustrates that media (in this case, a Bus) implement interfaces (CpuAPI). This medium interface implements two methods. These methods can change values (as denoted by the keyword "update") or read values (as denoted by the keyword "eval"). As with processes, media can be parameterized as well.

Using Quantity Managers to annotate cost

Quantity managers in METROPOLIS are similar to aspects in aspect-oriented programming [KLM+97] languages. Quantity managers can be used to assign costs to operations in the architectural platform. The cost can be in terms of any useful quantity, such as time, power, or access to a shared resources. The quantity manager collects all requests for annotation and determines which requests are to be satisfied and which ones need to be blocked.

An example of this type of annotation is shown in Figure 4.4. In this example, the *L_Exec* operation in a media is annotated with cost of *EXEC_CYCLE* cycles. First, the relevant events *beginEvent* and *endEvent* are identified which correspond to some thread executing the beginning and end of the operation respectively. The first event (*beginEvent*) is annotated with the current time according to the quantity manager whereas the second event (*endEvent*) is annotated with the time of the first event plus the cycle time. *gt* refers to the global time quantity manager.

```
L_Exec {@
    event beginEvent = beg(getThread(), L_Exec);
    event endEvent = end(getThread(), L_Exec);
    {$
      beg { gt.RelativeReq(beginEvent, 0); }
      end {
       currentTime = gt.getQuantity(beginEvent, LAST);
       gt.AbsReq(endEvent, currentTime + EXEC_CYCLE);
      }
    $}
    ... // code for this operation
@}
```

Figure 4.4: Annotating costs for operations with quantity managers

Definition 4.3.3 Quantity Manager - *Quantity managers act as schedulers. They are used to define scheduling policies which are used to satisfy constraints. They are passive objects but run functions when constraints need to be satisfied. Quantity managers control the execution of process. Quantity managers have ports which are hooked to state media to communicate with the processes they schedule. Quantity managers operate during the second phase of simulation separate from the processes and media.*

As is shown in Figure 4.5 there are four functions which a quantity manager must implement. The *request()* function generates a quantity request for a particular event. This function adds the event to a list of "pending" events. As can be seen in the code, two arguments are required. One is the event to

92

```
public interface QuantityManager extends port {
    eval void request (event e, RequestClass rc);
    update void resolve();
    update void postcond();
    eval boolean stable();
}
```

Figure 4.5: METROPOLIS Quantity Manager Example Code

request and the other is a class object which will aid in that request by providing information about the system. *resolve()* is used to resolve the existing quantity requests. This can be seen as the scheduling step. This pulls an event from the "pending" queue. *postcond()* is used to clean up the state of the quantity and the quantity requests. It is at this point that events are annotated. *stable()* indicates the success of the quantity resolution process and is used to determine when the simulation can switch phases.

Definition 4.3.4 StateMedia - *A special media type used for communication between processes and quantity managers. It passes the state of the process to the quantity manager and returns to the process the results of scheduling.*

```
public interface StateMediumSched extends Port {
    eval process getProcess();
    eval ArrayList getCanDo();
    .... (other support functions)
    update boolean setMustDo(event e);
    update boolean setMustNotDo(event e);
}
```

Figure 4.6: METROPOLIS State Media Example Code

Figure 4.6 shows the "setMustX" functions which enable or disable a particular event (and thus control which processes can proceed). *getCanDo()* returns an array of events upon which to begin to schedule. *getProcess()* returns the process associated with this state medium. There are other support functions not shown here but are discusses in [Tea04].

Definition 4.3.5 Port - *Ports are special interfaces which declare methods which can be used through ports. The methods themselves are implemented by media. In this way ports declare a set of function prototypes. These functions are called by processes or other media connected to the implemented media via ports.*

```
public interface CpuAPI extends Port{
    public eval void busRequest(int processID);
    public update void driveData(int addr, int data);
}
```

Figure 4.7: METROPOLIS Port Interface Example Code

Figure 4.7 illustrates that interfaces extending ports are simply the function prototypes which will later be implemented in media.

Definition 4.3.6 Netlist - *A netlist is a collection of meta-model objects, their ports, and the connections between them. This is instantiated with a variety of mechanisms including, connect(SrcObject, SrcPort-Name, DestObject) and addcomponent(NodeObject, Netlist Object).*

Definition 4.3.7 Scheduled Netlist - *A connection and parameterization of architecture elements in* METROPOLIS. *These include processes and media. Objects in this netlist generate events which need to be scheduled.*

Definition 4.3.8 Scheduling Netlist - *A connection and parameterization of quantity managers and state media in* METROPOLIS. *These objects receive events from the scheduled netlist and perform the resolve() function.*

The scheduling netlist contains elements primarily involved in the simulation engine. The scheduled netlist is where the architecture services are located and indicates which components are actually going to be captured and eventually used to create a description for a programmable platform.

Definition 4.3.9 Top Level Netlist - *A netlist which is only composed of sub-netlists and is itself not part of any higher level netlist. This is typically the combination of both the scheduled and scheduling netlist.*

Figure 4.8 illustrates a METROPOLIS architecture model. In this case there are meta-model processes which represent tasks (T_1 to T_N). These tasks will ultimately be mapped one-to-one with processes in the functional model. These tasks trigger the use of services. In METROPOLIS, architecture services are collections of media. In the scheduled netlist, the processes (called mapping processes) are squares and the media are ovals. Shown are a CpuRtos service, Bus service, and Mem service. Examples are given showing potential interface calls on ports. Each service has a corresponding quantity manager (diamond) in the scheduling netlist which communicates to the process through a statemedia object (small circles). Also shown is global time. Global time manages the logical time of the simulation. This is a

94

quantity manager which manages the annotation of events with physical time quantities. If this model were to be simulated, the result of simulation would be an estimate of the physical time that would be required. The quality of the model is often measured as the accuracy between this estimate and the actual value of the implementation.

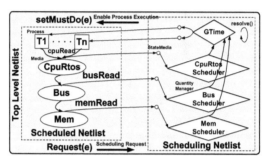

Figure 4.8: METROPOLIS Architecture Netlists

Another view of a METROPOLIS architecture model is shown in Figure 4.9. This shows a graphical representation of both the scheduled and scheduling netlists of a METROPOLIS architecture model. This figure also includes a characterizer database which will be discussed in Chapter 7. On the left hand side the scheduled netlist is shown and it illustrates how each MicroBlaze service (media) is connected to a task (process). These tasks drive the simulation. It also shows media-to-media connections between MicroBlazes and FIFO based communication channels.

Mapping with synchronization constraints

Synchronization statements are used to intersect the behaviors of the functional and architectural models by constraining events of interest from each to occur simultaneously. Along with simultaneity, we can also control the values of specific variables that are in the scope of these events. This type of synchronization can be used to restrict the behavior of architectural processes to follow that of the functional processes to which they are mapped.

An example of a function that would emit these synchronization constraints is shown in Figure 4.10. In this example, the function takes as arguments the two processes that are to be mapped together – a functional process and an architectural process. First, the beginning of the read operation is identified for both processes and recorded as the events *e1* and *e2*. The two events are synchronized

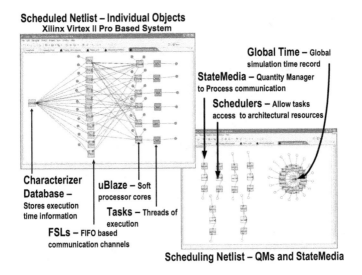

Scheduled Netlist – Individual Objects
Xilinx Virtex II Pro Based System

Global Time – Global simulation time record

StateMedia – Quantity Manager to Process communication

Schedulers – Allow tasks access to architectural resources

Characterizer Database – Stores execution time information

uBlaze – Soft processor cores

Tasks – Threads of execution

FSLs – FIFO based communication channels

Scheduling Netlist – QMs and StateMedia

Figure 4.9: Graphical METROPOLIS Architecture Representation

together and two variables in the scope of these events are constrained to be equal. In this example, the number of items read by the functional process is constrained to be the same as the number of items read by the architectural task. By changing the arguments to this function, various mappings can be realized and evaluated relatively easily.

4.3.3 Limitations of METROPOLIS

In this chapter, lessons learned from the METROPOLIS framework will be applied to drive future requirements for the METRO II framework. The case studies that have been carried out are from the multimedia [DZMSV05] and automotive [ZDSV+06a] domains, as described in Chapters 8 and 9, have used METROPOLIS for modeling. Additional case studies including [DRSV04] and [DDSV06a] have also utilized this framework.

While validating the core ideas of our approach, these case studies also revealed three main limitations of the METROPOLIS framework: design import, handling of quantity managers, and the description of mapping. In this section, these limitations will be addressed by describing the proposal for

```
void mapPair(process f, process a) {

    event e1 = beg(f, f.read);
    event e2 = beg(a, a.read);
    ltl synch(e1, e2: numItems@e1 == numItems@e2);

    event e3 = end(f, f.read);
    event e4 = end(a, a.read);
    ltl synch(e3, e4);

    ... // similar code for write() and exec() services
}
```

Figure 4.10: Synchronizing events to realize mapping

the next-generation METRO II framework. One of these goals is supporting automated design space exploration with a service-based approach, more compatible with the design flow presented next in Chapter 5.

The generality of the MetaModel language [Tea04] leads to difficulties both for users and framework developers. For users, learning a new language is usually difficult, especially if the execution semantics are complex. From the framework developer's point-of-view, a new language requires vast amounts of support. Designing compilers, debuggers, and simulators for a new language is quite time-consuming.

The two-phase execution semantics of the MetaModel language requires that interactions with quantities must be explicitly represented. So, the simplifying assumptions made in domain-specific languages cannot be made for the MetaModel. Moreover, since quantity annotation requests are interwoven with the behavioral code, the request-making statements cannot be encapsulated into separate libraries, and the specification task is therefore complicated.

The MetaModel allows both denotational and imperative specification. Mapping is specified as synchronization constraints between events from the functional and architectural models. The two main limitations relate to the granularity at which mapping is specified as well as the restrictions on variables associated with the mapped events. First, mapping can only be specified with event-level synchronization constraints. Since there is no built-in mechanism to agglomerate events, this must be implemented by the system designer within the mapping code. The ability to export events in a structured manner would address this limitation. Second, arbitrary local variables in the scope of events can be used within the constraints. This is an encapsulation failure and results in designs that are difficult to debug or reuse.

By focusing on the key value-added features of METROPOLIS, and addressing these limitations, we plan to make METRO II an IP-integration framework with enhanced support for PBD activities.

4.4 The METRO II Design Framework

The second platform-based design framework to be discussed is METRO II [DDM$^+$07], which is the successor to METROPOLIS [BHL$^+$03]. METRO II was also developed at UC Berkeley beginning around 2006. METRO II was created specifically to perform IP import better than METROPOLIS (via a more lightweight, wrapper syntax rather than the *Metropolis Meta Model* specification language). In addition, it expands the two phase simulation semantics of METROPOLIS into a three phase system to better enforce the separation of concerns ideology. METRO II is based on the concept of *events*: components generate events by means of imperative code (written in SystemC, for instance), events are annotated with performance metrics of interest, and events can be related together using declarative constraints or imperative code. With such an event based framework, we can express a variety of MoCs, formally capture system behavior, and provide support for design space exploration. It is important that using an event-based framework does not significantly reduce simulation performance or efficiency. Later in this work, we investigate this issue further in the case studies of Chapters 8 and 9.

The following sections introduce the reader the the goals of METRO II, the components of the framework, and the mapping and execution semantics employed.

4.4.1 Goals of the Framework

The three main goals that form the basis of the second-generation METRO II framework are based on the limitations of METROPOLIS described in Section 4.3.3. The three features are:

1. *Heterogeneous IP Import.* IP providers create models using domain specific languages and tools. Requiring a singular form of design entry in a system-level environment requires complex translation of the original specification into the new language while making sure that semantics are preserved. If different designs or different components within the same design can have different semantics, heterogeneity has to be supported by the new environment.

2. *Behavior-Performance Orthogonalization.* For design frameworks that support multiple abstraction levels, different implementations of the same basic functionality may have the same behavioral representation but different costs. For instance, different processors will be abstracted into the same

programmable component. What distinguishes them is the performance vs. cost trade-off. More-over, not all metrics are considered or optimized simultaneously. It should be possible to introduce performance metrics during the design process, as the design proceeds from specification to implementation.

3. *Mapping Specification.* Mapping relates the functional and architectural models to realize the system model. Specification of this mapping must be carried out such that there is minimal modification to the functional and architectural models themselves. In addition, the mapping specification must be compatible with design flow presented in Chapter 5 to facilitate automated design space exploration.

The following subsections describe these three requirements in more detail.

Heterogeneous IP Import

Heterogeneous IP import shapes the nature of METRO II to be primarily an integration environment. There are two main challenges that have to be addressed: wrapping and interconnecting IP.

First, IPs can be described in different languages and can have different semantics that can be tightly related to a particular simulator. Importing the IP entails providing a way of exposing the IP interface. The user must have the necessary aids to define wrappers that mediate between the IP and the framework such that the behavior can be exposed in an unambiguous way.

Secondly, wrapped components have to be interconnected. Even if the interfaces are exposed in a unified way, interconnecting them is not usually a straightforward process. Data and the flow of control between IP blocks must be exposed in such a way that the framework has sufficient visibility.

Behavior-Performance Orthogonalization

The specification of what a component does should be independent of *how long it takes* or *how much power it consumes* to carry out a task. This is the reason why we introduce dedicated components, called *annotators* to annotate *quantities* to events.

A distinction has to be made between quantities used just to track the value of a specific metric of interest and quantities whose value is used for synchronization. For instance, time is used to synchronize actions and it is not merely a number that is computed based on the state evolution of the system. For quantities that influence the evolution of the system, special components, called *schedulers* are provided by the glue language. Schedulers are used to arbitrate shared resources.

The separation of schedulers from annotators allows for simpler specification and provides a cleaner separation between behavior and performance. As a result, instead of two-phase execution as in METROPOLIS, the execution semantics become three-phase.

Mapping Specification

Following the PBD approach, we want to keep functionality and architecture separate. The implementation of the functionality on the architecture is achieved in the mapping step. In order to explore several different implementations with minimal effort, the design environment needs to provide a fast and efficient way of mapping without modifying either the functional or the architectural models. The main problems to tackle are related to the specification of mapping itself and the execution semantics of the two models.

The mapping specification needs to be specified in such a way that it can easily be modified in order to facilitate design space exploration without touching either the functional or architectural models. To accomplish these goals while remaining compatible with the design flow advocated in this work, mapping needs to be able to easily manipulate references to services.

The execution semantics of mapping in METROPOLIS relied on special "mapping" processes that were instantiated in the architecture. These mapping processes were completely nondeterministic in their usage of architectural services. The mapping specification was a one-to-one association between these processes and the functional processes. The functionality and the architecture were then executed concurrently with synchronization constraints present between them. Arbitrary variables in the scope of the synchronized events are allowed to be referenced by the mapping.

In the METRO II framework, we would like to reduce the complexity by not requiring special processes in the architecture only for mapping purposes and sequentially executing functionality and architecture if possible. Also, access to variables in mapping needs to be strictly regulated in order to maintain IP encapsulation.

4.4.2 METRO II Building Blocks

To simplify the designer's task of specifying models that conform to the three-phase semantics to be described in Section 4.4.3, different types of objects are defined in METRO II. First, components - the primary object for imperative specification - are described. Then, the different types of ports and connections in METRO II are described. After this, the specialized METRO II objects - constraints, mappers, annotators, and schedulers - are covered.

Figure 4.11 illustrates the major Metro II components. We attempt to use the iconography here throughout the work. A subtle but useful distinction to be made is between *required* and *provided* ports regarding their relative direction in relation to the component they are associated with. Required ports point outward while provided ports point inward.

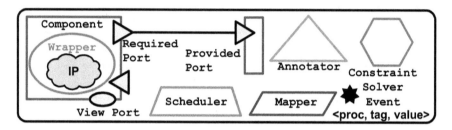

Figure 4.11: Overview of Metro II Design Components

Components

A *component* is an object which encapsulates imperative code in a design, either functional or architectural. Components interface with other components via zero or more ports. There are two descriptions of component composition: *atomic components* and *composite components*. An atomic component is a block specified in some language and is viewed by the framework as a black box with only its interface information exposed. A composite component is a group of one or more objects as well as any connections between them.

An atomic component with zero ports is shown in Figure 4.12. The IP encapsulated by the component is interfaced by means of a *wrapper*, which translates and exposes the appropriate events and interfaces from the IP. This is a major aid in the import of external IP into METRO II.

In Figure 4.13 a composite component is shown. This illustrates the hierarchy available with components as well as how connections can be made between components.

In addition to atomic and composite components, another designation given to components is *active* or *passive*. Active components have a thread of control and "actively" make calls on their ports. Passive components do not have a thread of control and respond "passively" to calls on their ports. Figure 4.14 provides an illustration of these two concepts.

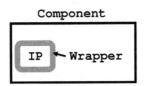

Figure 4.12: Atomic Component

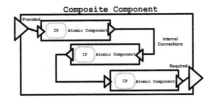

Figure 4.13: Composite Component

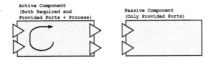

Figure 4.14: Passive and Active Components

Ports

There are two types of ports that components may have: coordination and view ports. Coordination ports are used for two-way interaction with other components by using events. View ports, on the other hand, may only expose internal events to the outside.

A *coordination* port is used to interact with other components. Each coordination port is associated with a set of methods. This collection of methods is called an *interface*. A *method* is a sequence of events, with a unique begin/end event pair. Variables in the scope of the begin event are method arguments. Variables in the scope of the end event are return values.

By setting constraints between events associated with coordination ports of different components, the execution of these components can be coordinated. Coordination ports are divided into three types based on the type of interaction: required ports, provided ports, and view ports.

Required Ports

Required ports are used by components to request methods that are implemented in other components. Connections are made only between a required port and a provided port.

For required ports, a component proposes a begin event and associates values with the proposed event that represent the arguments of the method being requested. When the proposed event is executed, control transfers to the component at the other end of the connection, which owns the provided port. The component waits for the end event to be executed and obtains the return values from the method. The method is executed in the same process as the caller.

As can be seen, only active components will have required ports since a thread is required not only to make the initial request on the port but also to execute the method.

Provided Ports

Provided ports are used by components to provide methods to other components. As stated before, connections are permitted only between a required port and a provided port. For provided ports, no separate process exists in the component to carry out the provided method. Instead, the component inherits the process from the caller component and executes the events in the provided method using that process. After the method has been executed, the process proposes the end event.

Passive components are only contain provided ports whereas active components can have both required and provided ports.

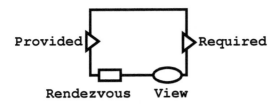

Figure 4.15: Component with 4 ports

View Ports

A *view port* exposes some of a component's internal events to the outside world. These events are read-only, i.e., they cannot be blocked by outside world. View ports cannot be connected to other ports.

These ports can be useful for several reasons. Firstly, when calls are made on required ports, this represents a request for scheduling (as will be shown in Section 4.4.3). This will ultimately cause the process to be suspended. View ports however will not cause such a preemption. This can be useful for learning about the state of the system or for debugging purposes. Interleaving the use of view ports with other ports will not change the overall synchronization of the system. Secondly, the use of view ports can make a designer's intention more clear. This is useful to make designers more readable.

A component with required, provided, rendezvous, and view ports is shown in Figure 4.15.

Syntax Example: Components, Interfaces, and Ports

Before moving on to the discussion of the remainder of the METRO II concepts, this section illustrates a snippet of METRO II code with many of the pieces discussed thus far. A reader component in a typical producer-consumer design example is shown here. There is an interface designed for the reader along with a reader METRO II component with one required port. It has a process associated with it so it is *active* and simply attempts to read five times on its port.

```
/*
This is a Metro II interface called ''i_func_receiver''
It contains one method called ''receive''
It has two arguments of type void * and unsigned long
*/

M2_INTERFACE(i_func_receiver)
{
```

```
public:
    M2_TWOARG_PROCEDURE(receive, void *, unsigned long);
};

/*
Here is an active component (Reader)
It has one process and thread which runs the ''main'' function
It has one required port called ''out_port''
*/
M2_COMPONENT(Reader)
{
public:
  //required port using ''i_func_receiver'' interface
  m2_required_port<i_func_receiver> out_port;

  sc_process_handle this_thread;

  SC_HAS_PROCESS(Reader);

Reader(sc_module_name n) : m2_component(n)
{
    SC_THREAD(main);
}

void main()
{
        this_thread = sc_get_current_process_handle();
        int array[3];

        for (int i=0; i<5; i++)
        {
           cout << "reader begins iteration " << i << endl;
           out_port->receive(array, 3 * sizeof(int));
           cout << "reader reads value: " << array[0] << " "
           << array[1] << " " << array[2] << " ends iteration " << i << endl;
        }
   }
};
```

Connections

Connections between coordination ports are the primary means of component interaction. One-to-one port connections are allowed between a required port and a provided port. Provided ports do not need to be connected, but each required port must be connected to a corresponding provided port.

The syntax for a connection in METRO II is shown here. This illustrates the instantiation of three components. The first is a passive component which implements a communication channel. The second two components are reader and writer components (the reader was shown in Section 4.4.2). The connection statements have the following syntax: M2_CONNECT(component, req_port, component, prov_port).

```
c_double_handshake c("rendezvous");
Writer w("Writer");
Reader r("Reader");

M2_CONNECT(r, out_port, c, read_port);
M2_CONNECT(w, out_port, c, write_port);
```

Constraints and Assertions

Constraints are used to specify the design via declarative means (as opposed to imperative specification which is used in components). Assertions are used to check whether the rest of the design conforms to given requirements. Both constraints and assertions are described in terms of events: their execution, the values associated with them, and their tags. The events referenced by constraints or assertions must be exposed by means of coordination or view ports. Depending on the logic used to describe them, constraints can be enforced either by the base model or the scheduling phases of execution. Similarly, assertions may also be checked by monitors either in the base model or in the scheduling phase.

An example of a constraint in METRO II is shown here. In this case it is a rendezvous type constraint. When the constraint is instantiated it requires two pointers to events. These pointers, m1 and m2, typically are associated with two different components. The constraint solver of METRO II will first call the "isSatisfied()" function. If this returns true, then the "solveConstraint()" function will be called. This function changes the state of the events which ultimately effects the progression of the components. The relation between event state change and process state change is discussed in Section 4.4.3.

```
class m2_rendez_constraint : public m2_constraint
{
  protected:
    m2_event *_m1, *_m2;

  public:
    m2_rendez_constraint(m2_event* m1, m2_event* m2)
        : m2_constraint(M2_RENDEZ_CONSTRAINT)
```

```
    {
        _m1 = m1;
        _m2 = m2;
    }

bool isSatisfied()
{
    return ((((_m1->get_status() == (char)M2_EVENT_PROPOSED)
    || (_m1->get_status() == (char)M2_EVENT_WAITING))
    && ((_m2->get_status() == (char)M2_EVENT_PROPOSED)
    || (_m2->get_status() == (char)M2_EVENT_WAITING))));
}

void solveConstraint()
{

 if (isSatisfied()) {
        _m1->set_status((char)M2_EVENT_PROPOSED);
        _m2->set_status((char)M2_EVENT_PROPOSED);
    }
    else {
        if (_m1->get_status() == (char)M2_EVENT_PROPOSED) {
            _m1->set_status((char)M2_EVENT_WAITING);
        }
        if (_m2->get_status() == (char)M2_EVENT_PROPOSED) {
            _m2->set_status((char)M2_EVENT_WAITING);
        }
    }
}
};
```

Mappers

Mappers relate functional methods to architectural services. The most common usage of mappers is to transform or add values to the parameters of architectural methods. For instance, a functional method may have two arguments, while the architectural service has a third argument which the functionality is unaware of. The mapper object bridges the two together and provides the third argument.

An example of a mapper is shown here. This mapper is called "receive_mapper" and is used to map the consumer in a producer-consumer design example to a processing element, p. During mapping when the receive method is called by the functional model with two arguments, the mapper's *out_port* will call the architectural model's *receive* method which has 3 arguments. Also shown in the figure are the instantiation of the mapper along with how the mapper is connected between the functional model and the architectural model.

```
//mapper definition
class receive_mapper: public m2_mapper<i_func_receiver, i_arch_receiver>
{
public:

receive_mapper(sc_module_name name) : m2_mapper<i_func_receiver, i_arch_receiver>(name)
{}

void receive(void * data, unsigned long len)
{
out_port->receive(data, len, 1000);
}
};

//instantiation
receive_mapper r_mapper("receive_mapper");

//mapping between ports
M2_MAP(r, out_port, r_mapper, p, read_port);
```

Annotators and Schedulers

In METROPOLIS, the scheduling of events along with performance annotation was carried out with a special component called a quantity manager. It is difficult to have a general mechanism to handle both scenarios since different design styles are used specify both. In METRO II, these two aspects will be separated by using *annotators* and *schedulers*.

Annotators are objects that write tags to events. Each tag is determined in terms of the event, the event's values, and any parameters supplied to the annotator. Only static parameters are permitted for annotators, which may not have their own state.

The instantiation of a physical time annotator is shown here. *r.read_event_end* and *w.write_event_end* are events associated with a reader and writer component respectively. These two events are added to a list of events to be considered for annotation. In addition, a table indexed by these events is created along with assigning time units required for execution (1 and 2 units respectively). This list and table are then added to the annotator object itself. If these events are present during the second phase of execution, their tags will be updated accordingly.

```
// setup physical time annotator
```

```
std::vector<m2_event *> ptime_event_list;
ptime_event_list.push_back(r.read_event_end);
ptime_event_list.push_back(w.write_event_end);

std::map<const char*, double, ltstr> ptime_table;
ptime_table[r.read_event_end->get_full_name()] = 1;
ptime_table[w.write_event_end->get_full_name()] = 2;

m2_physical_time_annotator* ptime = new
m2_physical_time_annotator("pt_annotator",ptime_event_list,&ptime_table);
register_annotator(ptime);
```

Schedulers are objects that can disable proposed events based on their scheduling policy. After the annotation phase has completed, the scheduling phase begins. Based on the scheduler's local state, the proposed events, and their values and tags, scheduling occurs which can lead to the disabling of some proposed events.

A logical time scheduler is annotated and events which should have an effect on the logical time of the system are added in the following code snippet. During the third phase of execution, any events proposed which are in this list will have the ability to contribute to the advancement of the logical time of the simulation.

```
// setup logical time scheduler
m2_logical_time_scheduler* ltime = new m2_logical_time_scheduler("lt_scheduler");
ltime->add_event(r.read_event_beg);
ltime->add_event(r.read_event_end);
ltime->add_event(w.write_event_beg);
ltime->add_event(w.write_event_end);
register_scheduler(ltime);
```

4.4.3 METRO II Execution Semantics

Like METROPOLIS, the semantics of the METRO II framework will be centered around the connection and coordination of components. The execution semantics discussed here are involved in the simulation of a system for design space exploration. The two key aspects of the METRO II execution semantics are events and processes.

The key concept underlying METRO II is an *event*. An event is a tuple $< p, T, V >$ where p is a process, T is a tag set, and V is a set of associated values. An event denotes an action taken by a process (p). Events may be associated with annotations (T) and state (V). Annotations correspond to quantities in the design, such as time or power. State includes variables that are in the scope of an event.

METRO II has a three-phase execution semantics. In order to discuss this semantics, two other concepts must be introduced: process states and event states.

In Figure 4.16 the states that an event can have are shown. Events can be inactive, proposed, and annotated. All events begin as inactive. As the self loop shows, they can remain inactive indefinitely. When a method call on a required port generates an event it becomes proposed. It then will be annotated. If the event is then deemed appropriate to enable (via a variety of scheduling decisions) it will transition to inactive again.

Each process in Metro II has two states: *running* or *suspended*. Processes execute concurrently until an event is *proposed* on a required port of the component containing the process or until they are blocked on a provided port. At this point they transition to suspended. Once the event is enabled or the internal blocking is resolved, the process will return to running.

Based on this treatment of events, the design is partitioned into three phases of execution. In the first phase, processes propose possible events, the second phase associates tags with the proposed events, and the third phase allows a subset of the proposed events to execute.

1. **Base Model Execution.** The base model consists of concurrently executing processes that may suspend only after proposing events or by waiting (blocking) for other processes. A process may atomically propose multiple events – this represents non-determinism in the system. After all processes in the base model are blocked, the design shifts to the second phase. The execution of processes between blocking points is beyond the control of the framework.

2. **Quantity Annotation.** In the second phase, each of the proposed events is annotated with various quantities of interest. For instance, a proposed event may be annotated with local and global time tags. New events may not be proposed during this phase of execution. In this way, events and the methods they correspond to can be associated with cost.

3. **Scheduling.** In the scheduling phase, a subset of the proposed events are enabled and permitted to execute, while the remainder remain suspended. Events are enabled according to schedulers and constraint solvers. These enabled events then become *inactive* again while simultaneously allowing their associated processes to resume to the *running* state. At most one event per process is permitted to execute. Once again, new events may not be proposed during this stage. Scheduling may be based on the resolution of declarative constraints or on imperative code.

A collection of three completed phases is referred to as a *round*. Figure 4.16 illustrates the process states, event states, and the three phases in the execution semantics. Self loops on the inactive and annotated states illustrate that multiple rounds may pass without an update to a particular event's state.

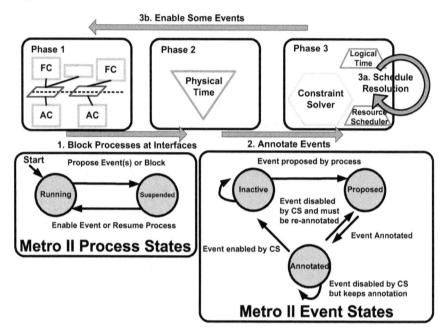

Figure 4.16: Metro II Three Phase Execution Semantics

Table 4.2 illustrates the relationships between events and phases. In the first phase (base) events can be proposed and their values can be read or written. In the second phase (annotation), tags can be read and written and values can be read. In the final phase (scheduling), events can be disabled and their tags and values can be read. The semantics have been carefully designed so that event manipulation adheres to our separation of concerns methodology. This is very helpful not only in debugging simulation but also in making sure that the framework functions efficiently.

Table 4.3 indicates the METRO II elements and their characteristics. This details the presence of threads as well as the ability to manipulate events, tags and values. Also it indicates if there is hierarchy.

Phase	Events		Tags		Values	
	Propose	**Disable**	**Read**	**Write**	**Read**	**Write**
Base (1)	Yes				Yes	Yes
Annotation (2)			Yes	Yes	Yes	
Scheduling (3)		Yes	Yes		Yes	

Table 4.2: Phase/Event Relationships

Table 4.3: METRO II Elements and their Characteristics

Type	Threads	Events	Tags	Values	Hierarchy
Component	0+	Generate	R/W	R/W	yes
Adaptor	0+	Generate	R/W	R	yes
Annotator	0	Propagate	R/W	R	no
Scheduler	0	Disable	R	R	yes

Events, and by extension, services, may be annotated by quantities of interest. Quantities capture the cost of carrying out particular operations and are implemented using quantity managers. *Annotators* are special components that provide annotation services. *Schedulers* are similar to quantity managers, but instead of a quantity they provide scheduling and arbitration of shared resources. *Adaptors* modify tags and provide interfacing between different models of computation. Depending on the MoC used and the needs of the design, different annotators and schedulers can be used.

4.4.4 METRO II Mapping Semantics

A key feature of METRO II is the ability to separately specify the functional and architectural models. The two are then *mapped* together to produce a system model with performance metrics. Mapping is realized by adding constraints between events from the functional model and events from the architectural model.

This section will present the three options for the execution semantics of mapping in METRO II. Each subsection will contain two components. The first is a table which summaries the options. This includes the correspondence between events in the architectural and functional model as well as any assumptions that the proposal makes. Secondly a small description of the option's operation is provided as well.

112

For Options 1-2, the *call graph* of the mapping is shown in Figure 4.17. Option 1 is the first call graph shown, option 2 follows, and option 3 is the last. For options 1-2 the structural view (upper right of the figure) is a connection between required ports in the functional model and provided ports in the architectural model. For option 3, the mapping structure is different and is between required and required ports. This is shown in Figure 4.18.

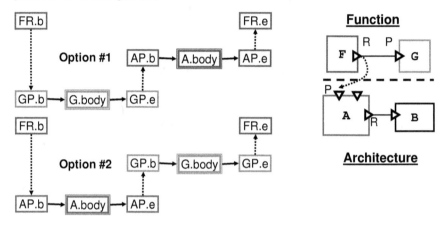

Figure 4.17: METRO II Options #1 and #2 Mapping Semantics

Option 1: Execute Functionality then Architecture (FTA)

The first option is a sequential option in which the functional model begins execution before the architectural model. Some of the highlights of this option are captured in Table 4.4.

Figure 4.17 shows both a structural and "call graph" view of mapping in the first option. The ports in these and future diagrams are specified with the first letter of the component they belong to. Also, ports are designated as "R" or "P" if they are required or provided. "b" and "e" designate the begin and end events respectively. These designations can be combined. For example "FPe" would indicate the end event of component F's provided port.

Figure 4.17 shows the mapping structure of a system using this option. The functional model contains a method call to G from P. The mapping of this method call occurs by assigning events proposed by FR to events proposed by AP. This is considered a required port to provided port mapping structure.

Figure 4.17 also shows the "call graph" of the system. Boxes with single line borders are events.

Option	Execution Order in Simulation	Mapping Structure (Func ↔ Arch) Port	Event Correspondence	Requires Blocking
1	Functionality then Architecture	Required ↔ Provided	FR_b, GP_b GP_e, AP_b AP_e, FR_e	Yes

Goals
The initial option to get a functioning framework.
Prevent unnecessary architecture execution by ensuring functional correctness/progress first.
Unique Assumptions
Architectural model can not directly influence functional model execution.
Functional model must resolve non-determinism in the absence of an architecture.
Architecture components which are mapped are passive.

Table 4.4: Option 1 Overview

Boxes that have two line borders are code blocks that may or may not contain events. The arrows indicate program flow (from left to right). If an arrow is dashed it means that two events connected to it are treated as a single event by the framework. The functional component F calls a method from component G. This is mapped to the architectural component A, which further uses architectural component B when providing the service.

Execution in this option occurs as follows: component F contains a process. This process is responsible for proposing event "FRb". "FRb" corresponds to "GPb" (in G). Once these events are enabled, "G body" (the code body of the function call to G) can now execute. Upon completion, "FPe" (in G) will be proposed. This event corresponds to "APb" in the architecture. The architecture body "A body" can now execute and culminate with the proposal of "APe". As shown, "APe" corresponds to "FRe" which completes the execution.

As shown, mapping of methods is carried out by invoking the mapped architectural service in the process of the caller AFTER the corresponding functional method has completed execution.

Option 2: Execute Architecture then Functionality (ATF)

The execution semantics of mapping involves executing mapped architectural services before their functional counterparts. When a mapped method is invoked by a functional process, the begin event of that method is initially proposed, and a phase change is permitted to occur. If this event is enabled, then the architectural service executes first, immediately followed by the invoked functional method. After this, the end event of that method is proposed, with a subsequent phase change. Both the functional method and the architectural service are executed by the functional process; there are no special mapping processes. Additionally, both the functional method and the architectural service may block internally while waiting for other processes.

The functional method is parameterized with arguments and has a return type. The architectural service is also parameterized, but the return value is not used. The correspondence between the architectural service parameters and the functional service parameters is specified at compile-time.

This proposal is in some regards the opposite of the first proposal. It is summarized in Table 4.5.

Option	Execution Order in Simulation	Mapping Structure (Func ↔ Arch) Port	Event Correspondence	Requires Blocking
2	Architecture before Functionality	Required ↔ Provided	FR_b, AP_b AP_e, GP_b GP_e, FR_e	Yes
Goals				
Allow the architecture model to resolve non-determinism in the functional model.				
Provide a more intuitive semantics of how a architecture resource driven system would behave.				
Unique Assumptions				
Whenever the mapped functionality "internally" blocks, the corresponding architecture "internally" blocks.				
If the architectural model "internally" blocks, the functional model does not have to "internally" block (but can).				
Architecture components which are mapped are passive.				

Table 4.5: Option 2 Overview

Figure 4.17 for the previous option shows the "call graph" for execution between the functional and architectural models. Basically, the functional methods need to be completed before corresponding architecture services start. However, in some cases this approach might not be able to reflect all the situations in the mapped system.

For instance, let's consider a shared FIFO example. Option #1 cannot assure that the architectural ordering decision impacts the functional execution, since the function methods will finish before

Option	Execution Order in Simulation	Mapping Structure (Func ↔ Arch) Port	Event Correspondence	Requires Blocking
3	Concurrent Functionality and Architecture	Provided ↔ Provided	FR_b, AP_b, FP_b FR_e, AP_e, FP_e	No

Goals
More closely follow a platform-based design philosophy.
Introduce concurrent architectural and functional execution.
Remove any suspension assumptions between models
Unique Assumptions
Finer granularity of execution through resource (i.e. bus, cpu, etc) access protocols.
Requires more opportunities for coordination between functional and architectural models.
Architecture components which are mapped are active.

Table 4.6: Option 3 Overview

the architecture is invoked. Therefore, the shared FIFO example may not work as expected with option # 1 if one wants to use the state of the architectural FIFO to block functional processes (i.e. it is full). Essentially, functional non-determinism cannot be resolved by the architecture. Such operations may be desirable when the architecture is better able to perform given the opportunity to make decisions based on its state (free resources for example). This also removes some scheduling burden from other areas of the system.

This second option remedies this problem by requiring that architecture services should be completed before the corresponding function method starts. The new "call graph" is shown in Figure 4.17. This proposal shares the same mapping structure as option # 1.

Option 3: Execute Functionality and Architecture Concurrently (FAC)

The third option is summarized in Table 4.6. There is a consensus that the MetroII environment is rooted in the Platform-Based Design methodology [SV02], where the functional model and the architectural model meet in the middle with a set of well-defined services as the binding contract. To the architectural model, the middle point represents what services it can provide to implement certain functionalities, or to estimate the implementation cost. To the functional model, the middle point describes its need of services to achieve its entire function. If we look at the design scenarios, the services that are exposed at the middle point include *execute, read_fifo and write_fifo*. Therefore, the architecture model has to provide at least those services. As the three proposals exhibit, there are multiple possibilities in terms of which ports to map. In fact, the syntactic difference does not really matter.

What matters is the role of the mapped architectural component and its relationship to the components on the functional side. Imagine on the functional side, the source component calls *write_fifo* that is provided by FIFO1. No matter which part in the connection (the required port, the provided port, the connection) is mapped to the architecture, we expect the architectural service at some point to perform *write_fifo*. In that sense, the architectural counterpart corresponds to FIFO1, where both of them react to *write_fifo* request and do the job. If we can agree on this correspondence, then any mapping syntax will work. i.e. on the functional side, the required port, the provided port and the connection each represents a pair of events; on the architectural side, the service is also represented by a pair of events. Then mapping establishes another pair of correspondences between the two pairs of events. However, from the methodology point of view, where we emphasize the meeting point between functional and architectural models, mapping connections or provided ports from the functional side seem to be better choices.

When running the functional and the architecture models together[1], we would like the mapped services on both sides to finish simultaneously, because this will provide the most information about how an architectural model implements a functional model. However, there are concerns about the fact that suspension of processes on either side would prevent the entire mapped system from progressing. This is primarily caused by the semantics mismatch of the services from both sides. By carefully designing the consistent services, we should be able to make the mapped system work even with blocking behaviors on either or both sides.

The mapping structure and "call graph" for the 3rd proposal are shown in Figure 4.18. Notice that in option # 3, provided ports in the functional model are mapped to provided ports in the architectural model as well. This is different from the previous two option. Also in the "call graph" it is shown that "correspondence" points must be created in the form of "protocols" in order to create a more granular operation at the event level in each model. An example of such a "protocol" will be shown in more detail in the hand traces for proposal # 3.

F and *G* are two components in the functional model, where *F* is making a method call on its required port *Req* to *G*'s provided port *Prov*. In this example, the architecture is represented by components *A* and *B*, and the provided port of *G* has the same interface as the provided port of *A*. In this

---

[1]Note that we can also run the functional model first, recording the service demands, then drive the execution of the architectural model. But this eliminates behavior where the feedback from the architectural model would affect the execution of the behavior model.

case, we can say that component *G* has been *mapped into* component *A*. For simplicity, assume that the interface of ports *F.Req*, *G.Prov*, and *A.Prov* contains only one method. The mapping between *G* and *A* is realized by placing rendezvous constraints on the *begin* and *end* events associated with this method, as shown at the bottom of Figure 4.18. Starting at the bottom left of the figure, one sees the initial event proposal of *F.Req.b*. Moving to the right, the other events are proposed in turn. Left-right arrows indicate causality while vertical arrows indicate the presence of a constraint.

Within the framework, these rendezvous constraints are handled in the same way as any other event constraints during phase 3 of the execution semantics. Mapping uses the same infrastructure as the rest of the system and therefore the simulation is not burdened with another set of semantics for mapping.

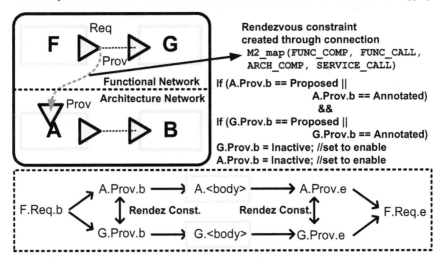

Figure 4.18: METRO II Option #3 Mapping Semantics

4.4.5 METRO II Implementation

An initial implementation of the METRO II framework has been carried out in SystemC 2.2. The framework has been tested under Linux, Solaris, and cygwin.

The infrastructure is summarized in Figure 4.19. The *sc_event* and *sc_module* classes from SystemC are leveraged directly to derive the corresponding *m2_event* and *m2_component* classes. A method is characterized by begin and end events. Multiple methods are wrapped together into an interface, which

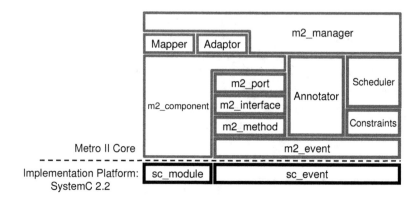

Figure 4.19: Implementation of Metro II

is associated with ports. Components contain possibly multiple ports. Mappers are a special type of component which translate arguments between functional methods and architectural services. Annotators directly annotate events, while constraints are defined over them. Adaptors translate between different models of computation. Schedulers enable certain events after carrying out constraint resolution. The manager coordinates the execution of objects in all three execution phases.

4.5   Conclusions

The Platform-based design methodology imposes a number of requirements on system-level design frameworks. METROPOLIS represents the first complete attempt at such a framework. To address the limitations of METROPOLIS, in this paper we identified three main features that must be enhanced and described how the next generation METRO II framework will support them. The aim is to develop a framework that supports the import of heterogeneous IP, facilitates behavior-performance orthogonalization, and eases design space exploration. This is achieved by building an integration framework based on events with three separate phases of execution.

We are currently implementing the three options related to the mapping semantics of the METRO II environment, and developing further case studies to exercise its capabilities. Future work involves the continued development of the METRO II infrastructure, especially as it relates to adaptors as well. Additionally, we are trying to integrate other tools such as Simulink and Modelica so that continuous time systems can be simulated very easily. Also of interest is the tradeoff between the use of constraints in METRO II to perform synchronization versus built in mechanisms of the underlying SystemC framework. There exists much work to be done and we have just touched the tip of the iceberg regarding the framework's potential.

Chapter 5

Design Flows

"To design is to communicate clearly by whatever means you can control or master." –
Milton Glaser, Graphic Designer

Ask any chef and they will tell you that it is not enough to only have the finest ingredients and the best cooking utensils. Even if the menu selections and wine are paired perfectly and the ambiance just right, if the recipes and techniques to cook the food are not in place, the end result will be a disaster. The same is true for embedded system design. It is not enough to have the concepts of abstraction and modularity and the implementation targets of programmable platforms if the design flows are flawed. As vital as the individual components are, it is only by unifying them with a rigorous design flow that we will truly offer techniques to tackle the many challenges in this area.

What now remains is to demonstrate how the contributions outlined in previous chapters (programmable platforms, platform-based design, and modeling frameworks) combine to create a design flow to accomplish the desired outcomes of accurate and efficient design space exploration. This chapter presents an example of design flows which are used later in the work along with illustrating a naïve design flow.

This chapter will present a naïve design flow and how specific flows for functional design, architectural service modeling, and mapping are superior to this and other approaches. These superior flows will be used throughout the remainder of this book.

5.1 Chapter Organization

The chapter begins by providing a naïve design flow in Section 5.2. This flow provides a point of comparison for the other flows introduced throughout this chapter. Section 5.3 provides a description of related work in this area. This is broken down by RTL and system level flows. Section 5.4 provides the needed background and definitions. Section 5.5 kicks of the discussion of the proposed design flows. Specifically Section 5.6 describes a flow for functional modeling and mapping. Section 5.7 describes a flow for architectural service modeling. Section 5.8 discusses the tradeoffs which are made in each of these flows. Section 5.9 provides an overview of the multimedia and distributed systems which will be investigated later in this book. Finally Section 5.10 provides conclusions and future work.

5.2 Naïve Design Flow

Before presenting the improved design flow approaches to be elaborated on in this work, a naïve approach will be presented as an example of how embedded system design is often done and to clearly illustrate the advantages of the proposed approach.

A typical simulation and synthesis design flow which minimally attempts to use ESL ideas (abstraction and modularity) may proceed as follows (Figure 5.1):

1. Create an abstract and modular architecture service design in a system level design environment. This will be accomplished in an environment supporting various models of computation and mapping strategies in the best case. In the worst case only one model of computation will be available and the functional and architectural modeling efforts will not be distinct.

2. Estimated data is used to annotate the simulation. This data may come from best practices, back of the envelope calculations, data sheets, or area based timing information. A final design is chosen from a set of simulations based on this data.

3. Once a design decision is made during design space exploration, one creates a "C" model (or equivalently a high level language description which is sequential in nature) *manually* which should represent the abstract system. This is needed since the abstract system has no automated path to synthesis.

4. Create an RTL model *manually* from the "C model". This is done since RTL has a synthesis path to implementation and industry expertise exists with designers who routinely perform this transformation.

5. Finally, from the "golden" RTL model create an implementation.

As Figure 5.1 shows, just because the initial design is *abstract and modular* it does not guarantee accuracy or efficiency. In fact, one must take explicit steps to ensure such characteristics. Weaknesses are found in all areas of the naïve design flow. These include but are not limited to error prone manual steps and lengthy design iterations. This design flow is currently tolerated because the level of complexity in today's designs is such that the methodology gap introduced in Chapter 2 can be overcome with the iterations seen in this flow at a cost low enough to justify continuing this path. However, this will not be the case as the iteration time grows and design times shrink. Additionally the "length" of the iterations in terms of designer teams involved and processing steps will grow as well.

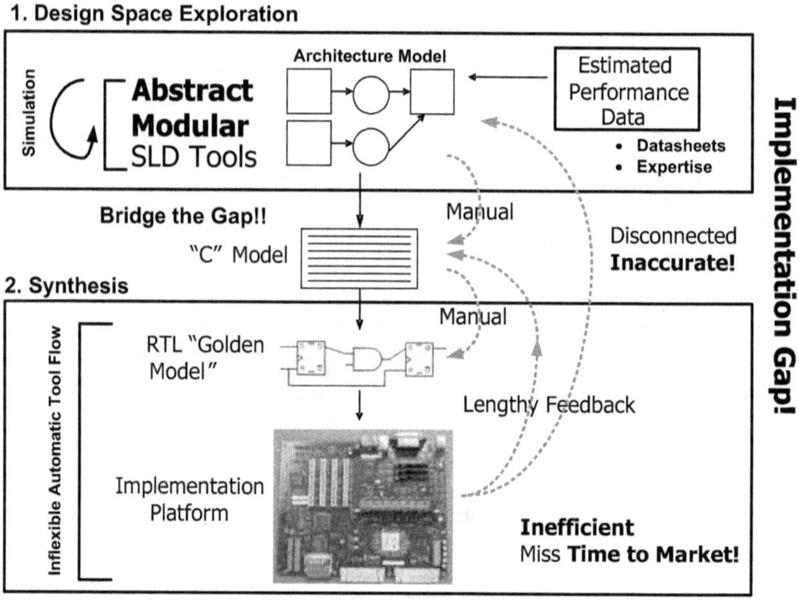

Figure 5.1: Naïve Design Flow

5.3 Related Work

When discussing related work for design flows it is important to describe the context in which they are presented. Each design flow will have certain goals and requirements when put in the context by which it is intended to be used. There are two contexts which are important to examine for this work.

RTL flows typically start with an HDL style language and target synthesis and low level (gate level) verification. *System Level* flows start with higher level programming languages and target design space exploration and functional verification.

RTL level flows are traditionally very top down. They start with an HDL style description written in VHDL or Verilog. This is a mixture of what the design should do along with structural notions of how it should be done. These descriptions proceed through a process called *logic synthesis* whereby the functionality described in the HDL is translated into a *netlist*. This netlist then becomes a collection of Boolean algebraic expressions. These expressions can be manipulated by a number of exact and heuristic optimization stages to get a *gate level* netlist. The gate level netlist then undergoes a process known as *physical design* which includes assigning standard cells (a physical schematic) to the gates (technology mapping), deciding on the physical location of logic blocks (placement), the connection of signals in the design (routing), and then the optimization of the design for area (compaction). The end result is a design specification (i.e. GDSII) for fabrication by a semiconductor manufacturer. These flows are well established and described in [HS02], [HS00], and [DGK94]. The majority of these flows are fully automated and require little input from the designer. The community at large is reliant on the correctness of these flows and has a high level of confidence in them.

System level flows approach the design problem in a variety of ways. They differ in their ultimate goal (simulation, synthesis, verification) and also their ideology (top-down, bottom-up). In [CGJ⁺02] a complete flow for systems specified as finite state machines is described. The xPilot tool from UCLA [CCF⁺05] begins with high level descriptions written in C or SystemC and transforms them to RTL. This RTL will then go through a flow as described earlier. Another type of flow begins at the same entry point, but instead of generating RTL, the code is mapped onto a mix of prefabricated and generated hardware. Tensilica is a company that epitomizes this flow [RL04], and has been successful at selling highly customizable processors that are tailored to customers' workloads. In [GK05] the authors describe a variety of flows to build Application Specific Instruction Processors (ASIPs). Finally, another type of system-level flow focuses only on validation and does not consider a path to implementation. Specification capture usually takes place in a formally defined language which can be understood by modeling checking tools. System properties are then specified within these flows and checked by the model checking tools. SMV [McM93] and SPIN [Hol97] are two model checking tools that enable such flows.

Tradeoffs can be made between architecture models based on information they provide regarding the cost of their selection. These costs can be performance, power, area, etc. These costs need to be accurate while allowing the architecture models to be *abstract and modular*. These are the challenges outlined in Chapter 1 and will be addressed in this chapter. Figure 5.2 shows qualitatively how "this work" compares with other architecture modeling styles in terms of relative accuracy (how simulation compares to actual implementation) and relative efficiency (how easily complex systems can be captured). The other styles compared are based on the classification given in [Don04] regarding TLM modeling styles. This illustration clearly places this work in the context of the existing approaches.

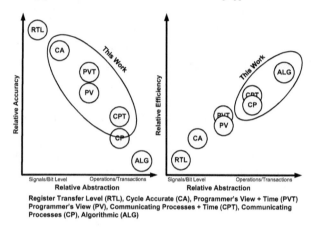

Register Transfer Level (RTL), Cycle Accurate (CA), Programmer's View + Time (PVT) Programmer's View (PV), Communicating Processes + Time (CPT), Communicating Processes (CP), Algorithmic (ALG)

Figure 5.2: Proposed Service Style Versus Existing Service Styles

5.4   Background and Basic Definitions

This chapter and this work in general requires that many terms be defined. These terms often have been used in other work using different language and in different contexts. This section is an attempt to reduce ambiguity. The terms here are meant to specifically highlight concepts developed and leveraged by this work. The language used in these definitions is meant to strike a balance between being too generalized but at the same time not forcing unnecessary formalisms.

Abstraction (a concept), behaviors (a result), and functions (a specification) can only be physically realized (i.e. implemented) if they are put in the context of a model with a relation to a device or collections of devices in the physical world. The first such idea is a *platform*. Generically we can think of architecture platforms with the following two definitions (from [BPPSV05]):

Definition 5.4.1 An Architecture Platform *consists of a set of elements, called the* **library elements**, *and of* **composition rules** *that define their admissible topologies of connection.*

Definition 5.4.2 *Given a set of library elements D_0 and a composition operator $\|$, the platform closure is the algebra with the domain $D = \{p : p \in D_0\} \cup \{p_1 \| p_2 : p_1 \in D \wedge p_2 \in D\}$ where $p_1 \| p_2$ is defined if and only if it can be obtained as a legal composition of agents in D_0.*

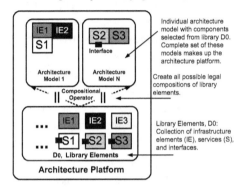

Figure 5.3: Architecture Platform Composition and Creation

Figure 5.3 demonstrates the definitions related to platforms. This illustrates a set of library elements and examples of the architecture models that can result. This is especially important for the work discussed here since the library of elements represent smaller architecture service IP models for programmable platforms and the collection of these elements creates a *platform instance*.

Definition 5.4.3 Architecture Model - *an architecture platform instance. Of the possible platforms that can result from a collection of library elements, one particular selection is an architecture model.*

System level event based services are of particular interest for two reasons. First the level of abstraction directly attacks the level of complexity currently seen in designs today. Additionally, it allows rapid design space exploration. Event based frameworks are useful since a wide variety of models of computation can easily be framed using events. Events also have the ability to carry the service cost along with them in the form of an event "value". This makes such frameworks very flexible. Examples of such frameworks were shown in Chapter 4.

Definition 5.4.4 Service - *a library element with a set of related interface functions and a cost. A service* *is a tuple $<f, c>$ where f is a set of interface functions and c is a set of costs. Services are the building blocks of an architecture model. All services are library elements but not all library elements are services. Library elements may provide infrastructure for creating an architecture model but not be visible to the functional model through interfaces or may not have costs. These two aspects however are requirements of a service.*

When creating an architecture service, there are two primary issues that must be resolved. The first issue is choosing the level of abstraction at which the service should be created. This answers the question at what level of granularity will the services be offered and how the components which comprise the service can interact. The second issue is what is the underlying semantics of the service. How will they synchronize? How will they communicate? How are they scheduled? These questions are in regard to inter- and intra-service relationships. In this section, these issues will be addressed specifically at the system level using an event based semantics. In system level models, the level of abstraction is at the transaction level or higher. This answers the first question. The semantics that will be discussed here are those which use events during simulation for a variety of issues including synchronization, annotation, and communication. This answers partially the second question.

A system level service is a collection of components (library elements). Each of the components have interfaces which expose their capabilities to other components. A service requires that at least one of the interfaces is exposed to the functional model or to other services. These interfaces are called *provided interfaces*. Interfaces between components making up the service alone are called *internal interfaces*. A service can be composed of a single or multiple components. If more than one component has a provided interface, then it is considered having *multiple interfaces*. Multiple interfaces allow for a service to have more than one cost model.

Definition 5.4.5 Interface *- a set of operations included in a service which can be utilized externally. These can be collections of functions (C style methods) or transactions.*

A service ultimately then corresponds to a set of event sequences generated by collections of components with various sets of interfaces with various costs. This execution therefore represents one possible behavior of a system. Here is the taxonomy of service types at the system level (as shown in Figure 5.4):

- **Single Component, Single Interface** (SCSI) - a service composed of a single component. The provided interface is the only interface provided to the functional description or other services. There is only one cost model provided with this service which is accessed through the single interface.

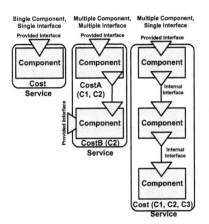

Figure 5.4: Architecture Service Taxonomy

- **Multiple Component, Single Interface** (MCSI) - a service composed of multiple components. Only one of the components has a provided interface. The presence of multiple components allows for a more complicated cost model, hierarchical composition of services, and hierarchical interfaces. Only one cost model is provided to the functional description.

- **Multiple Component, Multiple Interface** (MCMI) - a service composed of multiple components with multiple provided interfaces. This configuration allows for multiple cost models along with the advantages of a MCSI configuration.

Notice that single component, multiple interface (SCMI) services are not present because this scenario does not make sense given the fact that there are no internal interfaces to allow for various inter-service component interactions (and hence generate different cost models). Note that SCSI services can return different cost values based on the parameters provided to the interface. They are just restricted to one cost model.

Services can also be active or passive. An active service is a service which can generate interface calls. This can be thought of as a component which has an executing thread. A passive service is one which responds to interface calls. Naturally this response could in turn cause it to trigger an interface call itself.

Architecture topologies can be formed as well using collections of these types of services. Ul-

128

timately it is these topologies which form the architecture model. For example there are two primary styles. The first is a branching structure. This allows for services that use all types of service categorizations. This is illustrated in Figure 5.5 of the left. A branching structure is one in which services are connected in such a way that a service may interact with any number of other services. For example a bus service can interact with two or more computation services and a memory service. A ring structure on the other hand only allows for single interface services (SCSI, MCSI). This is illustrated in Figure 5.5 on the right. A ring structure can be useful for certain networking topologies. Also this structure often simplifies the scheduling problem as well as an analysis of its execution. Of course it is possible to have mixed topologies in which various aspects can be classified as either branching or ring.

Definition 5.4.6 System Level Architecture Modeling - *a collection of services at the transaction level or higher of abstraction. These services can be classified as SCSI, MCSI, or MCMI. Additionally the topology of the system can be defined as branching, ring, or a hybrid of the two.*

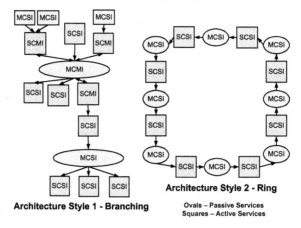

Figure 5.5: Composing Architectures Using Services

Figure 5.6 provides sample constructions of SCSI, MCSI, MCMI services built from METRO II components. Each of the services are either composed of one or multiple METRO II components as denoted by their classification. In addition, all components which constitute a service are encompassed by a "wrapper". This wrapper becomes the boundary of the service upon which the provided interfaces and the cost of the service are defined. This wrapper will provide a consistent global interface to all other services to facilitate their connection. These connections will create the architecture model itself. Additionally each service is provided with a scheduler and annotator. The scheduler will be used in the third phase of the service's execution and the annotator provides the cost model for the service in the second phase. Service "provided interfaces" of the wrapper are connected to METRO II "provided ports" of select components. One set of METRO II "required ports" is visible at the wrapper interface to allow the service to take advantage of other services if need be.

Each service classification (SCSI, MCSI, MCMI) differs in how many METRO II "view ports" are provided to the wrapper. View ports can be used to observe the operation of the service. These ports will be useful in creating structures to verify properties of the service (and hence architecture). In the case of SCSI and MCSI, there will only be one view port provided. This port corresponds with the component which connects its provided port to the wrapper's provided interface. In the case of MCMI, each component with provided ports serving as provided interfaces will have its view port present at the wrapper level. Additionally in MCSI and MCMI, "rendezvous constraints" are required to synchronize the components. In MCSI, rendezvous constraints between all components with provided/required port relationships are created. In MCMI, rendezvous constraints are created between components with provided/required port relationships provided that at least one of the two components does not contribute to the provided interface.

Both branching and ring architecture styles can be created using these service types. Whereas METROPOLIS required that active services have processes and passive services have media, METRO II does not have this distinction. All services are only composed of components. This potentially leads to more flexibility in specification or dual operating mode services (a switch that indicates if the service is passive or active). METROPOLIS required that mapping tasks be provided with the architecture model as active objects by which the functional model can be mapped to. This is not needed in METRO II. A provided interface itself can serve as this function using the interface of the METRO II component's provided port. More on METRO II can be found in Chapter 4.

For the sake of this work an architecture model has to perform the following tasks:

- **Capture the desired services for the given abstraction level**. For example at the logic gate level of abstraction, an architecture model must capture the number of inputs and outputs it is responsible for, as well as potentially capturing the interactions within the component during calculation. For the DCT example, it again must capture the inputs and outputs. The behavior internally during computation will be much more complex however. Note that an architecture model does not have to capture functionality. That is the job of the functional model. For example the architecture

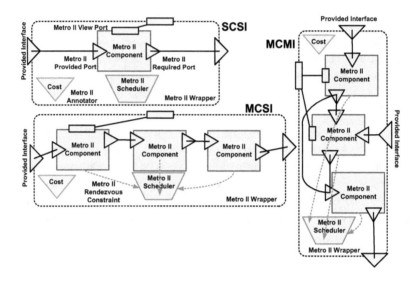

Figure 5.6: Architecture Service Model Proposal in METRO II

model of a logic gate does not need to calculate the outcome of "A AND B". It only needs to model the services involved in such a computation.

- The second aspect of architecture modeling is providing a *Cost* associated with the service. This cost will be associated at the granularity of the operations recognized in the architectural level. For example, the AND logic gate model may simply be annotated with the information that the cost of such an operation is 2 time units (whatever those units may be). However the DCT operation may not have a fixed cost. Its overall cost will depend on the type, order, and number of internal operations that are modeled within the DCT operation. This may depend on the state of the DCT, the types and size of its operands, or even the temperature of the device if that is so modeled.

Definition 5.4.7 **Cost** *is the consequence of using a service. Typically for embedded architecture models, cost is thought of as power, execution time, area, etc. Typically this is a physical quantity. These physical quantities are of interest during design space exploration. Cost can be a function of various variables or conditions such as input type, count, size, state of the system, etc.*

An event is the fundamental concept in the framework of solutions described. In METROPOLIS for example, an event represents a transition in the action automata of an object under simulation. An event is owned by the object that exports it and during simulation, generated events are termed as *event instances*. Events can be annotated with any number of quantities (i.e. costs). In addition, events can partially expose the state around them and constraints can then reference or influence this state.

Definition 5.4.8 Event - *logically an event instance denotes system activity. In this thesis, formally an event is a tuple $< p, T, V >$. p is the process which generated the event and therefore the event is associated with it. T is a set of tags. Tags are used to assign partial or total orders to events and correspond to performance annotations. Finally V is a set of values of the event. Values can be used to hold information regarding parameters and return types.*

Events can be used to transport costs throughout the system. This occurs through a process called *annotation*.

Definition 5.4.9 Annotation *is the assignment of a tag to an event. These annotations will typically be considered together at the conclusion of architecture execution in order to determine various metrics by which to evaluate the application running on a particular architecture.*

The overall cost of a given execution is the collection of a set of *transactions* that make up a system simulation.

Definition 5.4.10 Transaction - *a collection of service interface calls. Transactions can also be one service call which generates other service activity without explicitly calling their interfaces. This grouping is done to add abstraction and redirection into designs.*

Definition 5.4.11 Atomic Transaction *a transaction which only explicitly calls a service interface. This service must complete (the events generated are annotated and terminate) before another service can begin.*

The counterpart to an architectural model is the functional model. A key concept introduced in Chapter 3 was the separation between functionality and architecture. If a design flow is going to stay true to this separation these two model types must exist.

Definition 5.4.12 Functional Model - *a model which utilizes services to perform operations. A functional model itself has no notion of the cost of those services.*

For the sake of this work, a functional model has to perform the following tasks:

- It needs to utilize services in such as way to satisfy the specification of the system. The functional model is required to correctly capture the algorithm, protocol, or operation that the designer intends for it.

- It must perform required functional transformations on the data processed. For example a functional model of an adder must correctly produce the value of 5 if 3 and 2 are added. It cannot make any assumptions regarding the architecture's role in this process.

- It must adhere to the semantics of the model of computation it has been specified in.

The ultimate goal of this work is to allow design space exploration through simulation. In order to perform this simulation, mapping is required and the system must then be executed.

Definition 5.4.13 Mapping - *an assignment between behaviors in the functional model and services in the architectural model. Mapping can be "many-to-one". This allows "many" functional behaviors to "one" architectural service. For example a DCT and FFT behavior can be mapped to a single abstract service dealing with signal processing.*

For the sake of this work, a mapping has to perform the following tasks:

- It must bind all the required services from the functional model to compatible corresponding services from the architecture.

- It must ensure that the cost metrics obtained as a result of the specified mapping meet all system requirements.

Definition 5.4.14 System - *a complete mapping of functional model behaviors to architectural services.*

Definition 5.4.15 Execution *of a system is a set of architecture service interface functions invoked during the process of simulating an application. This results in a collection of costs as well which can be deemed the results of this execution. These costs are then used to evaluate the potential of the system model.*

Throughout this work the word "behavior" will be used. This is often an overloaded term. In this case it means the following:

Definition 5.4.16 Behavior - *a possible execution of a collection of services. How this execution is measured varies from system to system. An architecture model can be viewed as a set of behaviors. These*

execution sequences should be defined at an observable location such the memory contents, input and
output ports, or communication points (buses, switch, etc).

The next section (5.5) will detail two types of system level design flows. The first discusses functional modeling and mapping while the second details the creation of architecture services. These sections will discuss both the procedures and the components needed (using the frameworks from Chapter 4).

5.5 Proposed Design Flows

5.5.1 Functional and Mapping Flows

The approach taken in this work is to customize the platform-based design methodology to support our objective: enabling automated mapping for parallel heterogeneous embedded systems. Platform-based design advocates an initial separation between functionality and architecture, and a distinct mapping step to realize the system. The main problems tackled in this design flow are twofold: choosing how to model the functionality and architecture and then mapping the two together.

The problem of mapping between arbitrary functional and architectural models can be solved with two broad strategies. The first strategy attempts to bridge the gap between dissimilar models. The algorithms and techniques applied by this strategy are usually specific to the models of computation and abstraction levels employed by the models being mapped. This strategy has the capability to produce very good results for specialized problems, but there is little applicability to a broad class of systems. As a result, when either the functional or the architectural model changes, a new approach may have to be developed and reuse is difficult.

The second strategy initially transforms the models into a "common modeling domain" (CMD) where both the semantics and abstraction levels are compatible. The automated mapping approaches are then tailored to the common modeling domain. The advantage of this two-step approach is flexibility. Many models can be transformed into a common modeling domain and leverage the automating mapping flow.

The latter approach is adopted in this work. The development of the functional and mapping design flow in the remainder of this work is outlined in Section 5.6.

5.5.2   Architecture Service Flow

Architecture service modeling is the process of creating an environment to represent and expose *services* that can be used to implement functionality. Services represent capabilities of the underlying architecture upon which the design will be eventually implemented. These services are exposed to the designer and a correspondence can be made between the functionality present in the application model and the services exposed (this is a mapping). This service based environment is then used to investigate the performances that potentially can be obtained by using collections of these systems. This methodology requires the explicit separation of the functional model and the architecture model. This separation exists in the Platform-Based Design methodology (described in Chapter 3) and is required throughout this work. These sections will detail the design flow for creating architecture services of this type, provide details of programmable service models that have been created (based on the Xilinx IBM CoreConnect architecture), discuss extensions made to accommodate mapping and preemption, and outline the key features of this style of design space exploration.

5.6   Functional and Mapping Design Flow

The uniqueness of the functional and mapping design flow presented in this work is that it provides a unified way to view mapping problems from multiple domains. This allows the non-domain-specific aspects of the design flow to be combined into a common design framework. For instance, many manual design activities such design import, simulation, and debugging [CDH+05] can be handled in a common way. Based on this flow and the lessons learned from case studies, we have re-evaluated the goals from METROPOLIS and started development on the METRO II framework [DDM+07]. METRO II aims to support heterogeneous IP import and provide a more structured means for specifying performance annotations and mapping.

The approach taken in this work is based on the platform-based design (PBD) methodology [KMN+00]. PBD advocates a meet-in-the-middle design process, with the functional portion of the design (what the design does) constituting the top-down part and the architectural portion of the design (how this is carried out and at what cost) constituting the bottom-up part. An explicit mapping step binds these two parts together to realize the system model. Platform-based Design was discussed in more detail in Chapter 3.

To meet the challenges of deploying applications on heterogeneous multiprocessor embedded platforms, a synthesis – or automated mapping – approach is needed. Not only does such an approach decrease design time, but it also enables a correct-by-construction approach which reduces verification effort.

Structured modeling techniques are necessary to facilitate tractable synthesis. Our flow [ZDSV06b] within PBD is therefore focused on structured modeling between functionality and architecture. The 4-stage flow involves both modeling (functionality and architecture) and mapping. Modeling is based on the concept of common modeling domains (CMDs). CMDs are defined in terms of services. Mapping can be carried out using a variety of exact and heuristic techniques. Initially, we believe that mathematical programming [NS96] approaches are useful to consider since they are extensible, provide bounds on solution quality, and leverage advances in general-purpose solvers. The design flow is validated by applying it to studies from the multimedia and distributed system domains in Chapters 8 and 9 respectively. Functionality, architecture, and mapping for these systems are modeled within the METROPOLIS [BHL+03] framework.

Common modeling domains ensure that the semantics and granularities of the services being mapped match. First, concerns specific to parallel embedded systems modeling will be enumerated in Section 5.6.1. Next, these concerns are integrated within the functionality-architecture separation within PBD by using the concept of a common modeling domain. Mapping is described in terms of used and provided services in Section 5.6.3 and illustrated with a small example. A discussion of the modeling tradeoffs for both functionality and architecture is given in Section 5.8.

The flow is summarized in Figure 5.7. Stage 1 first finds the common modeling domain (CMD) between the functional and architectural models. Here, the architectural model is derived from the flow described in Section 5.7. Next, Stage 2 is concerned with transforming both models into the appropriate CMD. Mapping can then be formulated as a covering problem and solved in Stage 3. Further configuration of the system, i.e. assigning the architectural parameters, is carried out in Stage 4.

The focus of this research is to apply this flow to a number of heterogeneous programmable platforms from the multimedia and automotive domains. For the multimedia domain, the focus will be on stages 1, 2, and 3. For the automotive domain, the focus is on stages 1 and 4.

The reason that modeling domain tradeoffs are interesting to explore is that there is a lack of agreement on how heterogeneous parallel platforms should be programmed, evidenced by the large amount of experimentation for new MoCs for these systems. For lower abstraction levels such as hardware design, a single MoC – synchronous circuits – was sufficient for a large majority of designs, and considering such tradeoffs would have been less useful.

Also, these tradeoffs are feasible to consider due to the existence of several design frameworks that support many different MoCs. In particular, case studies from the multimedia and automotive do-

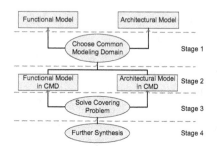

Figure 5.7: Common Modeling Domain Design flow

mains using different MoCs have been modeled in the METROPOLIS [BHL$^+$03] framework. These case studies will serve as the primary vehicles for further development of this flow.

5.6.1 Parallel Systems Modeling

Parallel embedded systems modeling can be carried out at different levels, depending on which aspects are explicitly captured. The modeling can be broken up into five levels, loosely based on the breakdown suggested in [ST98]. The levels are categorized by the modeling aspects which are made either explicit or implicit. Four levels are considered in this categorization, as summarized in Table 5.1.

1. The first aspect is whether or not *parallelism* is made explicit in the model. If parallelism is implicit, then the model has no concept of parallel execution, and parallelism must be extracted automatically to enable it to run on multiple resources. With explicit parallelism, processes that run concurrently have already been identified.

2. *Allocation* determines if computational resources are bound to the tasks/processes in the model. If allocation is explicit, the designer must specify which processing element each process executes on.

3. *Communication* fixes the realization of messages that are sent between processes. This involves determining the encoding of each message as well as the communication resources in the platform to which they are mapped.

4. *Performance* denotes whether the relevant cost metrics can be directly calculated or not. Models with explicit performance can evaluate cost metrics analytically, whereas implicit performance

Level	Parallelism	Allocation	Communication	Performance
1	Implicit	Implicit	Implicit	Implicit
2	Explicit	Implicit	Implicit	Implicit
3	Explicit	Explicit	Implicit	Implicit
4	Explicit	Explicit	Explicit	Implicit
5	Explicit	Explicit	Explicit	Explicit

Table 5.1: Modeling of Parallel Systems

metrics must be calculated via simulation. Different performance metrics are used for different systems. Predictable performance analysis is key to developing automation techniques.

In embedded systems programming, industry practice is to describe the system at level 1, in the form of untimed sequential C/C++ code. The latter stages of modeling are carried out manually. Typically, performance constraints for the system can be verified only at level 5. Since the transition from level 1 to level 5 is time-consuming and hard to verify, the design process becomes more complex. This roughly corresponds to the naïve design flow described in Section 5.2.

Note that the current industrial practice is well suited for general purpose uniprocessor systems. Levels 2–4 are absent from the general purpose uniprocessor flow, and there are less stringent (perhaps only average case) performance constraints on the system as a whole. For general purpose parallel systems, levels 2–4 are present, but the constraints to verify at level 5 are still lax, or even completely absent. This laxity is more tolerant of manual design flows. However, for parallel embedded systems, all the problems need to be addressed.

In order to remedy the problems, we need to automate the transformation of models between levels. Ideally, specification would be at level 1, and automation would enable transformation to level 5. However, this is difficult for a number of reasons. First, the transformation between levels 1 and 2 involves the automated extraction of parallelism, which is notoriously difficult. Secondly, in general, determining allocation, communication details, and analyzing performance in an automated fashion is infeasible. To tackle these concerns, the methodology has two main restrictions. First, the transformation from level 1 to level 2 is not automated, and still left to the designer. Second, system specification at level 2 can only be carried out in a specific manner. Namely, a correspondence must exist between the services that are used by the functionality and the services provided by the architecture.

Type	Effort	Runtime	Bounds	Portability
Deterministic Heuristics			X	X
Approximation Algorithms	X			X
Randomized Algorithms		X	X	
Mathematical Programming		X		

Figure 5.8: Automated Mapping Techniques

Given these type of systems to consider, we can now more fully define the mapping flow we will use.

5.6.2 Mapping Flow

"I have an existential map; it has 'you are here' written all over it." – Stephen Wright, Comedian

Mapping is the process of associating the functional and architectural models together such that the services used by the functionality are bound to those provided by the architecture. As mentioned, mapping involves allocating functional processes to architectural resources and configuring the parameters for architectural services. The allocation of process to resources can be seen as a covering problem, where each process may be "covered" by at most one architectural resource.

Mapping must ensure that the functional performance constraints are satisfied and that the relevant metrics are optimized. For embedded systems, satisfying performance constraints typically requires worst-case analysis. Optimization requires the ability to analyze a number of points in the design space, either explicitly with simulation or implicitly with analysis. Due to the complexity of modern multiprocessor embedded systems, our approach emphasizes automated approaches to mapping.

Figure 5.8 summarizes four techniques that can be used for automated mapping: heuristics, approximation algorithms, randomized algorithms, and mathematical programming. An "X" in a grid square denotes that the approach does not address the specified issue. Even though heuristics are relatively easy to create and can often provide solutions quickly, they do not provide bounds on solution quality. More importantly, they are brittle with respect to changes in the problem assumptions. For instance, partial solutions and side constraints are typically difficult to add to most heuristics without sacrifices in effectiveness.

Approximation algorithms do provide bounds, but the analysis applied to produce these bounds is even less resilient to problem changes. Consequently, even though heuristics and approximation algorithms excel at clearly defined problems, their applicability is limited within a design flow where platform-specific constraints are needed.

Randomized algorithms, such as genetic programming and simulated annealing are flexible to changes in problem assumptions. However, they do not provide bounds on solution quality and are typically computationally expensive.

An alternative is to use mathematical programming (MP) techniques. In MP, the system is represented with parameters, decision variables, and constraints over the parameters and decision variables. An objective function is defined over the same set of variables. Generic solvers can be utilized to find the optimal solution. The complexity of finding the optimal solution depends upon the variable types as well as the form of the objective function and constraints.

Mathematical programming techniques are much easier to customize, since application or platform-specific constraints can be added as required. Branch-and-bound solution techniques for MPs provide lower and upper bounds on the desired cost functions at each step of the solution process. This allows us to trade off solution time and quality. The main difficulty with using MP approaches lies in finding a formulation that is sufficiently accurate to capture the behavior of the system and yet remains amenable to efficient solving. A straightforward formulation may be inefficient but easy to understand, whereas an efficient formulation may be difficult to verify. Even though this is the same tradeoff that is made for the design itself, the terseness of mathematical programming formulation makes this much less of an issue.

5.6.3 Services

Operationally, the functional model consists of a set of *processes* $\mathcal{P}$ that execute a sequence of *services* S_F provided by different *components* C. The architectural model consists of a set of *resources* $\mathcal{R}$ which provide *services* S_A, each of which may be *parameterized*.

Mapping assigns processes from the functionality to resources in the architecture. This is a many-to-one mapping. Whenever a process $p \in \mathcal{P}$ executes a service $s_f \in S_F$, it is bound to a single resource $r \in \mathcal{R}$ invoking a service $s_a \in S_A$. s_a and s_f must be identical. Verifying this requirement is beyond the scope of this work, it is assumed that the functional and architectural designers agree on the definition of services. As a first cut, services can be the provided interfaces of the architectural model, thus ensuring at least structural/syntactic compatibility. Any parameter(s) π required for s_a are determined

140

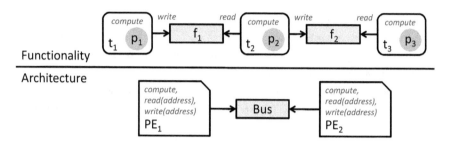

Figure 5.9: An Example of Service-based Mapping

by the mapping based on the component $c \in C$ in which the service s_f is executed. Note that service parameterization is not permitted for functional services, only for architectural ones. Also, parameters are determined statically based on the mapping, the architectural parameters cannot be determined during the execution of the functional model.

The semantics, or model of computation, depends on the set of services used by the functionality as well the sequence in which these services are used by the processes. The architecture determines the performance of the system by calculating a cost for each service. Functional performance constraints or performance metrics can only be verified after mapping has been carried out, not before.

Example

A simple example, shown in Figure 5.9, illustrates this approach to mapping. The functional model consists of three processes: p_1, p_2, and p_3, as well as three tasks: t_1, t_2, and t_3, and two FIFOs: f_1 and f_2. The processes can execute up to three services each: *read*, *write*, and *compute*. *read* and *write* services are executed by the processes within the FIFO components, while the *compute* services are executed within the task components.

The architecture consists of two resources (PE_1 and PE_2) connected with a bus. Both resources provide all three services required by the processes in the functionality. The *read* and *write* services are parameterized in the architecture with an address argument. Note that the amount of data written or read is missing from these services. This is because that amount is implicit in the definition of the service. For example, if the cost of reading two items is different than twice the cost of reading a single item, a new service should be introduced to differentiate these two situations.

The mapping for this system is described in Table 5.2. It indicates the mapping for each used

$< p_1, compute, t_1 >$	$\longrightarrow$	$< PE_1, compute, \emptyset >$
$< p_1, write, f_1 >$	$\longrightarrow$	$< PE_1, write, 0x0FFA >$
$< p_2, compute, t_2 >$	$\longrightarrow$	$< PE_1, compute, \emptyset >$
$< p_2, read, f_1 >$	$\longrightarrow$	$< PE_1, read, 0x0FFA >$
$< p_2, write, f_2 >$	$\longrightarrow$	$< PE_1, write, 0x0CC0 >$
$< p_3, compute, t_3 >$	$\longrightarrow$	$< PE_2, compute, \emptyset >$
$< p_3, read, f_2 >$	$\longrightarrow$	$< PE_2, read, 0x0FFD >$

Table 5.2: Mapping for example system

service to the associated provided service. The mapping has allocated p_1 and p_2 to PE_1, and p_3 to PE_2. The mapping has also determined the parameters for the *read* and *write* services. These parameters refer to the location in memory where the FIFO communication is mapped. Note that f_1 is realized in the local memory of PE_1, while f_2 is realized in the local memory of PE_2. Presumably, 0x0CC0 corresponds to the memory-mapped address of the remote memory for PE_1 while 0x0FFD is the same location in the local memory space of PE_2.

Specification of mapping may be assigned within *mapper* objects in Metro II, as described in Section 4.4.2.

5.7 Architectural Service Design Flow

"Hardware: the part of the computer that can be kicked" – Jeff Pesis

A key question which must be answered in this work is: "what is a service?". It is important to remember that architecture modeling efforts can span many abstraction levels. For example a basic logic operation "A AND B" can be implemented as a 2-input AND gate or it could also be implemented as a N-input AND gate where two of the inputs are A and B and the other $N-2$ inputs are tied to "logical 1". In either case an *AND* service would be exposed but each case may have a different cost associated with it. Another example at the other end of the abstraction spectrum is a Discreet Cosine Transform (DCT). This operation can be carried out on a model of a general purpose processor or a dedicated HW DCT. Again each would be a *DCT* service but the former may have a higher execution time cost than a dedicated HW block. There are many different ways of modeling architectures to this end. A model can be a single entity or a collection of smaller entities that make up a larger system. Essentially a service has an *interface* and a *cost*. Services were defined formally in Section 5.4 and a high level picture is shown

142

in Figure 5.4.

While it is clear that architecture modeling is possible, naturally it is important to answer why it should be done. Primarily architecture modeling at the system level is done so that system designers can see the effect of design decisions prior to implementing the systems. This process is done primarily through simulation. The more abstract this process can be, hopefully the faster simulation will be and the less designer effort that will be required. Hence abstraction must be maintained. Simulation must naturally also be accurate or it is not a useful exercise as the implementations will not have a correspondence to the simulation.

5.7.1 Design Flow Overview

Traditionally architecture and functional models have been merged. For example, synthesizable RTL design implicitly ties what the system *does* along with the physical structures that will *implement* it. Other systems begin with a functional description and this description morphs into hardware and software through a series of refinement steps. While this process can be automated, to some extent there does not exist descriptions of the system which are purely functional or purely architectural. As a result, when a designer wants to reuse the functional specification using a new architecture, either a new design must be written, or minimally rolled back to the most abstract version. In any event, much design and verification work will be lost. This work's approach looks to eliminate this inefficient and potentially error prone process. One of the critical design activities in a platform-based design flow is the creation of architectural services. Services to map to functionality are a first class citizen in PBD.

However, in the event that this functional/architecture separation does not exist, architecture modeling is difficult to define, has a broad set of interpretations, and classically results in modeling structural or topological details of an electronic system. Examples of these alternate styles were covered in Section 5.3 (related work). There are a number of tradeoffs possible by using these alternate styles. If one is not careful the naïve flow may emerge. For this work however, it is assumed that such a separation exists and the ultimate goal of the modeling effort is for design space exploration via simulation. These sections will demonstrate why this style was chosen, how it was implemented, and how it ultimately achieved the architecture modeling goals outlined in Chapter 1.

It is immediately clear that the naïve approach is not acceptable and will not achieve the goals desired by this work. The proposed approach improves upon the naïve approach in the following ways:

1. The proposed flow replaces a generic, abstract modeling approach with a fundamentally solid, architecture service based modeling style. This is focused on programmable platforms and uses an

FPM based environment (as defined by the taxonomies in Chapter 2). Services are at the transactional modeling level using an underlying event based semantics.

2. The proposed flow replaces estimated data in simulation with characterized data from real programmable platforms. Characterization itself will be described in detail in Chapter 7.

3. The proposed flow replaces manual translation from the more abstract design to implementation with a "correct-by-construction" automatic method. This is discussed in Section 5.7.5.

4. The proposed flow provides refinement verification techniques that close the implementation gap while still allowing highly abstract design space exploration. Chapter 6 will explore the concept of refinement in more depth.

These improvements are made possible by focusing on transaction level representations of programmable architectures and leveraging an event based simulation environment (i.e. the METROPOLIS and METRO II design environments). Figure 5.10 shows the techniques to be discussed in this work as a whole and provides an ordered step by step explanation of the process. In the figure, on the left is shown the architecture space of the ASV triangles outlined in Chapter 3. The design flow is expanded on the right. It begins with the selection of architecture services from a library of elements. From these elements a system level design, transaction based architecture model is created. Once the final model has been chosen based on simulation results, it is fed to another process to create an actual file for programming a device.

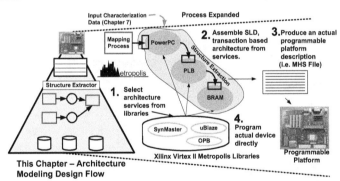

Figure 5.10: System Level Architecture Modeling in the Proposed Flow

These sections will illustrate how to model architecture services at the transaction level so that modularity and accuracy will be maintained. This requires support for a variety of architecture topologies, service exposure levels, and extensions for mapping. This process is clearly illustrated in a platform-based design environment.

5.7.2 Xilinx Architecture Modeling Exploration Example

In order to put the ideas proposed in this chapter to test, system level architecture service models were created based on the Xilinx Virtex II platform FPGA in METROPOLIS. An FPGA was selected since one set of services can be arranged in a wide variety of configurations. Whereas a static architecture only has one configuration, an FPGA has configurations only limited by the size of the configuration fabric and its topology. Building a library of programmable components will allow a designer to express many systems with maximum flexibility. These are some of the reasons mentioned earlier in Chapter 1 during the discussion of programmable platforms. The services (components) chosen were based upon those that could easily form embedded systems and that were well defined and characterized. To this end, the IBM CoreConnect [IBM99] based IP blocks were examined.

The architecture service models created can be categorized around the basic service type they represent.

- **Computation** - PowerPC, MicroBlaze, Synthetic Master, and Synthetic Slave - (4 services total)

- **Communication** - Processor Local Bus (PLB), On-Chip Peripheral Bus (OPB), BRAM, Fast Simplex Link (FSL) - (4 services total)

- **Coordination** - PowerPC Scheduler, MicroBlaze Scheduler, PLB Scheduler, OPB Scheduler, BRAM Scheduler, Bridge Scheduler, FSL Scheduler, and a General Scheduler - (8 schedulers total)

- **Hybrid Services** - Mapping Process, OPB/PLB Bus Bridge - (1 process, 1 service total)

Each service listed behaves as the device is described in its datasheet specification. PowerPC and MicroBlaze services are MCMI services. PLB and OPB are MCMI services. BRAM and FSL are SCSI services. Synthetic master and slave devices are used to represent dedicated peripherals created in the programmable fabric. For example if a designer wishes to create a dedicated hardware block, they would create the functionality and encapsulate it with the appropriate synthetic component. The synthetic components possess the interface of the PLB, OPB, and FSL and can be used with each if needed. Both master and slave devices are MCMI services.

In addition to the core architecture modeling concepts outlined previously, two key aspects were maintained:

- **Transaction Level Interfaces** - this requires that the interfaces provided by the services (media) are at the transaction level and that they corresponded syntactically to the methods that would be invoked in the process of executing the functional model. Each service transaction is denoted as "complex" or "atomic" as well. Each used event based semantics for synchronization.

- **Netlist instantiation and parameterization identical to the implementation IP** - the black box model of the IP is identical to the parameters used to instantiate an architecture object in the scheduled netlist. This black box "signature" can be obtained from the Xilinx IP implementation and generation tools such as CoreGen.

Transaction level interfaces not only are important to maintain the system level of abstraction, but they were also very easy to map to the functional model. Examples of the transaction level interfaces are:

Task Before Mapping: These function prototypes are what the mapping process task will export to the functional model. Additionally it will export the service functionality and affinity. The parameters the prototypes require should be assigned by the functional model and the method should correspond to one or more internal interface functional calls.

```
Read (addr, offset, cnt, size)
Write(addr, offset, cnt, size)
Execute (operation, complexity)
```

A "Read", "Write", or "Execute" service connected to the mapping process can be SCSI, MCSI, or MCMI. In this work it is typically implemented as a MCSI. The interface itself is provided to tasks through either an RTOS or CPU service. The services will make use of multiple components potentially such as caches, buses, or memory elements.

Task After Mapping: This is the result of mapping when the parameters are provided from the functional model. The "operation" field in the execute function is provided during mapping thanks to the mapping process. Complexity of the execute function is provided by the functional model mapped to the architecture service. Complexity itself is determined by the designer of the functional model and becomes a variable in the service cost model.

```
Read (0x34, 8, 10, 4)
Write(0x68, 4, 1, 2)
Execute (add, 10)
```

Computation Interfaces - These interfaces are the same interfaces which are exposed to the functional model through mapping tasks but their implementation will be much different. Whereas the mapping tasks are concerned with determining the parameters of the interface calls, computation interfaces actually have to implement the services and most importantly the cost models.

```
Read (addr, offset, cnt, size), Write(addr, offset, cnt, size),
Execute (operation, complexity)
```

Computation services are MCMI type services in the majority of cases. Most computation services at the system level are still composed of multiple components with multiple interfaces (and hence costs) for those services. In the event that the model has a very coarse granularity they may be SCSI.

Communication Interfaces (Buses) - These interfaces will utilize services to translate read and write requests into sequences of atomic transactions. The interfaces listed here can be combined in a variety of ways to form a number of bus protocols.

```
addrTransfer(target, master)
addrReq(base, offset, transType, device)
addrAck(device)

dataTransfer(device, readSeq, writeSeq)
dataAck(device)
```

Bus based communication services are MCSI services with the bus often itself being the single interface point. They may be MCMI in the event that they represent a hierarchy of buses.

Communication Interfaces (FSL) - Unlike bus services, FSL interfaces only need read and write capabilities since a FSL acts as a FIFO.

```
Read (cnt, size), Write(cnt, size)
```

FSL services are those which interact with buffer based communication. These are SCSI where the component is a simple buffer and they only have a single interface. These are used often in ring topologies as illustrated or in dataflow applications.

These interface prototypes are shown to give the reader a feeling for the types and level of abstraction provided by the services. In the following sections, the actual interfaces for the components modeled will be shown.

5.7.3 Xilinx Vertex II Pro Execution Estimation

In order to begin to estimate the performance of the architecture service models in METROPO-LIS, performance numbers for various operations must be determined for particular architecture instances. These operations should correspond to services that can be requested by the mapping process (task) in a given architecture model. These estimates are the *cost* of the services. The services requiring estimates will be described in the appropriate sections to follow.

The Xilinx Virtex II Pro was chosen due to its flexibility. It is the combination of FPGA fabric along with embedded PowerPC units. This flexibility allows for static architecture configurations along with custom implementations. This allows one device to represent many architecture models. Using this platform will allow for rapid, meaningful performance estimation across many architecture models. Additionally models can be quickly compared to their implementation counterparts.

There are many issues with this estimation method as will be demonstrated in Chapter 7. In fact that chapter will go to great lengths to show why this method is not desirable. However it is included as it is important to show how such a process may be carried out. It is important that this estimation process occur to see if a more robust characterization method offers an advantage.

The services that must be annotated with an execution metric are in three areas and are as follows:

1. CPU services - these will ultimately be represented on the *PowerPC* embedded core and *MicroBlaze* soft core which are available in the Virtex II Pro.

   - The interfaces of interest are *cpuRead( ), cpuWrite( ), execute( )*. These interfaces will result in event based requests which potentially access the bus and then an external memory. Therefore they should represent uncacheable loads and stores.

2. BUS services - these will be represented by *CoreConnect Processor Local Bus (PLB)* and *On-Chip Peripheral Bus (OPB)* requests. In addition there are FSL write and read interfaces that will not be discussed explicitly here.

   - The interfaces of interest are *busRead( ) and busWrite( )*. These are event based requests to the PLB and OPB and will include both the address and data tenure phases.

3. Memory services - these are SelectRAM+ (BRAM) requests which will be characterized by *SelectRAM+* operations which are event based requests as well.

- The interfaces of interest are *memRead() and memWrite()*. These will be read and write operations which are fully synchronous for the SelectRAM+. This information was used to develop a general BRAM model since it is more robust, portable, and scalable than the static estimation data available for more complex memory models such as DDR or other SDRAMs. Also BRAM is very prevalent in Xilinx devices and very close to the configurable fabric which aids in performance.

The following sections detail the various components modeled in METROPOLIS and each culminate with a performance estimation for each interface operation. The information for estimation is gathered from [Xil03b], [IBM99], and [Xil02].

PowerPC

The PowerPC core on the Xilinx Virtex II Pro is the PPC405 RISC CPU. This is a five stage pipeline, 32 bit processor. There are several basic guidelines regarding instruction execution.

- Instructions execute in order

- Assuming cache hits, all instructions execute in one cycle

  – With the exception of divide, branch, MAC, unaligned memory accesses, and cache control instructions.

Figure 5.11 provides details on the PowerPC model created for METROPOLIS. Included in this figure are the parameters, ports, and interfaces implemented by this object. This same style of illustration will be shown for each of the services described in this chapter.

Since the load and store instructions do not "assume cache hits" they will take more than one cycle. For the purposes of the initial architecture service models, the CPU functions that need to be estimated are the read (load) and write (store) instructions. There are loads and store instruction for data in *byte, halfword, and word* formats. The format desired is expressed as the "size" argument shown in the function prototype. In addition, there are various addressing modes and side effects that can be associated with each data size request. However, neither the size of data transferring, address mode, or side effect have any effect on the cycle count within the load and store family of instructions (thanks to the strict RISC regularity). Tables 5.3 and 5.4 show the wide variety of loads and stores that need to be given performance numbers.

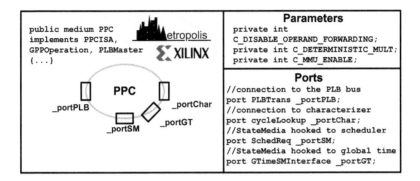

Figure 5.11: METROPOLIS PowerPC Model

An uncacheable load instruction will incur penalty cycles for accessing memory over the PLB. Assuming the PLB is at the same speed as the processor and that the address acknowledge is returned in the same cycle that the data cache unit asserts the PLB (OPB), the number of penalty cycles will be *6 cycles with operand forwarding* and *7 cycles without operand forwarding*. The architecture service models in METROPOLIS do not explicitly include operand forwarding so a load will take **7 cycles**.

The PowerPC data cache unit has a queue so that store instructions that miss in the data cache appear to execute in a single cycle. These services are constructed assuming aligned memory access and no usage of the stwcx (conditional store; takes 2 cycles). Therefore stores will take **1 cycle**.

Table 5.5 gives the final analysis of the interfaces' estimated performance. All instructions assume aligned accesses.

MicroBlaze

In addition to the PowerPC processor, an architecture service model was created for the MicroBlaze processor. The MicroBlaze is a soft processor core which is created in the FPGA fabric. Whereas there are only 2 to 4 PowerPC cores available to Xilinx Virtex II Pro, one can fit a much larger set of MicroBlazes on a die. This allows for interesting, highly concurrent architecture topologies. When the designer wishes to construct a netlist using these components they are restricted only by the size of the overall device and not a static number (as is the case with the PowerPC). Another way in which this device contrasts the PowerPC is that it connects to the OPB bus and FSL units as well (not the PLB).

lbz	lha	lmw
lbzu	lhau	lswi
lbzux	lhaux	lswx
lbzx	lhax	lwarx
	lhbrx	lwbrx
	lhz	lwz
	lhzu	lwzu
	lhzux	lwzux
	lhzx	lwzx

Table 5.4: PowerPC load instructions

stb	sth	stmw	stw
stbu	sthu	stswi	stwbrx
stbux	sthbrx	stswx	stwcx
stbx	sthux	stwu	stwux
	sthx		stwx

Table 5.3: PowerPC store instructions

Interface	Assumptions	Cycle Count
cpuRead()	*Any load instruction without operand forwarding*	7 cycles
cpuWrite()	*Any store but stwcx*	1 cycle
execute(int inst, int comp)	*Valid inst field*	(1 * complexity) cycles

Table 5.5: PowerPC Service Performance Estimation Summary

The MicroBlaze is a 32-bit Harvard architecture processor. Its base architecture has 32 registers, ALU, shift unit, and two levels of interrupts. This is a DLX-style microprocessor with a 5-stage pipeline in which most instructions complete in one cycle. The processor can operate at speeds up to 210Mhz on the Virtex 5. Optional configurations include a floating point unit, barrel shifter, divider, and multiplier. It also interfaces with a high speed, local memory bus (LMB). The METROPOLIS model is shown in Figure 5.12.

Table 5.6 provides execution time estimates for the the MicroBlaze. Note that the function prototypes here are pseudocode and not what is actually provided in the actually meta-model code for the element. Typically what differs is the list of arguments. These are left off in order to keep the table size manageable. These typically include IDs, control arguments, or addresses.

Synthetic Masters and Slaves

Synthetic master and slave services are used to represent custom made programmable functionality created in the device fabric. The difference between a master and slave device is the way in which

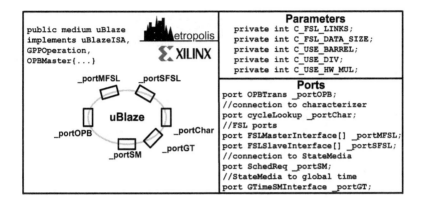

Figure 5.12: METROPOLIS MicroBlaze Model

Interface	Assumptions	Cycle Count
cpuRead(int bus)	*Bus Dependent*	1(LMB), 7(OPB) cycle
cpuWrite(int bus)	*Bus Dependent*	1(LMB), 2(OPB) cycle
fslRead(int size)	*Transfer Size*	(1 * size) cycles
fslWrite(int size)	*Transfer Size*	(1 * size) cycles
execute(int inst, int comp)	*Valid INST Field*	(1 * complexity) cycles

Table 5.6: MicroBlaze Service Performance Estimation Summary

it interacts with the bus (PLB or OPB) it is attached to. A slave can only respond to requests whereas a master can generate requests. In the terms of the services in this work, a slave is a passive service and a master is an active service. Figure 5.13 illustrates the METROPOLIS service model.

The estimated execution times for bus and FSL communication interfaces of a synthetic service are the same as the MicroBlaze service costs. The PLB access time for a synthetic service is the same as the PowerPC service. However, the execution time is a function of what function is being computed, its complexity, and the port that it is being accessed from. The port being accessed has differing overhead for a master device as opposed to a slave device. The equation for execution time is $inst * complexity + PortAccessOverhead$ where $0 < inst \leq 1, complexity \geq 1$, and $2 \geq PortAccessOverhead \geq 0$.

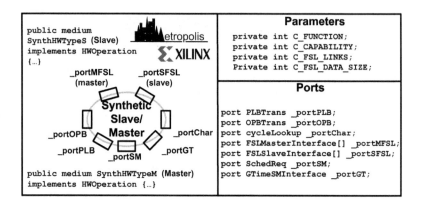

Figure 5.13: METROPOLIS Synthetic Master/Slave Model

CoreConnect Buses

The CoreConnect environment provides three buses. The Processor Local Bus (PLB), the On-Chip Peripheral Bus (OPB), and the Device Control Register (DCR) Bus. This discussion begins with the PLB which is where the PowerPC will reside in the majority of designs. The PLB is used to make requests to memory elements or other peripherals. The OPB which is primarily used with the MicroBlaze, will be discussed next. The DCR was not modeled because the investigations involved with this work did not require it.

The PLB is the connection provided to the PowerPC cores which gives them high speed access to peripherals. It has separate 32-bit address and 64-bit data buses. It is a fully synchronous bus which supports multiple master and slave devices. Read and write transfers between master and slave devices occur through the use of PLB bus transfer signals. Each PLB master has its own address, read-data, and write-data buses. Slaves have a shared but decoupled interface. Figure 5.14 illustrates aspects of the PLB bus model in METROPOLIS.

The PLB bus transactions consist of multiple address and data tenures. The address tenure has *request, transfer, and address* phases. The data tenure has *transfer and acknowledge* phases. Begin by assuming that there are only one master and one slave on the bus. In the event that a requesting master is immediately granted the bus and the slave acknowledges the address in the same cycle, then all three address tenure phases happen in 1 cycle for a total of 3 cycles. The data tenure phase requires n cycles for

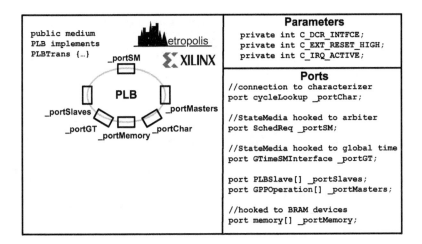

Figure 5.14: METROPOLIS PLB Model

the transfer phase where n is the number of 32-bit words transfered and then 1 cycle for the acknowledge phase. This is a total of n+1. Combining the data and address tenures results in **4+n** total cycles. It is understood that one master and one slave is a gross oversimplification and it will be shown to have its disadvantages when compared to the characterized process described in Chapter 7. The OPB has a more sophisticated estimation scheme than the PLB but its accuracy also ultimately pales in comparison as well to the characterized method.

Table 5.7 provides the final PLB bus estimation numbers. The "Size" argument in the functions is translated to the number of 32-bit words transferred, n.

Interface	Assumptions	Cycle Count
busRead(int size)	*Single Master, Single Slave on Bus*	4+n cycles
busWrite(int size)	*Single Master, Single Slave on Bus*	4+n cycles

Table 5.7: PLB Bus Service Performance Estimation Summary

The OPB is a low speed interface for the PowerPC. It was modeled however since it is available to the MicroBlaze soft cores as a master interface (which the PLB is not for the Xilinx ML310 board used in the experiments in Chapters 8 and 9). It is a fully synchronous bus which is intended to work at a lower level of hierarchy as compared to the PLB. It supports separate 32-bit address and data buses. It accesses slave peripherals through the PLB-to-OPB bridge.

Figure 5.15 illustrates aspects of the OPB bus model in METROPOLIS. It is similar to the PLB in most respects (ports for example) but implements different service interfaces on its ports and has different parameters.

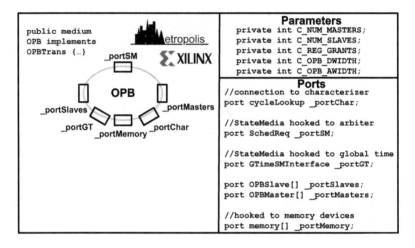

Figure 5.15: METROPOLIS OPB Model

Based on IBM's OPB Bus Functional Toolkit [IBM03], three scenarios were supported for OPB operation. These scenarios formed the basis of the performance estimation data. The first scenario is a synchronized, unlocked, multiple master memory access (SUMMA). In this scenario, there are two or more masters and one slave device. Each master wishes to access the this slave. It is assumed that one master receives access to the slave first, completes its transaction, and then notifies the second master that it can now proceed. This notification is why this scenario is denoted "synchronized". Since the masters work together in this scenario, the transfer time is **2nm+nm** (3nm) where n is the number of 32-bit data words transfered and m is the number of masters which wish to transfer. $2nm$ is the set up (request and grant) for each transfer of each master. nm is the transfer cycles themselves for each master.

The second scenario is a locked, multiple master memory access (LMMA). This assumes that once a master obtains the bus it is "locked" which will prevent other masters with higher priority from accessing the bus. This is a less cooperative scenario as compared to SUMMA. This increases the overhead of obtaining the bus from 2 to 4 cycles (assuming an additional request and grant phase). Therefore, the transfer time is **4nm+nm** (5nm). Again n is the number of 32-bit words and m is the number of master devices involved in the transaction. *4nm* is the set up (request and grant times two) for each transfer of each master. *nm* is the transfer cycles themselves for each master.

The third scenario is a burst read or write using bus lock and sequential addresses (BRWLSA). This scenario is for a single master and slave with bus parking disabled, round robin arbitration, and the bus locked for the entire transfer. Since the addresses of the burst are sequential, the OPB can work more efficiently. It does not need to go through a request and grant addressing phase for each transfer. Since the bus is locked, it does not need to worry about multiple masters interrupting the transfer. Since bus parking is disabled and round robin arbitration is assumed, other masters should have access to the bus in such a way that fairness is preserved and starvation avoided. The transaction time is **2 + n** where n is simply the number of 32-bit words transfered during the burst along with the two extra cycles for the initial request and grant phases. Table 5.8 provides the final OPB bus estimation numbers.

Scenario	Assumptions	Cycle Count
SUMMA	*m Masters, Single Slave, Synchronization, n words*	3nm cycles
LMMA	*m Masters, Single Slave, Locked bus, n words*	5nm cycles
BRWLSA	*Single Masters, Single Slave, Locked, Burst, Seq. Addr.*	2+n cycles

Table 5.8: OPB Bus Service Performance Estimation Summary

SelectRAM+ (BRAM)

The memory chosen to profile for performance estimation is the SelectRAM+ memory which is prevalent on the Virtex II Pro device. This is a dual port RAM which comes in 18Kb blocks. Each of its two ports can be independently configured as a read port, write port, or read/write port. Depending on its configuration as single port or dual port, various different memory partitions are available. In order to access the memory, there is one read operation and three write operations (write_first, read_first, and no_change). Operation is synchronous and behaves like a register in that address and data inputs need to be valid during a set up time and hold time window prior to a rising edge of a clock edge. Data output changes as a result of that same clock edge. SelectRAM+ was chosen since it is very easy to profile, prevalent on the device, and easy to model.

SelectRAM+ is often called block RAM or BRAM because of how it is available in cascaded blocks along the FPGA configurable logic blocks (CLBs). This makes them available to implement deeper or wider single- or dual-port memory elements. In the largest Xilinx Virtex II Pro device (XC2VP125) there are 18 columns of BRAM for a total of 10,008 Kbits.

BRAM interfaces were exported up to the functional model to simplify the creation of basic systems. Parameters required are the enable (EN), write enable (WE), and Set/Reset (SSR) signals. Also BRAM was very easy to use in creating actual implementations for comparison with the simulations. Often times, simple communication with BRAMs made for very effective dataflow systems based on the MicroBlaze and FSL components.

Figure 5.14 illustrates the BRAM bus model in METROPOLIS.

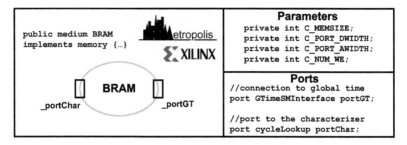

Figure 5.16: METROPOLIS BRAM Model

The read operation for BRAM uses only one clock edge. If the read address is provided by that clock edge the stored data is loaded into the output latches after the RAM access interval has elapsed.

The write operation as mentioned could be in one of three forms. The default mode is *write_first* where the data is written to the memory then that data is stored on the data output (as opposed to *read_first* where the "old data" is sent to the output while the new data is stored). *No_change* maintains the content of the output register throughout the the the new operation.

In order to get the latency estimates of the memory, one can use estimates from the application note [Xil03a] in which FIFO units are created using SelectRAM+ blocks and operate at 200Mhz or with a latency of 5ns. Table 5.9 summarizes the memory interfaces used. These estimates are for each 32-bit read or write. *memWrite* accepts a integer mode argument between 1 and 3 which indicates its operating mode.

157

Interface	Assumptions	Cycle Count
memRead()	*200Mhz System Clock*	1 cycle initiate and˜5ns of latency
memWrite(int mode)	*200Mhz System Clock*	1 cycle initiate and˜5ns of latency

Table 5.9: Memory Service Performance Estimation Summary

CoreConnect Quantity Managers

Each of the services mentioned have a corresponding METROPOLIS quantity manager (scheduler) associated with them. Each of these quantity managers implements the request, resolve, postcond, stable functions and their operational semantics as described. The resolve function is specific to each quantity manager and reflects the device it is intended to interact with.

Figure 5.17 illustrates some aspects of the Quantity Manager models in METROPOLIS. This figure differs from the earlier figures in this chapter in that instead of showing the parameters of the model, the interfaces are shown. Parameters belong to models in the scheduled netlist. Not only will those parameters be used to configure the simulation, they will also be used to create a programmable device description to be used during synthesis (this is shown in Section 5.7.5). To this end parameters are not of importance for scheduling netlist components. However, what is of interest are the interfaces which are called to schedule components.

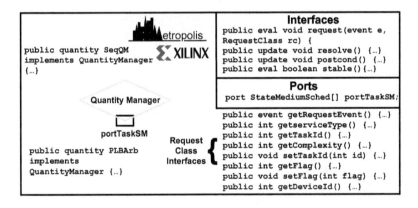

Figure 5.17: METROPOLIS Quantity Manager Model

The first set of interfaces are the request(), resolve(), postcond(), and stable() functions discussed in Chapter 4 Section 4.3.2. Notice that request requires two arguments. One is the event that is scheduling is requested for. The second is a request class. A request class consists of a separate set of interfaces and variables.

The request class' set of interfaces shown are: getRequestEvent(), getserviceType(), getTaskId(), getComplexity(), setTaskId(int id), getFlag(), setFlag(int flag), getDeviceId(). These are "getter" and "setter" style functions that allow information to be gathered regarding the service to be scheduled. These interfaces involve access to the generated event, type of service, complexity of requested service, id information, and synchronization flags.

5.7.4   Architecture Service Extensions

During the development of architecture service models in a system level design environments many issues need to be addressed. Many of these issues are explicitly discussed in this chapter (transaction level modeling, estimation techniques, event based simulation, etc). This section is going to focus on two issues which were not natively supported in the METROPOLIS modeling environment and hence specific solutions needed to be created. They are highlighted since their solutions are highly generic and easily supported to other environments (SystemC [Ini07] for example). The first issue relates to *preemption* and the second to *mapping*. The first issue is concerned with capturing the correct behavior of the architecture being modeled and supporting the appearance of architecture task level concurrency and interruption. The second issue looks to provide a more efficient path to automatic mapping of functional and architecture models. Implicitly the first issue deals with accuracy and the second issue with efficiency.

Preemption Extensions

In the course of creating an architecture it becomes clear that some *services* are naturally preempted. Examples scenarios are a CPU context switch or Bus transactions. Preemption must allow for one thread of execution (METROPOLIS process) to relinquish control of a service, the system must save the state of that execution, and the the service must allow for the new thread to use the service. Additionally, the simulation must make sure that the measured simulation time not only reflects the execution of the operation but also the overhead that would be required to perform such a transaction.

In an event based architecture service, there is no way to preempt a single event. Architectures services which can be preempted therefore must be a series of events which are related together to form a transaction. Some transactions can be preempted and others can not. Transactions should identify as to which group they belong to. It becomes also clear that this will require the notion of *Atomic Transactions*. A atomic transaction is one which cannot be preempted. In many cases this will be a single event but it is possible that it can be a collection of events as well. Atomic transactions were defined in Section 5.4.

Preemption can be dealt with in METROPOLIS quite simply. Prior to dispatching events to a quantity manager via the *request()* method, decompose transactions (using a "decoder") in the scheduled netlist into non-preemptable chunks (the atomic transactions). There must be infrastructure which maintains the scheduling status with an FSM object (counter) and controller. Figure 5.18 illustrates the preemption process. There are several stages involved in the process:

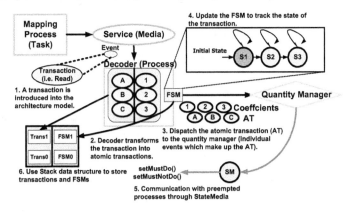

Figure 5.18: Architecture Extensions for Preemption

1. A transaction is introduced into the architecture model. This is done by the mapping process which is mapped to the tasks in the functional model. This is a single event generated by a task process in the scheduled netlist.

2. The transaction proceeds to a *decoder* object (process) connected to the service media. This decoder must perform several tasks:

    - Identify if the transaction is atomic or not. This is done through a table lookup or by a transaction argument detailing its status.

- In the event that it is not atomic, decompose it into atomic transactions (A, B, C in the figure). Each atomic transaction is typically made up of transactions from SCSI services but it does not have to be, it just will raise the notion of atomic to contain more components. Often atomic transactions are defined around provided interfaces of services.

- Augment the atomic events with information regarding architecture execution. This is a *co-efficient* which will be used to ensure that each atomic event takes a fraction of the total execution time for the entire transaction. Coefficients can be created dynamically using simple floating point operations.

- Introduce events to represent the overhead associated with the preemption type (1, 2, 3 in the figure). These events will also have a coefficient value. How to generate these events, how many of them, and their cost is determined by the decoder.

- Create a finite state machine. The number of states is a one-hot encoding based on the number of atomic transactions. Each atomic transaction now has a partial ordering assigned to it. It is a partial ordering since atomic transactions may issue nondeterministically.

3. Dispatch (request()) the atomic transactions as normal to the quantity manager. Here they will go through the standard resolve(), postcond(), stable() iterations.

4. Update the FSM to track the state of the transaction as a whole. When no transition can be made, the transaction is considered complete. In the event that a new FSM has been created by a preemptive process, you should push the existing FSM requests on a stack and pop them off as other FSM finish. This is assuming a LIFO preemption policy.

5. Use statemedia to communicate with processes using the setMustDo() and setMustNotDo() functions. The preempted process will be blocked whereas the preempting process will be allowed to proceed.

6. There is a decoder assigned to each service which supports preemption. In the event that preemption occurs, the FSM and bundled atomic events are pushed onto a stack (LIFO object). Each decoder (and hence services supporting preemption) has its own stack. Once the preemption is done, the stack is popped and execution continues. Once the FSM reaches the final state, the information for that transaction is discarded and the stack can be popped again.

This approach can be improved to some extent by a more compact FSM encoding but at the level of abstraction required by transaction level modeling, there will rarely be more than ten states in any transaction. In addition the time required for the overhead of a preemption is unique to each service and must be provided by the designer.

Mapping Extensions

A unique aspect of programmable platforms is that they allow for both SW and HW implementations of a function. For example there may be a soft processor model (Xilinx MicroBlaze for example) which can perform a SW routine for DCT. Additionally there may be a dedicated DCT block in the programmable logic fabric. At some point one may want to explore automated mapping of functionality to services. A service which provides general purpose processing can handle a wide variety of functionality mappings whereas a service for a specific HW component can only offer one type of functionality. In is not appropriate to map functionality to any architecture block. If this were done, services would not be available to the functional netlist and minimally the simulation would halt, if not fail altogether. Not to mention that not all architecture blocks efficiently perform all operations. Not knowing the service capabilities severely limits the ability to do intelligent mapping. Therefore there is the need to express which architecture components can provide which services and with what *affinity*. Affinity refers to how well the service can provide the desired operation. For example, an ASIC service providing an "ADD" service will have a high affinity to provide this service if it has a lower cost (execution time perhaps) than a general purpose processor software service "ADD". This information is used for mapping of functionality to architecture models. Mapping can employ greedy, task specific, or other strategies to maintain the best average affinity rating over all mapped tasks. An example of the information provided to the mapping network is shown in Figure 5.19.

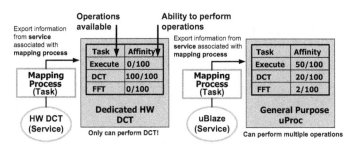

Figure 5.19: Architecture Extensions for Mapping

In Figure 5.19 there are two processes (tasks) connected each to their own media (service). The service on the left is a HW Discreet Cosign Transform (DCT). The service on the right is a MicroBlaze soft processor model (think of this as a general purpose computation service). The mapping processes are equipped with two functions. One function can be queried to return all of the services that it has access to. In this picture, the task on the right has access to a service which can provide execute (generically), DCT, or FFT services. The task on the left only has access to a DCT service. The tasks know this information regarding available services since each service reports itself and its capabilities to the task. Additionally the second function assigns an affinity to each service. This is also reported to the task by the service. This process is done statically initially but can be updated at runtime by the performance of the simulation. Affinity is a relative value, but in the illustration it is shown as a score out of 100. As is shown, the task on the left can only perform DCT (since it is tied to a dedicated HW block). In practice it would not even have other task operations shown as available. However, the task on the right can do all three operations, including DCT (albeit with a lower affinity).

The models developed in this work provide a service interface (getCapabilityList()) which returns the affinity and operations of the service in a hash table. It is the responsibility of the mapping network to use this information to efficiently map functionality to architecture. The algorithms for automated mapping which utilize this information will not be covered in this chapter.

5.7.5 Synthesis Path for Architecture Services

One of the goals of the proposed flow presented in this work is to find a way to take architecture service models and produce output which can be used in various synthesis flows. This process has been termed, "narrowing the gap". A desirable outcome of the Xilinx modeling effort is the production of a file for the programmable tool flow which does not suffer from the translation gaps present in the naïve flow. This process will ensure that the the architecture topology created not only matches that of the model used in METROPOLIS simulation but also that it has the same parameters which effect the simulation.

Because of the enforcement of parameterized IP like service construction, Xilinx Microprocessor Hardware Specification (MHS) file generation is automatic. It consists of the following steps which are illustrated in Figure 5.20:

1. **Assemble the scheduled netlist** - This step consists of making the connections between architecture elements the designer is interested in simulating. This is naturally part of the design process and is required for simulation. The elements in this netlist should be selected from the provided METROPOLIS library of Xilinx elements.

2. **Provide parameters for the architecture component instance** - These are required by the constructors of the architecture elements themselves. Examples of these are shown in Figures 5.11, 5.14, and 5.16.

3. **Simulate the architecture** - This step requires running the parameterized architecture model mapped to a functional model. This is the design space exploration process.

4. **Decide on the architecture model which meets your goals** - This is the outcome of the design space exploration method. The rest of this process will use the model chosen in this step.

5. **Run "Structure Extractor" script** - This script works by traversing the scheduled netlist. It identifies components, their parameters, and their connections. The final result is the production of a MHS file.

6. **Take resultant file and feed to Xilinx EDK** - This process will produce an FPGA with the specified components, topology, and parameterization. All that need be done now is to provide the FPGA with the software aspects captured only in the functional model with the configuration determined through mapping.

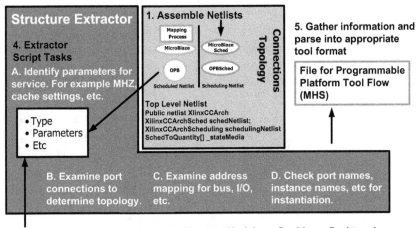

Figure 5.20: Automatic Xilinx MHS Extraction

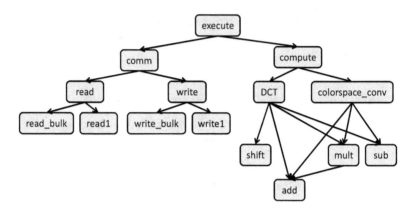

Figure 5.21: Example: Hierarchy of services

5.8   Tradeoffs

Functional and architectural modeling must balance the conflicting aims of accurate/efficient system performance vs. automated design space exploration. This is directly related to the granularity of services that are mapped. Service relationships can be captured in the form of a directed acyclic graph (DAG) where nodes represent services and the directed edges capture service containment. A service s_1 contains s_2 and s_3 iff s_1 can be decomposed into some sequence of invocations of s_2 and s_3. Note that since we do not deal with the definitions of the services, the construction of this graph is beyond the scope of this work.

An example of a service relationship graph is given in Figure 5.21. In this example, a set of services is hierarchically arranged from most to least abstract. The more abstract services, such as *comm* and *compute*, are invoked in a much simpler pattern than the leaf services. However, the leaf services can be more accurately characterized than the higher level services.

5.8.1   Functional Tradeoffs

The aim of functional modeling is to capture the behavior of the application and enable both behavioral and performance verification. These aims are often contradictory. Under the definition of services given in Section 5.6.3, the behavior of the application can either be captured with a large set of focused services that are used in a complex manner or a small set of generic services that are used in a straightforward manner.

Behavioral verification relates to the way in which the atomic services are used by the processes to carry realize the behavior. Therefore, coarse granularity services are preferable; as long as the services themselves have been pre-verified. With coarse-granularity services, the usage patterns are simpler, and more likely to fall into a verifiable pattern.

5.8.2 Architectural Tradeoffs

The choice of the CMD also has an important impact on the architectural modeling. Architectural modeling has to be carried out in such a way that the services which define the CMD are provided by the architectural model. For instance, for a dataflow CMD, blocking read operations must be provided by the architectural model. These services are typically provided with some measure of cost, which may not be captured in the CMD. For instance, the blocking read operation may have latency and energy costs associated with it in the architectural platform. The tradeoff to be considered is that the services associated with the CMD may be fairly expensive for a given architectural platform. So, the CMD may a priori restrict the maximum achievable performance of any system that can be implemented on that architectural platform. Ideally, the architectural services should be defined at a fine granularity to enable accurate and low overhead characterization.

The other choice in architectural modeling is to expose or hide the flexibility that may be present. For instance, a blocking read operation may either be carried out using limited local memory or a larger amount of global memory with different energy and latency costs. Exposing this choice from the architectural model may require more detailed modeling and unnecessarily complicate the mapping stage.

5.9 Systems Overview

The applicability of the design flow presented in this chapter to various systems is the proof of its usefulness. In this work, we focus on two classes of systems: those from the multimedia domain and those from the distributed systems domain. These case studies touch on all aspects of the design flow presented here, including functional modeling, architectural modeling, mapping, refinement, and characterization.

5.9.1 Multimedia Systems

Multimedia or streaming systems deal with computation carried out on streams of data. The model of computation (MoC) [JS05] most often considered for multimedia systems is usually a specialization of Kahn Process Networks, where actors consume data from input streams, carry out computation, and produce data on output streams. The five case studies in this domain explore all stages of the design flow, and are described in further detail in Chapter 8.

The first case study consists of deploying a JPEG encoder application onto the Intel MXP5800 architectural platform [DZMSV05]. The JPEG encoder is a multimedia application whose building blocks are used in many image and video processing algorithms. The MXP5800 is an imaging processor which is highly parallel and heterogeneous. The focus for this case study is choosing the appropriate dataflow MoC and abstraction level for modeling and then carrying out manual mapping. The case study demonstrates that the appropriate choice for MoC and abstraction level can allow designers to fully exploit the capabilities of such platforms while also enabling future automation.

The second case study [DZSV06] investigates the automated [DCZ+06] mapping of a motion-JPEG application onto a heterogeneous soft-core multiprocessor FPGA platform. Motion-JPEG is an extension of JPEG for video and is used in consumer electronics and video editing systems. The architectural platform consists of soft-core uBlaze PEs and custom hardware connected with point-to-point FIFOs on the Xilinx Virtex II FPGA platform. The automated mapping is carried out with a Mixed Integer Linear Programming approach that is shown to be both efficient and extensible.

The third case study deals with the deployment of a portion of the h.264 application [WSBL03] [JM9] onto a FPGA architectural platforms. This case study focuses on the deblocking filter portion of the application and considers the implementation on the Xilinx Virtex architecture platform.

Finally, the fourth case study looks at the User Equipment Domain of the UMTS protocol which is of interest to mobile devices [Pro04]. Design space exploration is carried out for an architectural platform composed of a diverse set of processing elements and helps demonstrate the modeling costs associated with the METRO II framework.

5.9.2 Distributed Systems

The distributed systems we consider are typically control oriented and have strict timing requirements. Their distributed nature means that they are more decoupled with respect to multimedia systems. One type of distributed system that we consider is an automotive system. Due to the intrinsic physical distribution of the sensors/actuators, automotive applications are deployed on multiple electronic control units (ECUs) connected with standardized buses. Again, all aspects of the design flow are addressed in the four case studies, described in more detail in Chapter 9.

The first case study looks at a new distributed microarchitecture for processors known as FLEET [Sut06] [CLJ$^+$01]. The main objective of this case study is to demonstrate an example of refinement verification, as described in Chapter6.

The second case study looks at the SPI-5 [GT01] packet processing system and attempts to configure it to meet certain performance requirements. The objective of the case study is to determine whether METROPOLIS can aid in the system configuration as well as to carry out automated refinement of different portions of the design.

The third case study in this domain is an automotive system case study. It involves reconciling the MoC used for the functional model with the MoC exhibited by the architectural platform [ZDSV$^+$06a]. We demonstrate that current industry practice can lead to problems such as message loss and priority inversion due to limited buffer sizes in the bus controllers. These problems are not evident in the original functional model, making correct-by-construction deployment difficult. To remedy these problems, we modify the functional MoC such that these problems can be diagnosed prior to mapping.

The fourth study develops automated mapping techniques that meet worst-case end-to-end latency requirements [DZN$^+$07] in the automotive domain. The three stages of the automated mapping problem for this domain are allocation, priority assignment, and period assignment. For the period assignment stage in mapping, and iterative approach to assign task and message periods has been developed and applied to two case studies.

5.10   Conclusion

A design flow is as important as the development of design frameworks and methodologies. This chapter has described design flows for functional modeling, architecture service creation and usage, and the mapping of the two together. These are vastly superior to a naïve flow presented. However, the real usefulness of this flow will only be demonstrated after looking at a number of case studies later in this work in Chapters 8 and 9.

Future work in the area of design flow development has to do primarily with automation and creating a "correct-by-construction" environment. In the case of the architectural design flow presented, there still exist a number of manual steps. For example, the analysis of design space exploration requires that the user look at the results of each individual simulation. Based on these results, another architecture configuration and mapping is often performed. Ideally given a particular objective (i.e. overall execution time), a simulation could be run in which depending on the results, another architectural model is created (using standard services from a library) and then the simulation is run again. Once the desired objective is met, not only could the design be exported for synthesis but also the transformations which were applied could be recorded and applied to future designs in the hopes that they would be of use.

The automation of the mapping process is naturally highly sought after. If given a functional and architectural model, ideally a mapping could be generated automatically which will meet a certain objective. This would include automatic generation of the formal objective function from high level specifications as well as identifying which aspects of the functional and architectural models need to be exposed for mapping in order to meet that objective.

Chapter 6

System Level Service Refinement

"Program construction consists of a sequence of refinement steps." – Niklaus Wirth, Designer of Pascal

Increasing abstraction as shown in Chapters 4 and 5 is a powerful tool in the fight against increased complexity. Platform-based design explicitly accommodates various levels of abstraction in what has been termed, "the fractal nature" of the design process [KMN$^+$00]. This technique is particularly useful for architectural *services* and the *platform instances* that result when functional descriptions are mapped to those services. Working at various levels of abstraction is useful also in many synthesis situations. For example, one may want to work at lower abstraction levels for the purposes of optimization. Logic minimization of a gate level netlist as opposed to a more structural netlist is an opportunity for optimization. Analysis of design decisions during design space exploration, design transformation from one model of computation to another, or the introduction of physical and implementation concerns, such as wire delays, are all additional reasons for abstraction. When deciding upon the initial abstraction level or the move to another abstraction level, it becomes critical that one can ensure that newly introduced models correspond to their more or less abstract counterparts. Therefore three issues become paramount:

1. What is the *behavior* that should be required to correspond between the abstract and refined systems?

2. How can that *behavior* be captured *efficiently* and *formally*?

3. How can *behaviors* once captured be compared?

This chapter will identify three strategies to classify, capture, and verify the behavior of system level architecture service models. The verification process will be involved in demonstrating that two systems at various levels of abstraction can be safely used in place of each other during simulation without damaging the functionality of the overall design.

Figure 6.1 highlights how refinement plays a part in the proposed design flow first outlined in Chapter 5. In this picture, one can see that refinement is intended to be coupled with the design space exploration process. It is used to ensure that as the design moves down abstraction levels, successively refined models maintain particular properties important to correct system level operation. Specifically structural modifications as well as component modifications should be verified.

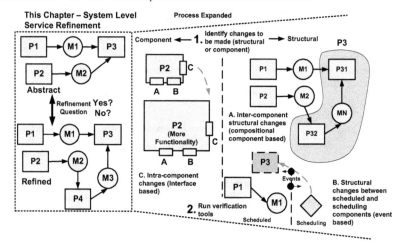

Figure 6.1: System Level Service Refinement in the Proposed Design Flow

In order to clarify refinement's place in the design flow, consider this example scenario. A designer using the METROPOLIS design methodology wishes to provide various architectural service instances upon which to map a functional description. These services could represent new processing elements (such as a CPU) or storage elements (such as memory). These architecture services may each be unique or each may be incremental additions to existing services as well. Those falling into the latter category are considered *refined services*. The system composed of both new and incrementally modified services is a *refined architecture*. This refinement is of interest since these changes represent a variety of intentions on the part of the designer. Refinement attempts to either preserve or introduce new properties to the architecture, raise or lower the abstraction level, or introduce or remove elements bringing it either closer to or further from the requirements of synthesis. Due to the effort often associated with creating entirely new models, refinements are often the most common architecture service modifications. Target architectures tend to be a family of architectures as opposed to entirely new designs. Additionally, testing effort is very valuable and it is desirable to repeat as little of this as possible when designing new systems. It is with all these factors in mind that this work examines three methodologies in which to introduce, categorize, create, and test architecture service refinements. Ultimately we will provide the results of these methodologies in Chapters 8 and 9.

6.1 Chapter Organization

Before beginning any in-depth discussion, Section 6.2 will provide the required background and definitions. This section is followed by an overview and classification of related work in Section 6.3

The remaining organization of this chapter is such that the reader is introduced to the three refinement methods in order of ascending specificity. In Section 6.4 an event based structure for refinement in illustrated. This method uses events to define system behavior and demonstrates how properties can be defined over these events. Section 6.5 builds on the previous section by demonstrating how traditional interface based refinement techniques used in the formal verification community can be utilized in a design environment such as METROPOLIS using events. Finally Section 6.6 shows how a very specific structure (labeled fair transition systems) can be effectively used to represent communication structures in systems by using events as well in compositional component based refinement. Finally, conclusions are provided in Section 6.7.

6.2 Background and Basic Definitions

While it is impossible to make this work completely self contained, it is the goal of this section to at least provide the intended audience with the necessary definitions to understand the majority of this chapter. When the definitions are not unique to this work (which is the case regarding much of the underlying theory) citations are given. It is important that the reader also examine the background and basic definitions provided in Chapters 2, 3, and 5 since they will not be repeated here and their understanding is often assumed.

To start any conversation which attempts to relate two or more systems to each other, the concept of *equivalence* versus *refinement* is critical.

Definition 6.2.1 Equivalence - *The property describing two systems which cannot be distinguished from one another when each is provided the same input stimulus or operating environment.*

For example two combination circuits such as AB or AB + ABC are equivalent since they have the same truth table. Two states in a FSM are equivalent if for any input sequence the set of observable output values which result as the FSM transitions does not differ.

Definition 6.2.2 Refinement - *The process of of removing behaviors of a system through the introduction or removal of components.*

Typically this process is removing over specification or nondeterminism in a design as it proceeds to implementation. An example is developing a USB device from the USB spec.

This difference between the refinement and equivalence definitions is important and should not be discounted. This work will be involved in verifying refinement (not equivalence) between two or more platforms. A system is refined by the existence of a refined architecture model (and its refined services) as defined:

Definition 6.2.3 Refined Service - *A service which provides a subset of the interface methods provided by its more abstract counterpart. This subset will result in fewer possible behaviors. This service may be composed of more or less components than the more abstract service.*

Definition 6.2.4 Refined Architecture - *An architecture model having one or more refined services.*

These definitions provide a sufficient starting point for the discussion to follow. Terms such as architecture, service , and system have been defined earlier as mentioned.

6.2.1 State Equivalence

State equivalence is a well defined concept. It is often applied to finite state machine optimization. The general notion is that two states are equivalent (indistinguishable) if upon applying any input sequence of any length to one state, the output sequence produced is the same as having started from the other state using the same input sequence. Groups of equivalence states are called *equivalence classes*. More formally:

Definition 6.2.5 State Equivalence - *Two states S_1 and S_2 are equivalent if for every possible input sequence X: 1) the corresponding output sequence $Z_1 = Z_2$ and 2) the corresponding next states $S_1^+ = S_2^+$.*

State equivalence is important because certain refinements can be defined loosely as requiring that every state in the refined model have an equivalent state in the abstract model.

6.2.2 Trace Containment

Trace containment is a more specific refinement definition requiring that behaviors be captured as trace sequences. This process will be used in the discussion of interface based refinement. Formally it can be described in the following manner taken from [AH03].

A model is generically defined as an object which can generate a set of finite sequences of behaviors, B. One of these possible finite sequences, B, is considered a trace, $\bar{a}$. Given a model X and a model Y, X refines the model Y, denoted $X \preceq^{Ref} Y$ if given a trace $\bar{a}$ of X then the projection $\bar{a}[ObsY]$ is a trace of Y. A trace, $\bar{a}$ is considered a sequenced set of observable values for a finite execution of the module. A projection of a trace, $\bar{a}[ObsY]$, is the trace produced on Module Y for the execution which created $\bar{a}$ over the observable variables of Y. An observable variable is one which can be read by the surrounding environment or other objects. The two modules X and Y are *trace equivalent*, $X \simeq^{Ref} Y$, if $X \preceq^{Ref} Y$ and $Y \preceq^{Ref} X$.

The answer to this particular refinement problem (X,Y) is YES if X refines Y and otherwise NO.

6.2.3 Synchronized Parallel Composition

Synchronized Parallel Composition is a concept used to create systems specified using sets of finite state machine based descriptions. The operation of these composed systems is described using what is called *synchronization*. Synchronization is the process of explicitly denoting the requirements for individual component state transitions based on the state of other components in the system. For example, a pedestrian walk signal can be activated if another component (traffic light) is in the "red" state. The advantage of this approach is that each individual component is relatively simple but the composition of the systems and corresponding synchronization can be quite sophisticated. Ideally the small component operation can be shown to be sound and therefore composition itself is sound if created following a set of requirements. These concepts will be useful for the third method proposed (compositional component based refinement) in Section 6.6. The definitions in this section are reproduced almost verbatim from [KL03a].

Let $Var = \{X_1...,X_n\}$ be a finite set of variables with their respective domains $\mathbb{D}_1,...,\mathbb{D}_n$. Let

AP be a set of atomic propositions $ap \overset{def}{=} (X_i = v)$ with $X_i \in Var$ and $v \in \mathbb{D}_i$. Let SP be a set of state propositions sp defined by the following grammar: $sp_1, sp_2 ::= ap \mid \neg sp_1 \mid sp_1 \vee sp_2$.

Definition 6.2.6 Interpreted Labeled Transition Systems (LTS) - *A interpreted labeled transition system S over Var is a tuple $<Q, Q_0, E, T, l>$ where:*

- Q is a set of states,

- $Q_0 \subseteq Q$ is a set of initial states,

- E is a finite set of transition labels or actions,

- $T \subseteq Q \times E \times Q$ is a labeled transition relation, and

- $l : Q \to SP$ is an interpretation of each state on the system variables.

Definition 6.2.7 Sum of Two LTSs - *Let $S_1 = <Q_1, Q_{01}, E_1, T_1, l_1>$ and $S_2 = <Q_2, Q_{02}, E_2, T_2, l_2>$ be two transition systems over Var. The sum of S_1 and S_2, written $S_1 \uplus S_2$ is $<Q_1 \cup Q_2, Q_{01} \cup Q_{02}, E_1 \cup E_2, T_1 \cup T_2, l_{12}>$ where l_{12} is defined by:*

$$l_{12}(q) = \begin{cases} l_1(q) \text{ if } q \in Q_1, \\ l_2(q) \text{ if } q \in Q_2, \end{cases}$$

Moreover, $\forall q_1. \ q_1 \in Q_1, \forall q_2. \ q_2 \in Q_2 \ . \ (l_1(q_1) = l_2(q_2) \Leftrightarrow q_1 = q_2)$.

Definition 6.2.8 Synchronization of n Components - *Let $S_1,...,S_n$ be n components. A synchronization Synch is a set of elements (α when p) where:*

- $\alpha = (e_1,...,e_n) \in \prod_{i=1}^{n} (E_i \cup \{-\})$, where - is a fictive action "skip"

- p is a state proposition on the component variables.

Definition 6.2.9 Context-in Component - *Let $S_1,...,S_n$ be n components. Let Synch be their synchronization. A context-in component S_i^c is defined by the tuple $<Q_i^c, Q_{0i}^c, E_i^c, T_i^c, l_i^c>$ where:*

- $Q_i^c \subseteq Q_1 \times ... \times Q_n$ with $(q_1,...q_n) \in Q_i^c$,

- $Q_{0i}^c \subseteq Q_{01} \times ... \times Q_{0n}$,

- $E_i^c = \{(e_1,...,e_i,...,e_n) \mid (((e_1,...,e_i,...,e_n) \text{ when } p) \in Synch) \bigwedge (e_i \in E_i)\}$,

- $l_i^c((q_1,...,q_n)) = l_1(q_1) \bigwedge ... \bigwedge l_n(q_n)$

- $T_i^c \subseteq Q_i^c \times E_i^c \times Q_i^c$ with

$((q_1,...,q_n), (e_1,...,e_n), (q_1^{'},...,q_n^{'}))$ in T_i^c iff:

- $((e_1,...,e_n) \text{ when } p) \in Synch$,

- $l_i^c((q_1,...,q_n)) \Rightarrow p$, and

- $\forall k.(k \in \{1,...,n\} \Rightarrow ((e_k = - \wedge q_k = q_k') \vee (e_k \neq - \wedge (q_k, e_k, q_k') \in T_k)))$.

Definition 6.2.10 Synchronized Composition of n Components - *Let $S_1,...,S_n$ be n components and Synch their synchronization. Let $S_1^c,...,S_n^c$ be their respective context-in components. The synchronized parallel composition of $S_1,...,S_n$ under Synch is defined by:*

$$\|_{Synch}(S_1,...,S_n) \stackrel{def}{=} \uplus_{i=1}^{n}(S_i^c)$$

Definition 6.2.11 Gluing Relation - *Let GI be a gluing invariant between SR and SA. The states $q_R \in Q_R$ and $q_A \in Q_A$ are glued, written $q_R \mu q_A$, iff $l_R(q_R) \wedge GI \Rightarrow l_A(q_A)$.*

6.3 Related Work

The idea of refinement and its verification is not new and is not limited to the notion of architectural services. In fact, much of the work in refinement verification is concerned with software design. While software design focuses on program correctness, this work is more focused on system functionality. Correctness assumes a desired result where functionality assumes the validity of several outcomes. This is a subtle difference but can be seen as two similar problems. The former wants to ensure that the system arrives at a particular state(s) whereas the later wants to avoid a particular state(s). Aspects of this are built upon the fact that there are "don't care" states and nondeterminism in architecture models. This section will provide an overview of the existing work regarding refinement verification of architectural services at the system level. This section will be used to highlight the unique contributions of our proposed approach and clearly define which aspects of this problem have been addressed. It should naturally be mentioned that there are many types of verification methods related to electronic system design. These include simulation based approaches, model checking [CGL93], symbolic simulation [BBS91], combinational equivalence checking, sequential equivalence checking [MS05], statecharts [Har87], and process algebras (CSP)[Hoa78], (ACP)[BK85], and Robin Milner's (CCS) for example. The work to follow in many cases uses concepts from these areas as a foundation.

Refinement verification work has been proposed in a number of forms. From these forms, refinement verification can be broadly categorized as *style/pattern based, event based, and interface based*. Additionally, in work by Gong et. al. [GGB97], there is a discussion of refinements as *control related, data related, or architecture related*. The first classification (control related) denotes that the execution sequence is to be preserved when the design is refined over multiple components. The second classification (data related) denotes that data accesses must be updated appropriately when the design is refined.

The final classification (architecture related) denotes the ability to perform changes to the communication structures between components (buses for example) which facilitate communication during the refinement process.

Table 6.1 uses these two groups of classification schemes to organize the approaches discussed in this section as well. Firstly each approach is grouped according to its place in the first categorization (style, event, interface, other). Then each approach is assigned a "+" (focused on), "-" (not focused on), or "?" (not applicable or known) in each area of the later classification regarding control, data, and architecture. The lack of support for these constructs does not indicate a particular weakness but rather serves to illustrate the intended purpose and scope of each tool. Additionally a very brief description is provided as well for each tool.

	Control	Data	Arch.	Description
Style/Pattern Based				
ForSyDe [SJ04] [RSSJ03]	+	-	+	Transformation rules
METROPOLIS	+	-	+	TTL vs. Yapi
Model Algebra [AG06]	+	+	+	Algebraic rules
Moriconi [MQR95]	+	-	-	Six design patterns
Virtual Prototyping [BHKR05]	+	+	+	Parameter and data streams
Event Based				
METROPOLIS	+	?	+	Tagged signal model
METRO II	+	+	+	3rd phase constraint solving
Rapide [LV95] [LKA$^+$95]	+	-	+	Event pattern based EADL
Interface Based				
METROPOLIS	+	?	+	Function calls on ports
METRO II	+	+	+	Required and provided services
Reactive Modules [AH99]	+	-	-	Hierarchical verification
Signal [TGS$^+$03]	+	+	?	Polychrony/Flow equivalence
SPADE [LvdWD01]/Sesame [PEP06] [vHPPH01]	+	+	+	Kahn process network based
SynCo [KL03b]	+	-	?	Compositional Components/LTS
Others				
Obj. Orient. in C2 [MORT96]	?	-	+	Explicit subtype relationships

Table 6.1: Refinement Verification Related Work Classification

Style Based Refinement Related Work

Style based refinement requires that rules be developed *a priori* defining what is considered refinement. Each *style* has associated with it a set of rules. Once those rules have been shown to be sound on one set of component style instances, the components can then be reused or substituted for many other components which use that same style. Often rules can be used to convert components in one style to another style. Styles which can undergo this transformation are often called *substyles*. Style based refinement is often called "pattern based refinement" as it categorizes styles as groups of compositional patterns. Valid composition of these patterns introduces corresponding compositional rules. A style based approach is shown in the work of Moriconi [MQR95]. In this work, six patterns are proposed for classifying the refinement of components, connectors, and interfaces. These patterns are batch sequential, control transfer, dataflow, functional, process pipeline, and shared memory. Example applications of these rules are restrictions placed on the architectures regarding variable types, access to variables, and ordering of variable access. A drawback of this approach is that all system instances may not fall into a given style thus limiting the types of systems expressed. Another pattern based approach is called "Virtual Prototypes" [BHKR05]. This work creates verification patterns from an algorithmic level description (considered their highest level of abstraction) and automatically applies these patterns to lower level models. These patterns can be viewed as streams. There are both data-in and data-out streams as well as parameter-in and parameter-out streams. The parameter-in streams set up the device under test and the data-in streams stimulate it. The "out" patterns detail what results are expected for each verification "in " pattern. A drawback of this method is that an algorithmic pattern must be created potentially for each device under verification. While some devices may share patterns, this process is potentially very user intensive. Finally Sander and Jantsch present ForSyDe [SJ04] [RSSJ03] which uses what it calls *Transformation Rules* in order to perform refinement. These rules denote specifically how one process network (the model of computation in ForSyDe) maps to another process network. It requires that they have the same input signals and the same number of output signals. These transformations can be either "semantic preserving" which do not change the meaning of the model, or "design decisions" which do. All of these transformations form a transformation library. While not necessarily a drawback, this work does not make an explicit separation of architectural services from the functionality of the system. For an overall assessment of style and pattern based approaches see [Gar96] which explains in more detail these types of systems. In general it states that style and pattern based approaches need to accommodate a large set of system instances to be useful, that style classification is useful if the styles are chosen carefully, and that refinement can be made more flexible if styles are accompanied by a set of properties of interest (those which should be maintained during refinement). While style/pattern approaches are discussed to give a complete survey of the field, a style/pattern based approach will not be presented in detail in this work but this technique is possible in METROPOLIS as shown in Figure 6.2 using the REFINE keyword. This is a very simple re-routing of the connections in a netlists. This re-routing requires that the port configurations of the components being swapped match.

There are naturally attempts to prove equivalence (not refinement) between models at various levels of abstraction. Algebraic approaches [AG06] appear promising. This is ultimately also a pattern type approach. This work describes systems in an algebra which consists of behaviors, channels, variables, interfaces, ports, and labels. From a set of roughly 7 rules, transformations can be made on models composed of those components. These rules themselves have been shown to be sound and thus the transformations made by these rules are sounds. Our methodology is amenable to such an approach as we can describe the architecture model (or a functional model) in this algebra. Work is currently being done in METRO II which will facilitate this process by creating libraries of the required components corresponding to the algebraic rules. Drawbacks of this approach are that it requires models to be described in one model of computation (dataflow), use specific components, and there is not a canonical representation for two equivalent architectures (i.e. there exist false negatives regarding two equivalent models).

An example of METROPOLIS style refinement involves the YAPI (Y-Chart API) [KES+00] and the TTL (Task Transaction Level) libraries which are provided as part of its distribution. These are both process network based FIFO libraries. In METROPOLIS, the YAPI library has unbounded FIFO-like elements while the TTL library attempts to be a refined version with *boundedfifo*, *yapi2TTL*, *TTL2yapi*, and *rdwrthreshold* elements. The boundedfifo simply is the storage mechanism now with a fixed size. The rdwrthreshold element acts as the coordination for access to this element. Finally, yapi2TTL and TTL2yapi are used for the refinement interface in the refined netlist similar to the example in Figure 6.2.

During the use of these elements in a multi-media application exercise in METROPOLIS, several bugs in the design were discovered. This drew attention to the fact that refinement checking is a crucial element as the design process becomes more complex and specifications are adhered to in an *ad hoc* manner.

Event Based Refinement Related Work

Event based refinement use events to define how architectures are related to each other. While this idea could be used to mimic a style or pattern based approach (one which uses event patterns for instance), in practice event based approaches allow a much wider variety of systems to be related. For example, an architecture at a high level of abstraction may require just one of its events to be related to a set of events in a model created at a lower level of abstraction. A bus read event in the abstract model may correlate to a request, ack, and read set of events in the refinement. Events also are typically part of the operational semantics of the architectures they are part of. Therefore, event based refinement can also be used to specify a specific operation or a restriction on system operation. For example a particular refinement may require the notification of an event which triggers one behavior of many possible behav-

179

```
//In Metropolis Netlist

/*Introduce to the netlist(this),
an object for refinement(ref_obj)*/
refine(ref_obj,this);

/*Redefine the connections
so that the refinement input
and outputs map to the abstracts ports*/
//''ref_obj'' is TTL, ''abs_X'' ports are YAPI's
refineconnect(this,src_connect(ref_obj,out),
    port(ref_obj,out),abs_out);
refineconnect(this,src_connect(ref_obj,in),
    port(ref_obj,in),abs_in);
```

Figure 6.2: METROPOLIS Style Refinement Example

iors. The abstract model would relax that notification requirement allowing more behaviors. An event based approach is shown in Rapide [LV95] [LKA+95]. Rapide is an executable architecture definition language (EADL). Rapide uses "event patterns" to relate architectures together during mapping. This should not be confused with patterns as previously defined however, as Rapide uses these patterns or *maps* to trigger events or generate events. The event based approach proposed in this work uses events to define properties. These properties could be seen as patterns defined over a set of events. These properties use events specifically belonging to components depending on their role in the simulation (either part of the scheduling or scheduled mechanism). This is different from Rapide which does not make such a distinction. One can not directly make a comparison between the two approaches outside of the fact that both require the notion of an event object defined as a tuple containing more than one field of data.

Interface Based Refinement Related Work

Interface based refinement is premised on the fact that most modeling systems encapsulate services through the use of *interfaces*. These interfaces often are a function of the language that the system is described in. SystemC and Java for instance have the idea of interface functions for classes. Often times however, the term interface denotes the legal interaction points between models and other models or models and their environment. Refinement which focuses on these points is interface refinement. Typically changes to the model which cannot be observed at the interface are not considered in this refinement style. Therefore interface based refinement introduces a notion of observability not necessarily implied in the other styles. The observability can be exploited in the event that a designer does not wish to have subsequent models be viewed differently (e.g. keep the same interface) while at the same time this observability can be an issue when design differences are desirable but difficult to push to the interfaces.

Interface based refinement methodologies are illustrated in the Reactive Modeling Language (RML) [AH99] and SPADE (System level Performance Analysis and Design space Exploration) [LvdWD01]. In RML a concept called *hierarchical verification* is employed. RML uses an object called a reactive model. These reactive models can be composed to form composite objects. Hierarchical verification requires that every finite sequence of observations resulting from the detailed module also be possible from the abstract model. This work is directly relevant to the proposed methods in this chapter and will be discussed in more detail. SPADE on the other hand uses Kahn Process Networks to describe functional models and a library of architectural building blocks. The functional model creates traces. These traces can be "accepted" by the architecture and contain information detailing which operation the architecture should perform. Trace transformation is the process by which functionality is assigned to available architecture resources. This transformation equates to forcing refinement traces to change to meet available architecture resources and topologies. This style of work has also been used in Sesame [PEP06] [vHPPH01] which builds on the SPADE framework. Refinement for these tools is a means to achieve a mapping. While mainly methodological in nature the difference between this approach and a tool like METROPOLIS is that SPADE starts with a functional description and works top down. METROPOLIS works both bottom up and top down, refining both architectural services and functional descriptions independently.

Two other interface approaches are SynCo [KL03b] and Signal [TGS+03]. SynCo is based upon the work of [KL03a] which is a compositional component based methodology. Transitions systems for both refined and abstraction systems are specified. States in those systems are then "glued" indicating which states are required to correspond to each other (this is a many to one mapping from refined to abstract model). Also synchronization mechanisms can be defined as well in order to create larger systems from individual LTS. SynCo will be used in this work. Signal on the other hand is a polychronous (i.e. multiclocked) design language. This language has the notion of *flow equivalence* between behaviors. This means that for two behaviors their signals hold the same values for the same order. This leads to *flow invariance* where an asynchronous implementation preserves flow equivalence.

Other approaches exist which can not be placed into one of the three previous classifications. These often consist of ad hoc or brute strength style approaches. For example in [MORT96] the authors demonstrate that object oriented (OO) subtype hierarchy type checking can be used to identify refinement. They investigate how concepts in OO programming languages can be used in C2, a component and message based system specification style. They find that by making subtyping explicit, identifying component substitution is possible. Also extending type checking mechanisms allows a richer set of architectural relationships to be expressed.

The proposed design flows in this work are a combination of event and interface based approaches. Specifically they most closely resemble the work of [AH99] and [KL03a] (both interface based). For example they incorporate concepts such as trace containment, labeled transition systems or control flow automata, and trace transformations. In fact the tools Mocha and SynCo are used in Chapter 9 on several cases studies to actually implement the methodologies presented. This work could be extended to style/pattern based approaches as well assuming a set of rules were created. Aspects of this process were performed in early METROPOLIS related projects using both TTL and Yapi channels as proposed by [LvdWDV01].

6.4   Event Based Service Refinement

Event based service refinement will be presented first. The presentation ordering was selected because of the approaches presented here, event based is the most general. Events can be leveraged to perform both interface and compositional component based verification as well (to be discussed). Event oriented frameworks have become popular with the increased interest in exploring design frameworks which allow specifying concurrent computation models. Lee and Sangiovanni in [LSV98] introduced the tagged signal model which demonstrated how an event based framework can be used to express a variety of models of computation. This characteristic has made them very flexible and gives them the ability to realize a wide variety of systems. Also event based models are portable because often they only assume the presence of events and do not make other assumptions about the framework which is implementing the events. Event based platform refinement is prefaced upon the following ideas:

- The design framework uses events to denote system activity and provide synchronization mechanisms. For example, imagine a basic producer and consumer example. The producer writes to a shared storage location. Upon doing so, it produces an event (production) signaling this. It then waits for the presence of another event (consumption). The consumer will use this notification (production) to realize that it it can now consume the data. Upon consumption it will signal this operation with an event (consumption) as well. This notifies the producer that it can safely produce again. This process continues indefinitely.

- Sequences of events (traces) can be captured to recreate or represent system behavior. For example a bus transaction has a fixed sequence of events as dictated by the protocol. A *request* event must proceed a *grant* event for example. If this sequence is not maintained then the system behavior has been violated.

- Event sequences can restrict or enforce behavior. For example often times a system has to make choices. A control statement (if, while, for, etc) often has conditions which allow the system to make a decision. Those conditions can use events as part of their evaluation. Allowing or restricting event appearance can be an effective mechanism to enforce behavior without changing the model explicitly.

- This enforcement or restriction has the ability to be a well defined, methodological refinement as will be shown in this work. Specifically examples will be shown in the METROPOLIS design environment in Chapters 8 and 9.

6.4.1   Proposed Methodology

In order to systematically refine a platform there must be a methodology in place which demonstrates the procedure for a designer to follow in order to perform various refinements. In following such a procedure, ideally one can enforce *by construction* which properties will hold between the abstract and refined model. In the event that the construction can not be provably correct, such a method will allow a property checking system to perform verification. One can therefore follow various procedures depending on which properties are of interest. Each section regarding a refinement style will begin with such a proposed methodology.

Within this section there are three refinement methodology proposals for event based service refinement. These three proposals will examine how event based platform refinement can be performed for two scenarios/goals:

1. Refinements between and within systems with changing *component-to-component relationships*. For example an architecture designer may wish to introduce a new bus, memory hierarchy, or processing component. These introductions will manifest themselves as new components. Alternately, one may wish to collapse services into a single component. This will result in the removal of components and existing components will therefore offer more services.

2. Refinements between systems with changing *component-to-scheduler/annotator relationships*. For example, it may become necessary to introduce components which act as arbitrators or controllers

which do not offer services directly to functional model components but rather only restrict the operation of existing components.

The initial refinement methodologies to be described are what this work terms *vertical* or *horizontal* refinement. These are both topological refinement techniques (the topology of the system is affected). Vertical refinement refers to the process of transforming relationships between components (scenario 1). For example in METROPOLIS this occurs in the scheduled netlist. This typically is done by targeting one particular process or media element and decomposing it into multiple media and process elements and then replacing that decomposed structure back into the model. This is geared toward changing the nature of the services and the interaction between those services the architecture provides.

Horizontal refinement refers to refinement which converts aspects of the model's scheduling mechanisms into components themselves in the scheduled mechanism (scenario 2). In METROPOLIS this requires that quantity managers from the METROPOLIS *scheduling netlist* move into the *scheduled netlist*. This represents refinements geared toward physical implementation.

Figure 6.3 illustrates a high level view of the various event based refinement styles to be discussed. This picture demonstrates that it is important to clearly separate the components in the model which provide services from the components which schedule these services. The number, type, origin, and order of events are the aspects which are modified by the refinement styles.

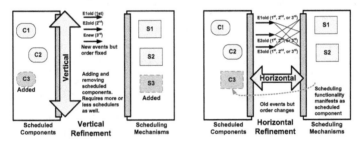

Figure 6.3: Event Based Refinement Proposal

For the rest of these sections let α be a set of components (objects which provide or use services). These are often processes in a METROPOLIS scheduled netlist. γ is a set of annotators or schedulers. For example quantity managers in a METROPOLIS scheduling netlist. Finally β is the overall behavior of the platform. β will mean something unique to each system. In this work β is a event trace.

Vertical Refinement

Vertical refinement is the notion that component-to-component relationship changes (scenario 1) are performed for three reasons.

- **Increase service interaction sequentially.** For example adding a cache hierarchy to a microprocessor model by physically stringing out a first and second level elements cache along with main memory. This modification is often done to reduce the number of processing elements (PEs) needed since the services can map to the same element.

- **Increase service interaction concurrently.** For example adding processing cores to a many-core architecture. This modification is done to provide performance gains over sequential execution. Also this can be done to expose parallelism for functional models to take advantage of during mapping.

- **Create coarser or more granular services.** While it could be said that these changes could be classified as one of the previous two reasons, this classification specifically occurs when the abstraction level changes. For example, migrating from task level modeling to transaction level modeling.

Definition 6.4.1 Vertical Refinement - *A manipulation to the scheduled component structure (netlist) to introduce or remove the number or origin of events as seen by the scheduling components (netlist).*

The term *vertical* comes from that fact that these changes are within the same domain (METROPOLIS scheduled netlist for example). It is not swapping aspects between netlists but rather moving within a particular netlist. Naturally this contrasts with horizontal refinement. Vertical refinement of an platform can be seen as a whole spectrum of refinement with the abstraction levels being defined as to what elements are passive (media for examples) and which are active (processes for example). One can change the number and types of processes in the scheduled netlist or one can change the number and type of media in the scheduled netlist. The primary method of vertical refinement in METROPOLIS is the addition of service media. This ultimately is the addition of architecture services at a different level of granularity compared to the abstract services provided initially. Other system design methodologies such as [SJ04] [RSSJ03] term this a *design decision* refinement since the behavior of the architecture will change.

Formally a vertical refinement is a transformation in the set of components (α) and annotators/schedulers (γ). Additionally behavior (β) may change:

$(V_1)\ \alpha_{refinement} = \alpha_{abstract} \cup \alpha_{additional}$

$(V_2)\ \alpha_{abstract} \subset \alpha_{refinement}$

$(V_3) \mid \gamma_{refinement} \mid \geq \mid \gamma_{abstract} \mid$

$(V_4) \beta_{refinement} \subset \beta_{abstract}$

V_1 requires that the refined system have all the components of the more abstract system and allows for additional components if needed. V_2 requires that the abstract components are a subset of the refined component set. V_3 requires that the number of annotators/schedulers in the refinement is greater or equal to the number in the abstract model. Finally, as with all refinements, V_4 requires that the behaviors of the refinement are a subset of the abstract model.

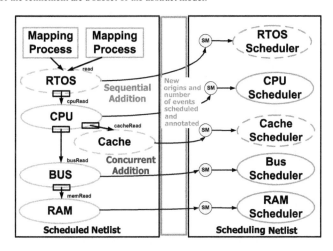

Figure 6.4: Vertical Refinement Illustration in METROPOLIS

Figure 6.4 is an illustration of how vertical refinement is carried out in METROPOLIS. An additional explanation of this vertical refinement is shown in Table 6.2. This example illustrates that the two subtypes of vertical refinement, sequential and concurrent, change the event traces. This change can be in the number/origin of events seen but not the overall ordering. In the left most column (labeled original), the sequence of events seen by the METROPOLIS scheduling netlist is shown. In a sequential, vertical refinement (second column) an RTOS is added. This introduces the new event RTOSREAD but the order amongst the events also in the original sequence is unchanged. The "concurrent 1" trace (third column) adds a cache. This adds an interleaved CACHEREAD but the order in which the other original events are seen is still unchanged. The same number of events do not appear since a cache hit is assumed.

The final column (concurrent II) is a cache miss which causes interleaving but does not eliminate the appearance of other events or change the organization amongst them. The events to notice are italicized throughout the table.

Original	Sequential	Concurrent 1	Concurrent II
E1 (CPURead)	*E1 (RTOSRead)*	E1 (CPURead)	E1 (CPURead)
E2 (BusRead)	E2 (CPURead)	*E2 (CacheRead)*	*E2 (CacheRead)*
E3 (MemRead)	E3 (BusRead)		E3 (BusRead)
	E4 (MemRead)		E4 (MemRead)

Table 6.2: Potential Vertical Refinement Event Traces

The vertical refinement methodology is explicitly shown in Algorithm 1.

Horizontal Refinement

Horizontal refinement is the transformation of scheduling (quantity managers in METROPOLIS for example) functionality into a scheduled component (a METROPOLIS process or media the scheduled netlist for example). This is the second scenario mentioned earlier. The spectrum of different horizontal refinements results from how many of the schedulers one moves and what portion/aspects of the schedulers are moved. Horizontal refinement is done in for two primary reasons:

- **In order to reduce the number of elements resolving quantities.** This potentially represents a way to speed up simulation. This can be accomplished by removing the number of events that need to be evaluated by the simulation manager.

- **Focus the scheduling effort more locally which reflects a more implementation based view.** This can be done in the event that the design environment can be targeted for synthesis.

Definition 6.4.2 Horizontal Refinement - *A manipulation of both the scheduled and scheduling components (netlists) which changes the possible ordering of events as seen by the scheduling components (netlist).*

The term *horizontal* comes from the fact that the changes made are from different domains. Objects once concerned with controlling the scheduling of components now become actual components which enforce that schedule through their behavior as components. In METROPOLIS this is a swapping

---

Algorithm 1: Vertical Refinement Process

    Input: Select Service, S, to refine vertically

    /*Decision made based on DSE results and performance desired   */

1 **if** $S == MCSI \mathbin{||} MCMI$ **then**

2   Add new Component, C_N;

3   **forall** *Components, $C_i \in S$* **do**

4   **if** *C_i interacts with C_N* **then**

5   Add(**internal**) interfaces to C_i to accommodate, C_N;

6 **else if** $S == SCSI$ **then**

7   Add new Component, C_N;

8   Add one **internal** interface to accommodate, C_N;

9   Reclassify Component as $MCSI$;

10 **else**

11   Add new Component, C_N;

12   Add quantity manager, QM_{New} /*C_N is a new stand alone Component   */

13   Classify the Component as $SCSI$;

14   Register the Service with the mapping process;

15 Reconnect the new topology

16 **forall** *Events, E, between Netlist$_{sched}$* **and** *Netlist$_{scheduling}$* **do**

17   Capture new behavior, B_{New};

18 **RETURN** B_{New};

---

of items from the scheduling to the scheduled netlist. [SJ04] [RSSJ03] terms this a *Semantic Preserving Transformation* refinement since it retains the overall behavior of the model.

Formally a horizontal refinement is a transformation in the set of components (α) and annotators/schedulers (γ). Additionally behavior (β) may change:

(H$_1$) $\alpha_{refinement} \neq \alpha_{abstract}$

(H$_2$) $\alpha_{abstract} \subset \alpha_{refinement}$

(H$_3$) $|\gamma_{refinement}| < |\gamma_{abstract}|$

(H$_4$) $\beta_{refinement} \subset \beta_{abstract}$

H$_1$ requires that the number and types of components in the refined model and the abstract model not be equal. H$_2$ requires that the abstract components be a subset of the refinement. This is also a requirement of vertical refinement. The number of annotators/schedulers must be greater in the abstract model as shown in H$_3$. H$_4$ requires the behaviors of the refined model to be a subset of the abstract.

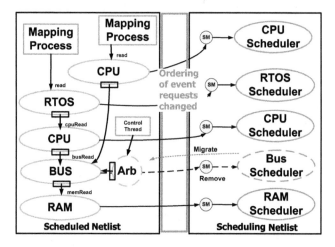

Figure 6.5: Horizontal Refinement Illustration in METROPOLIS

Figure 6.5 is an illustration of how horizontal refinement is carried out in METROPOLIS. This shows the migration of a bus scheduler which manifests itself as an arbiter component. The affect of this refinement on event traces is shown in Table 6.3. The left column shows the original trace (the event and which component generated it). The right column shows a possible trace of the refinement. Notice the second and third rows. Event E2 and E3 now are generated in a different order than in the original. This is a change which would not have been possible in solely vertical refinement. Horizontal refinement verification will require that the extent to which this re-ordering can occur be specified by the designer. This specification can be done typically by stating explicitly which sequences can not occur (a smaller set than allowed sequences). Typically boundary events are also specified denoting when this deviation can begin and when it should end.

Original *	Refined (Interleaved)
E1 (BusRead) → From CPU1	E1 (BusRead) → From CPU1
E2 (BusRead) → *From CPU1*	E3 (BusRead) → *From CPU2*
E3 (BusRead) → *From CPU2*	E2 (BusRead) → *From CPU1*
E4 (BusRead) → From CPU2	E4 (BusRead) → From CPU2

Table 6.3: Potential Horizontal Refinement Event Traces

The horizontal refinement methodology is explicitly shown in Algorithm 2.

Diagonal (Hybrid) Refinement

Diagonal refinement is a combination of vertical and horizontal refinement methods. The goal of any of these refinement methods is to *determine a set of properties that are held or not held depending on the refinement style*. Ultimately these properties will determine which refinement methodology is employed. One potential drawback of a diagonal refinement approach is that as more changes are made in parallel to a design, the more difficult (or impossible) it may become to determine the effects of the changes. The methodological recommendation is that for any one refinement, each stage only consist of a vertical or horizontal change followed by a verification of the relevant aspects before making any additional changes.

Event Based Properties

In order to make use of event based refinement methods, it is important to illustrate that properties can be defined over events. These events will be of use in describing architectural services. The ability to specify properties (event sequences) and verify these sequences is a key part of refinement. In Table 6.4 a simple set of event traces demonstrates how to represent resource utilization. The table is first broken into two sections. The left shows a "bad" *resolve()* function. The function resolve() is used in METROPOLIS during the scheduling phase by quantity managers which enable events. The other side

---

Algorithm 2: Horizontal Refinement Process

Input: Select Service, S, to refine horizontally

/*Decision made to affect scheduling of Services   */

1   $QM_{Old} \rightarrow$ creation of new Component, C_N which is **SCSI**;

2   **if** $S == MCSI \parallel MCMI$ **then**

3   **forall** *Components,* $C_i \in S$ **do**

4   **if** C_i *is required to interact with* C_N **then**

5   Add one **external** interface to C_i to accommodate, C_N;

6   **else if** $S == SCSI$ **then**

7   Add one **external** interface to C to accommodate, C_N;

8   Remove QM_{Old} /*S no longer requires a quantity manager   */

9   Add quantity manager, QM_{New} /*This is for C_N */

10   **if** $S == NULL$ **then**

11   Add quantity manager, QM_{New} /*This is a new stand alone Component   */

12   Classify the Component C_N as *SCSI*;

13   Register the Service with the mapping process ;

14   Reconnect the new topology

15   **forall** *Events, E, between* $Netlist_{sched}$ **and** $Netlist_{scheduling}$ **do**

16   Capture new behavior, B_{New};

17   **RETURN** B_{New};

---

shows a "good" *resolve()*. In this case a "good" resolve makes maximum use of the resources by scheduling events in such a way that resources are not idle. For example assuming events E1, E2, and E3 only use the CPU whereas event E4 uses the CPU, Bus, and Memory, it is ideal to let the Bus and Memory process the event E4 as soon as possible assuming that the events are independent. In the "bad" resolve() scenario the events are scheduled E1, E2, E3, E4. The P's represent phases at which the elements are idle. The italicized *Ps* in the "bad" resolve() illustrate phases in which resources are available but are not used. The difference can be seen in the "good" resolve which schedules E4 first on the CPU thereby enabling this event to be seen earlier in the other components. This property can be expressed later in such a way that allows the scheduling of events using the most resources first in the event that events are independent.

	Bad Resolve()	Good Resolve ()
CPU	E1, E2, E3, E4	E4, E1, E2, E3
Bus	P_4, P_3, P_2, P_1, E4	P_1, E4
Mem	P_5, P_4, P_3, P_2, P_1, E4	P_2, P_1, E4

Table 6.4: Resource Utilization Event Analysis

In Table 6.5 the latency of two different simulations are shown. Again one side illustrates a "bad" resolve() and the other side a "better" resolve(). Each resolve() side has two columns. The leftmost of these columns shows the events to be scheduled. The rightmost of these columns shows the event selected. In this case, "better" means that the average latency of events (time from generation to annotation of a particular event) is minimized. Each element is labeled with a number in parenthesis which illustrates which scheduling phase number is currently being evaluated. The italicized events illustrate key decision points in the table. Starting with the "bad" resolve() side a description of the table is thus: initially the CPU can choose between event E1 and E2. E1 is selected. In the next phase E3 enters the system so that E3 and the left over E2 can be selected. E2 is selected. The bus now can select E1 (what the CPU just passed on) or from an existing EX. EX is selected. In phase 2 the CPU can only chose E3. The bus now can select between the older E1 or E2. As shown by the italics, E2 is chosen. Then in the last stage, again E1 is passed over for E3. In this case E1 waits two phases longer than necessary. In the "good" resolve() trace this is not the case. It initially proceeds in the same way, but in phase 2 E1 is scheduled instead of E2 and in the final phase E2 selected ahead of E3. This scheduling would minimize the latency for each event.

The key issue for event based refinement is resolving what are the properties that are required

	Bad Resolve()		Good Resolve ()	
	Choices	Selected	Choices	Selected
CPU (0)	E1, E2	E1	E1, E2	E1
CPU (1)	E2, E3	E2	E2, E3	E2
Bus (1)	E1, EX	EX	E1, EX	EX
CPU (2)	E3	E3	E3	E3
Bus (2)	E1, *E2*	E2	E1, E2	E1
CPU (3)				
Bus (3)	E1, *E3*	E3	E2, E3	E2

Table 6.5: Latency Event Analysis

to hold between the abstract and the refined model. The first question is how do those properties manifest themselves as attributes of a model? For example if one is interested in the resulting latency of a process, what are the observable behaviors of the process which give them insight into this property? The second question is how do I capture and specify the properties? The third question is how are those attributes to be related between the two models? To begin to answer these three questions we introduce two definitions of a property.

Definition 6.4.3 MicroProperty - *The combination of one or more attributes (quantities) and an event relation defined with these attributes.*

Definition 6.4.4 MacroProperty - *A property which implies a set of MicroProperties. A MacroProperty is defined by the property which ensures the other properties' adherence (dominator property). The satisfaction (i.e. the property holds or is true) of the MacroProperty ensures all MicroProperties covered by this MacroProperty are also satisfied.*

Since the the implication does not commute there are MacroProperties which share Micro-Properties but they are not themselves the same. MacroProperties are also assigned a level (1 to ∞). The level indicates the length of the longest chain of implications the MacroProperty is responsible for. MicroProperties are by definition level 0.

Event Based Property Classification

This section will begin to discuss which properties can be specified during event based refinement. Right now the list is very sparse and high level. The majority of effort will now go into identifying these properties, their relationships, and how to check them.

One can categorize the properties as *structural, functional, and performance*. Platform-based design dictates that these be kept explicitly separate.

Examples of performance properties are:

- Latency - time for a task to complete. Given an appearance (start) time and a disappearance (end) time of an event, the latency is the positive difference between the two.

- Throughput - number of tasks completed per unit time. Given a period of time (t) and a number of completed tasks (T_A), throughput is T_A/t.

- CPI - cycles per instruction (request). This is simply an average to indicate system performance. For example, the goal of a basic pipelined microprocessor is CPI=1. In the superscalar microprocessor era, typically the inverse, IPC, is a more relevant metric.

- Jitter - random variation in a signal. For example given a periodic signal, the variation in the period or amplitude is an example of jitter.

Performance properties typically have to do with specifications regarding the desire for a certain level of performance. However sometimes these properties can actually be required for the correctness of a system. This is often true of safety critical or real time systems. In many cases performance properties are related to one another.

Examples of functional properties are:

- Mutex - mutual exclusion of a resource. This property can be realized by semaphores or shared memory/variables. In many model languages such as SystemC events are used to accomplish this.

- Data Consistency/Coherence - global data set contents must match and reading and writing ordering must be preserved. This property commonly is of interest in memory systems. Cache consistency and coherency are often maintained by such protocols like MESI (Modified, Exclusive, Shared, Invalid).

Functional properties typically have to do with maintaining the correctness of a system. Often they implicitly affect the performance properties of a system as well.

Examples of structural properties are:

- Memory Size - size of memory elements such as FIFOs. There exists both the physical memory as well as the virtual memory. These can be distributed or shared memories.

- ALU operand size - the size of the ALU operands (e.g. bits). In addition to operand size, operation type can also be important (floating point vs. integer).

- Datapath width - the size of the instruction and addressing datapaths. Datapath width may need to change in the presence of instruction level parallelism such as VLIW machines or the adoption of a new ISA.

Structural properties typically have to do with both the performance and the correctness of a system. They will interact with other structural properties as well as with functional and performance properties.

What will be of key importance is the way in which properties are related and categorized so that one can:

- Determine which properties are related and how. This can be defined over sets of Micro and Macro properties.

- Determine which refinements relate to which properties. These can be defined both explicitly (e.g. a list of required properties), construction (e.g. certain refinements automatically preserve properties), or implicitly (e.g. one property preservation requires the adherence of another).

In terms of grouping properties an initial attempt we have seen is in [NGMG98] which describes a method which uses the following terminology:

- Rule of Computation (CMP) - dictates how variable (stored) values are computed based on their old values as well other variables.

- Rule of Read Order (RO) - for any pair of read events x and y in a process, if x comes before y in program order, then x occurs before y.

- Rule of Write Order (WO) - for any pair of write events a and b in a process, if a comes before b in program order, then a occurs before b.

- Rule of Write Atomicity (WA) - writes become instantly visible to all processes instantaneously.

One can group properties by this method. For example the "mutex" functional property is a WA property. While Data consistency is a RO and WO property. This method can be used to examine Macro and Micro properties relationships.

Event Based Property Relationships

This section will describe how property relationships can be established. This is a key to the Micro and Macro properties discussed earlier. These relationships will be needed to check event based refinement in an efficient manner both in terms of its specification as well is its execution time. Here are several examples indicating the relationship and hierarchy of Micro and Macro properties. Future work outside of this book will be devoted to establishing these relationships more formally.

1. Data Consistency $\rightarrow$ Sufficient Space, Read Access, Write Access

   In the case of this relationship, if the MacroProperty "Data Consistency" (DC) is proven, it then implies the MicroProperties "Sufficient Space" (SS), "Read Access" (RA), and "Write Access" (WA). SS indicates that the data storage device itself has enough space. RA indicates that the storage device is allowing reads. WA indicates that the storage device is allowing writes. Notice that proving the MicroProperties SS, RA, or WA does not imply anything else at this point. However, assume that WA and RA were transformed into a MacroProperties such that:

   - Write Access $\rightarrow$ SS
   - Read Access $\rightarrow$ Data Valid

   In this case, SS is the same MicroProperty as described previously and "Data Valid" (DV) indicates that the data is marked as being valid in the storage device (i.e. during a cache or snooping update). If these MacroProperties are proposed and proven, then "Data Consistency" actually implies DV as well in addition to the other MicroProperties mentioned previously. It also will imply SS transitively through MacroProperty hierarchies and would not have to imply it explicitly. "Data Consistency" is a RO and WO property in terms of its grouping as well as a performance property in terms of its classification.

2. Data Coherency $\rightarrow$ Data Valid, Snoop Complete

   Notice that "Data Coherency" (DCo) implies DV. RA implies this MicroProperty as well and RA is implied by DC. However simply because DCo and RA share a MicroProperty they do not imply each other. Implication is a one way assignment. "Snoop Complete" (SC) indicates that the snoop process conducted by the memory controller is completed as part of the coherency protocol. DCo is a WA property by its grouping and performance property by its classification.

3. Data Precision $\rightarrow$ Sufficient Bits, SS

"Data Precision" (DP) implies that there are "Sufficient Bits" (SB) to hold the results. SB in turn implies that "No Overflow" (NO) is detected, and therefore the data is valid as well (property DV). Also required of DP is that there is sufficient space (property SS). DP is an example of a CMP group property and yet again it is a performance based property like the other properties discussed.

- Sufficient Bits → No Overflow
- No Overflow → Data Valid

The keys to these property relationships are: (1) There must be a method to prove the relationship, (2) the MacroProperties cannot be more expensive to check then the sum of their implied MicroProperty checking costs, and (3) the MicroProperties must be non-trivial. The relationships between the properties outlined are illustrated in Figure 6.6. Arrows from left to right indicate implications. The "leaves" (properties with no outgoing arrows) are MicroProperties while the others can be considered MacroProperties. This illustrates that properties can be classified as to which "level" they belong to. All MicroProperties are level 0 while MacroProperties are a level ≥ 1. MacroProperties at a higher level imply more properties overall and indicate how far each MacroProperty is topologically from the true MicroProperties. One possible heuristic selection as to which properties are proven first could be a "greedy" selection by level as to cover as many properties as possible. Another view is similar to logic minimization where MicroProperties are seen as minterms and MacroProperties as cubes.

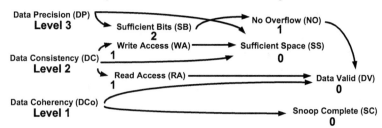

Figure 6.6: Macro and MicroProperty Relationships

Event Petri Net

Many systems (METROPOLIS and METRO II for example) progress through simulation in a series of phases. In each phase, simulation proceeds by enabling or disabling events. This enabling and disabling determines which active components (threads) will be allowed to make progress in the next phase. This scheduling behavior (which decisions can be made) can be captured formally. Capturing the behavior formally allows one to reason about the system and enforce system operation. These same structures will also be used to define Macro and MicroProperties relationships. Once both structures exist (one for the service behavior and one for the property relationships) they can be augmented together to either enforce or check adherence of the service to the property. A structure capturing this execution is presented here as an *event petri net*. Formally an event petri net is:

Definition 6.4.5 Event Petri Net (EPN) - *is a tuple* $<P, T, A, \omega, x_0 >$ *where:*
- P is a finite set of places,
- T is a finite set of transitions,
- A is a set of arcs, $A \subseteq (P \times T) \cup (T \times P)$
- ω is a weight function, $\omega: A \rightarrow \mathbb{N}$
- x_0 is a an initial marking vector, $x_0 \in \mathbb{N}^{|P|}$

The event petri net is defined as any normal petri net. The event petri net developed for the model (Model$_{EPN}$) requires that each service event of interest has a corresponding transition, t_{EN}. The firing of that transition denotes the occurrence (enabling, EN) of that event. The event petri net developed for the properties (Prop$_{EPN}$) requires that each property have a place, $p_{<Prop>}$. The property is satisfied when a token is present in its corresponding place. Prop$_{EPN}$ also is constructed in such a way that it is required that all transitions only fire once. Prop$_{EPN}$ begins with an empty initial marking vector. It is connected to the the Model$_{EPN}$ in such a way that when specific t_{EN} fire, they will produce the needed tokens eventually in Prop$_{EPN}$'s $p_{<Prop>}$ places.

For example, in Figure 6.7 the top half illustrates a sample Model$_{EPN}$. This model is for a basic CPU, Bus, and Memory system. The leftmost section corresponds to the CPU, the center to the Bus, and the rightmost to the Memory. The transitions are labeled with a name describing the function call which will cause the transition to fire (the initial transitions in this case) or as t_{EN} for the transitions indicating the enabling of specific events. Model$_{EPN}$ basically illustrates a producer/consumer structure. The CPU "execute" transitions are contained within the CPU petri net itself. However the other functions in the CPU interact with the Bus and Memory petri nets.

The lower half of Figure 6.7 is the Prop$_{EPN}$. The transitions are labeled numerically (for simple identification) and the places are labeled with the acronym used previously for each Micro and Macro-Property. Prop$_{EPN}$ is augmented with a set of transitions, t_{CN}, and a set of places, p_{CN}. Each transition t_{CN} produces the exact number of tokens needed such that each of the numeric transitions in Prop$_{EPN}$

198

fires exactly once. There are three property places, p_{DCo}, p_{DP}, and p_{DC} (from left to right at the bottom of the figure). Places "start1", "start2", "start3" receive the tokens from t_{C1}, t_{C2}, t_{C3} respectively.

The arcs into t_{CN} from p_{CN} are defined by the requirement of each property. In this case, they are defined by which enabling of events should constitute the satisfaction of a property. In Figure 6.7, t_{C1} which will generate property DCo is attached to the places related to the write transitions (p_{C1}, p_{C2}, p_{C3}). t_{C2} (DP) is only related to the "execute" function, t_{E3} through p_{C4}. The third and final property DC, is associated with t_{C3}. This transition requires bus read and memory read events t_{E3} and t_{E5} and places p_{C5} and p_{C6}. These scenarios are a simplified set of events to indicate the properties but should give the reader some indication of how this process occurs.

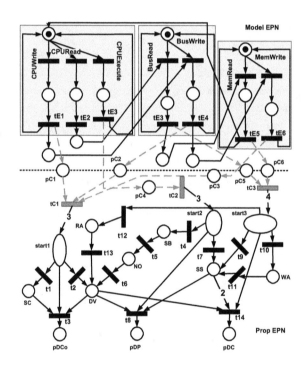

Figure 6.7: Event Petri Net Example

Event based refinement has introduced both vertical and horizontal refinement methodologies as well as an infrastructure to support those methodologies (Macro and MicoProperties and Event Petri Nets). Together these pieces will be used to reason about refinement. The next section will propose another methodology which will focus less on changing the structural interaction between scheduled services and their scheduling mechanisms and more on inter-component structural changes within the scheduled services themselves.

6.5   Interface Based Service Refinement

Interface based refinement denotes a method of verifying relationships between systems based on how they interact with other systems or the environment in which they are placed. These interactions occur at *interfaces*. This definition clearly defines which aspects of the system are required to be related. Interfaces become the only point at which system behavior is visible. Interfaces themselves in practice may be function calls, ports, visible variables, or any number of language dependent constructs. What makes this approach attractive is that it reduces the space of all possible behaviors to a fixed set and often requires no modification to the existing model. A drawback is that the designer often has to specify very clearly what the interfaces are and which interfaces require correspondence between two models. This section will detail how this work can be done at a high level followed by an explicit methodology in the METROPOLIS design environment.

6.5.1   Proposed Methodology

Thus far both *vertical* and *horizontal* refinement have been explained. Next *surface* refinement will be introduced. This term denotes system level refinement using interface based refinement. Figure 6.8 illustrates a proposal for this in an environment similar to METROPOLIS. This approach is called "surface" since interfaces can be viewed as the *surface* of potentially black-box components. All that can be assumed about the component is the number, name, and types of interfaces. The behaviors of interest therefore are the sequences in which these interfaces operate. This behavior may be the sequence of function calls made on or to these ports (as is the case in METROPOLIS), events generated on these ports, or even restrictions on what other components are attached to these ports. Interfaces which require services will be considered active interfaces whereas provided services are passive. What is of chief concern is how to capture interface activity.

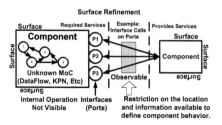

Figure 6.8: Interface Based Refinement Proposal

Syntactic Conditions

In order to automate the task of interface refinement verification, interfaces must be easily identified. Additionally, the two models being compared need to have interfaces which are easily and correctly identified as corresponding. This process can be facilitated by syntactic conditions in the modeling environment. These can include keywords, hierarchy, type checking, etc. While not a requirement explicitly of the modeling framework, there must be some way of indicating which interfaces are to play a role in the behavior of the system or component.

An example of syntactic conditions are given in [AH03]. This frames the refinement conditions in terms of the *reactive modules* [AH99] syntax and puts requirements on their variable structures for each model to be compared. These are the same types of syntactic conditions which will be used in this work. The similar syntactic conditions for METROPOLIS models are, given $X \preceq^{Ref} Y$, that $Y_{inputs} \subseteq X_{inputs}$ and $Y_{outputs} \subseteq X_{outputs}$. Essentially this simply requires that X have all of Y's inputs and outputs (if not more). This requirement could be viewed as simply a naming issue if one requires the same order and number of corresponding inputs and outputs for each model. In the methodology to be presented this requirement is the case (maintaining a strict order and naming style).

Trace Definition

As mentioned previously in the background and definitions, Section 6.2.2, a trace $\bar{a}$ is considered a sequenced set of observable values for a finite execution of the module. In the case of METROPOLIS, the key observable values that we are concerned with are *function calls to media*. This work will refer to a trace consisting of function calls to media as a $Trace_M$, where the "M" stands for "Metropolis". Due to the semantics of METROPOLIS, processes must communicate strictly via media. This restriction could exist in METRO II as well but since media are not present in this environment, it would be component-to-component connections. Ultimately the behavior of a process/component can be characterized by the sequence by which it makes these calls. Syntactically this results in an interface call attached to a particular port.

Definition 6.5.1 METROPOLIS **Interface Behavior** - *a sequenced set of observable values for a finite execution of the model, Trace$_M$. This sequence results from function calls on ports requesting (requiring) services.*

In order to characterize the METROPOLIS TRACE, *TraceM*, the key structure needed to be obtained from the model is the control flow automata (CFA) concerning the ways in which these sets of observable events can occur. Once this structure is created *State Equivalence* concepts such as *Bisimilarity* and *Similarity* [AH03] can be used to determine refinement. A *Trace$_M$* can be obtained by traversing this structure. This structure is described in Section 6.5.1. Before describing this structure however, one must select which set of interfaces should be considered for this structure. These sets are defined by what this work calls, *refinement domains*.

Refinement Domains

Naturally in a design there are many interfaces. However during refinement it may not be necessary or appropriate to consider all of these interfaces during refinement verification. Often components are composed in such a way that there is a single set of interfaces which sufficiently captures the behavior of the set of the components. These collections of components are termed, *refinement domains*, specifically:

Definition 6.5.2 Refinement Domain - *a collection of components C, ports P, and observed ports OP. Typically organized by component service. <C, P, OP> where OP ⊆ P.*

The refinement domain definition illustrates that only a subset of ports are involved in the interfaces to be verified. The components are typically organized into domains such as *computation*, *communication*, and *storage*. These organizations are constructed in such a way to minimize the number of observed ports needed.

Computation domains collect components involved in computation such as adders, multipliers, processing elements, etc. Interfaces typically have to do with the execution of specific services. An example of a refinement domain specifying interactions of interest in computation is when one changes from using an adder to multiply values to a dedicated multiplier. All that matters is the input and output interfaces, not the interfaces between the various adder or multiplier blocks.

Communication domains contain components such as buses, bridges, switch fabric, and buffers. Interfaces have to do with reading, writing, synchronization, or data movement. For example, two buses may be connected with a bridge. The interaction of interest is not between the buses and the bridge but rather that the end to end behavior is maintained. Hence bridge interfaces may be ignored.

Finally storage domains contain main memory, cache, or scratch pad storage. Interfaces have to do with loading and storing. Often with memory hierarchy one may add a new component (e.g. a cache). One is not interested in the cache's relationship with main memory but rather with the boundary between the memory system and the components which need the data stored there. One can restrict refinement only to those interfaces.

An example of refinement domains is clearly illustrated in Figure 6.9. This is an example based on a FLEET [Sut06] style dataflow system. The left hand side shows the original system. This system consists of two computation refinement domains (Adder and Producer based) and one communication refinement domain (Switch Fabric based). One unique component is assigned to each of the three domains. One can see which function calls each component can make next to the component itself. These include, "move.source.Adder", "prodLit()", and "Add(input1, input2)" for example. What is illustrated is a potential graph showing functional interaction in each system. A graph is composed of locations and arcs. The locations are states of the system as it proceeds through its execution. The transitions occur as function calls are made in each refinement domain. In the original system, the adder has two states. The first state waits for data. When data arrives, it can perform the addition and transition to the next state. The adder can then move the result to the switch fabric. The switch fabric can move the data back into the adder through a series of move instructions. The producer on the other hand waits to produce a literal value. Once this occurs, it transitions to its second state. From this state the produced literal can be passed through the switch fabric to the adder or back into the producer and used as a seed to produce another literal.

On the right hand side of the figure, the second system is shown. In this case a memory component is added to the system. This is an newly introduced storage refinement domain. Also a new computation domain is defined combining both the adder and the producer. These additions can be seen as refinements. Restrictions have been made regarding the source and destination behavior as well. For example, the producer can only be addressed through the adder now. The computation refinement domain now waits to "add" and then proceeds to the next state after the "Add(input1, input2)" function execution. This system now will transition to a state in which a literal value based on the addition operation is produced through the "prodLit()" function. The switch fabric now can route data to the memory component, or to the states concerned with adding. Notice that after the "add" function, the switch fabric can be reached directly. This effectively bypasses the producer states if desired.

As is shown, only the function calls which interact between domains (through observable ports; not within a domain) are part of the interface and will be used for refinement verification. This is illustrated as arcs cross refinement domain boundaries. These arcs are colored differently from the inter-domain arcs. The point of this example and figure is to show how refinement domain definitions can change which function calls can be seen through observable ports (OP). The function calls which pass between domains are placed in the figure between the domains (as opposed to within the domain) for clarity.

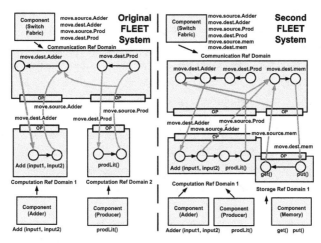

Figure 6.9: Refinement Domains in Interface Based Refinement

With refinement domains, interface traces, and the idea of syntactic conditions defined, a more detailed discussion of the actual surface refinement procedure can now be discussed.

Control Flow Automata in Metropolis

The key structure in this investigation is the Control Flow Automaton (CFA) representation of a METROPOLIS model. METROPOLIS has an *Action Automata* specification underlying it [BLP+02b] but this automata provides much more information than is required here and its structure is not suited to use in this refinement scenario. A CFA is defined as a very much like in [HJM$^+$02]. It is a tuple $<Q$, q_0, $\mathbf{X}$, Op, $\rightarrow>$.

Q is a finite set of control locations. These locations will be determined by the METROPOLIS model structure. q_0 is the initial control location, $\mathbf{X}$ is a set of variables, and Op are operations which denote: (1) function calls to media (2) a basic block of instructions starting (3) a basic block of instructions ending. This "ending" and "beginning" symmetry is taken from the *Action Automata* semantics. A basic block is defined in the traditional sense, meaning a section of code in which there is no conditional execution which could result in a different execution sequence. A basic block simply could be viewed abstractly as a function call assuming no conditional execution occurs within the function. It is for this reason that the start and end are denoted. This way, the CFA could be augmented with the body of the function call if desired, inserted inside the beginning and end portions.

An edge (q, Op, q') is a member of a finite set of edges and the transition relationship, $\rightarrow$, is defined as $(Q \times Op \times Q)$. A edge makes a transition based on the evaluated Op present, $q \rightarrow^{Op} q'$.

Ideally a CFA is created which represents the model and corresponding automata are created which represent the state of variables in the automata. These variable automata are used when decisions in the CFA depend on these variables. For example a model may have a loop which is checking the value of a particular variable. The CFA would have a variable, $v \in \mathbf{X}$, which has its own automata which can be queried as to the value of that variable to determine what edges can be transitioned. For the purposes of this work, these automata are not formally defined nor are they automatically generated. Figure 6.11 shows one possible representation that could be used to capture the incrementing of an integer with a functional range of 0 to 2.

Figures 6.10 and 6.11 demonstrate a code snippet and the resulting METROPOLIS CFA respectively as defined in this work. In Figure 6.11 there are two automata. The first automata is simply a hypothetical automata for the variable X. This automata is not actually created but it demonstrates what it would look like. This simply illustrates that X will begin in a state representing its value of 0 and proceed until it equals 2. It is even further simplified by not illustrating all the states (control locations) and edges which would result from begin and end events. The main automata has 10 control locations. Next to each control location is a description which indicates the number of the control location, the type of node which lead to its creation as it would be defined by the METROPOLIS abstract syntax tree (AST), and/or a description of the node in the event that it is not explicitly defined in the AST. The edge labels are as described with "+" or "-" indicating a start and end of a basic block.

Once a CFA is defined, a *Trace$_M$* is nonempty word $\overline{a}_{1...n}$ over the alphabet of Q control loca-

```
//sample code snippet
process example {
  port Read port1;
  port Write port2;
void thread(){
  int x = 0
    while (x<2){
      port1.callRead();
        x++;}
      port2.callWrite();
  }
}
```

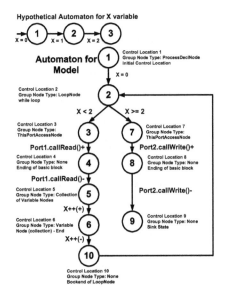

Figure 6.10: METROPOLIS Code Example

Figure 6.11: Resulting CFA for Code Example

tions such that $a_i \rightarrow a_{i+1}$ for all $1 \leq i \leq n$.

Naturally the potential for a CFA to be quite large is a concern. As will be illustrated in the description of the METROPOLIS backend (which generates CFAs) it is bounded by the nodes in the Abstract Syntax Tree (AST) created by METROPOLIS compilation which could be very large. However this can be reduced further by heuristic grouping of nodes to create control locations as will be shown in the section to follow.

CFA METROPOLIS Backend

The METROPOLIS design environment is designed around the concept of a *meta-model* as mentioned previously (Section 4.3). This allows for the initial model to be decomposed into an intermediate representation and then fed to a number of different tools called *backends*. This is demonstrated roughly in the structure shown in Chapter 4, Figure 4.1. As one can see, the model is parsed into an *Abstract Syntax Tree* (AST) and that AST is interpreted by the backends to generate another representation with semantics for another tool while maintaining some relationship to the original model. The creation of a backend to generate a CFA as described earlier (Section 6.5.1) was the primary tool flow of "surface" refinement as it currently functions in METROPOLIS.

The CFA backend traverses the AST and identifies the *nodes* of the AST. It is composed of two files:

- CFABACKEND.JAVA - top level METROPOLIS backend interface with file input/output functionality. This code interacts with the user and the METROPOLIS infrastructure.

- CFACODEGENVISITOR.JAVA - AST visitor functions and CFA construction mechanism. This code is the core of the tool and where 90% of its work is done.

CFABACKEND.JAVA is called when the backend is *invoked* and actually writes to various files the results of the AST node *visitor functions*. The file CFACODEGENVISITOR.JAVA actually contains the visitor functions. The visitor functions traverse the AST and determine what should happen at each type of node. There are over 160 different node types that can make up an AST. It is in these functions that the CFA structure is determined. In particular this is true when visiting what this work introduces as *Grouping Node Types* (GNT). Each AST node generates its own *location* structure, L. Groups of these structures belong to a *group location structure*, $\{L_1...L_N\}$. Each group location structure each contains *exactly* one node which is a member of the GNTs. These sets of group location structures with one unique node of the GNTs are what constitute a control location, Q, in the CFA. All of this information is stored in an internal list structure which can be traversed itself. It is this heuristic grouping which prevents the size of the CFA from being $O(AN)$ (where AN is the number of AST nodes in the model) and rather $O(GNT)$ (where GNT are the grouping node types in the model) which is substantially smaller in practice. In order to have this reduction, the GNTs are currently defined as:

- **Structure Nodes** - these include ProcessDeclNode, CompileUnitNode. These nodes capture the structure of the process description.

- **Control Nodes** - these include AwaitStatementNode, AwaitGuardNode, LoopNode. These represent control decisions which frequently result in branches in the CFA.

- **Variable Nodes** - these include ThisPortAccessNode. These nodes will be very important as these are the source of the interface function calls which ultimately define the behavior of a METROPOLIS system.

Also worthy of note is that the CFA internal structure can be created in one pass through the METROPOLIS model code. Therefore the running time it is $O(NV)$ (where NV is the number the nodes traversed by AST visitor functions) where $|visitor functions| \leq$ AST nodes types in code. There are restrictions currently on the types of METROPOLIS systems that can be handled by this backend. For example all processes much be single threaded with deterministic behavior.

CFA Visual Representation

The first and most trivial result of the CFA backend is a simple visual representation as shown in Figure 6.12.

```
Group: 3
Parents: 2
Types: 12
Inputs: in1
Outputs: #can be blank
Misc: #can be blank
Names: LoopNode
Cond Codes: 1
        |       |
        V       V
```

Figure 6.12: CFA Visual Representation

The visual representation is simply for debugging purposes and allows the user to see not only what the structure of the CFA is but also examine what individual AST nodes compose a control location. This information can be used to redefine any heuristics used to define what a GNT is and then observe the effects of the different heuristic choices for grouping. The *Group* field is an integer identification of what group this object is. In turn this corresponds to a control location, Q in the CFA. The *Parents* field is a collection of integers which define which groups are the parents of this group. *Types* is a set of integers which are associated with each node to identify its composition of individual AST nodes (as defined by the AST node types). Each AST node type has a unique integer "Type" value which makes up this list. The *Inputs* field denotes what input variables must be required to transition from this group. The *Outputs* field denotes which output variables will be present (i.e. go "high") when you transition from this node. *Misc* is used to hold such information as the occurrence of arithmetic nodes being visited (e.g. a PlusNode denoting a possible incrementing of a variable) or other information used to build the CFA. *Names* is simply a list of strings which indicate what types of nodes make up this group location (corresponding to the type field; easier for human debugging). And finally the *Cond Code* field indicates which type of conditional node was visited for the group (i.e. LoopNodes, AwaitStatementNodes, etc) and is internally defined to identify the branching structure of the CFA. The "arrow" like symbols are used where there are multiple children. This can be produced in one pass of the internal list structure of the CFA or $O(Q)$ (where Q is the set CFA control locations).

Finite State Machine Representation

The second more functional result of the CFA backend, is that it produces a Finite State Machine (FSM) representation of the CFA. The inputs to the finite state machine represent information provided by other automata to the CFA model (such as the variable automata) and the outputs are the function calls to media. This is formatted as a KISS representation. An example of KISS is shown in Figure 6.13.

```
#KISS File
.i 3 #input count
.o 4 #output count
.s 2 #state count
.p 2 #next state equations
#inputs current_state next_state outputs
010 s1 s2 0101
000 s2 s1 1010
.e
```

Figure 6.13: CFA FSM Representation

This format was chosen for two reasons: (1) It is easily produced from the internal list structure which also created the visual representation (2) it can be read by various tools such as SIS [SSL+92]. SIS in turn can produce other formats such as BLIF, PLA, EQN, etc. Of particular interest is BLIF (Berkeley Logic Interchange Format) whose close relative EXLIF can be read by FORTE [NLK03] as will be described in in a later section. Once the initial data structure is created by the backend, the algorithm to create a KISS file is as shown in Algorithm 3.

The running time of this algorithm is $O(2*(GL*IV + GL*OV))$. GL stands for Group Locations which are the CFA Structure Groups. IV and OV are input and output variable list sizes respectively. This computation is captured by the "for all" loop behaviors in Algorithm 3. Essentially one has to traverse the structure once to create the lists of inputs and outputs. Then you must traverse it again to actually

---

Algorithm 3: KISS Construction from CFA

    Input : CFA Data Structure, D

    Output: KISS File, K

    /*Create unique list of inputs **(step 1)**   */

1   **forall** *Group Locations*, $i \in D$ **do**

2   **forall** *Input Values*, $j \in i$ **do**

3   **if** $j \notin$ *Unique Input List, UIL* **then**

4   Add(j);

    /*Same procedure as step 1: Add Output Values $\rightarrow$ Unique Output List (*UOL*) **(step 2)**

     */

5   $\{...\}$

    /*Create the declarations section **(step 3)**   */

6   printf(".i %d", sizeof(UIL));

7   printf(".o %d", sizeof(UOL));

8   printf(".s %d", sizeof(D));

9   printf(".p %d", nstate_count) /*nstate_count = lines processed making the body (back

     annotated)   */

    /*Create the body **(step 4)**   */

    /*Input portion of KISS   */

10   **forall** $i \in D$ **do**

11   **forall** *elements*, $e \in UIL$ **do**

12   **if** $e \in i$ **then**

13   printf("1");

14   **else**

15   printf("0");

    /*Print information to describe the transition **(step 5)**   */

16   sprintf(current_group, child_group)

    /*Same procedure as step 4: Output portion of KISS using UOL **(step 6)**   */

17   $\{...\}$;

18   **Return** K;

---

generate the KISS file based on that information. Each line of KISS requires that you examine the input and output lists completely to see if they contain input or output at that location as well.

Reactive Module Representation

The third and final result of the *CFA backend* is a *reactive module* [AH99] file. This is a modeling language for describing the behavior of hardware and software systems. This file is produced as an additional benefit of the backend for three reasons: (1) It is very inexpensive to create a reactive module which models an FSM. (2) It allows for non-deterministic behavior which is not allowed by KISS models provided to SIS. (3) It can be read by tools such as *MOCHA* [AHM+98]. MOCHA allows a rich set of model checking algorithms to be run on the CFA model that are useful both for refinement and other verification tasks.

The first point mentioned for making this representation was that it was inexpensive to do from the FSM representation. Algorithms 4 and 5 give the algorithm to do so.

The process of creating a reactive module file can be done in one pass of the KISS file. The variable declaration initializations for the module are simply from the KISS input (.i), output (.o), and state (.s) declarations. The *init* command is simply another listing of the variables. The largest part of the file, the *update* commands, correspond one-to-one with each line in the KISS body. The running time of this process is naturally $O(L)$ (where L is the number of KISS Lines).

The second reason for using this representation, non-determinism, is inherent in the fact that multiple guards in a METROPOLIS *await* statement may be *true*. Also inherent is that the union of all guard commands does not have to equal the entire space of the inputs (i.e. a reactive module can be partially specified). Naturally, KISS currently has deterministic behavior so it will result in a reactive module with deterministic behavior. However, there is nothing preventing a reactive module from being produced from a KISS file which would not run in SIS. A CFA could be produced that has non-deterministic behavior simply with a modification to the backend.

The third and final reason, the verification tool MOCHA, will be discussed in its own section to follow.

FORTE Accommodations

Prior to the integration of Reactive Modules into the CFA Backend, this work was targeting a tool called FORTE. FORTE [NLK03] is a tool provided by Intel Corporation which is a collection of several tools. These are *Functional Language* (FL), *Symbolic Trajectory Evaluation* (STE), *FSM Logic*

Algorithm 4: Reactive Module Construction from KISS Description; Part 1

Input : KISS File, K

Output: Reactive Module File, R

1 RM R = new RM (<filename>);

/*Lists of the external, interface, and private variables (**step 1**)   */

2 **forall** $i \in UOL$ **do**

3   ivar [index1] = new interface_variable iv;

4   index1++;

5 **forall** $i \in UIL$ **do**

6   evar [index2] = new external_variable ev;

7   index2++;

/*D is the collection of FSM states   */

8 **forall** $i \in D$ **do**

9   pvar [index3] = new private_variable pv;

10   index3++;

/*Create new atom CFA (**step 2**)   */

11 printf("atom cfa");

12 printf(" controls ");

/*Control declaration   */

13 **for** $j = 0$ *to* $j < index1$ **do**

14   printf(ivar [j]); /*each interface variable   */

15 **for** $k = 0$ *to* $k < index3$ **do**

16   printf(pvar [k]); /*each private variable   */

17 printf(" reads "); /*Read Declaration   */

18 **for** $j = 0$ *to* $j < index2$ **do**

19   printf(evar [j]); /*each external variable   */

20 **for** $k = 0$ *to* $k < index3$ **do**

21   printf(pvar [k]); /*each private variable   */

Algorithm 5: Reactive Module Construction from KISS Description; Part 2

    Input : Variable lists and atom CFA from part 1

    Output: Reactive Module File, R

    /*The continuation starting from line 21 of part 1   */

    /*Initialize Module **(step 3)**   */

1 printf(" init ");

    /*All interface and private variables = false except first state variable   */

2 printf(pvar [0] = true);

3 **for** $k = 1$ *to* $k < index3$ **do**

4   printf(pvar [k] = false);

5 **for** $j = 0$ *to* $j < index1$ **do**

6   printf(ivar [j] = false);

    /*Update behavior **(step 4)**   */

7 printf(" update ");

    /*Pseudo Description   */

8 <For each line of the KISS representation, the guard is the appropriate input = true and that current state = true. The result is the next state variable = true and the appropriate outputs = true>;

9 **Return** R;

Data Model, and some circuit drawing tools. FORTE works on circuit descriptions of models. This is was a major factor in influencing the decision to reduce the CFA into a FSM representation originally.

Once a model has been created as a KISS file, that KISS file is given to the SIS tool. SIS is used to create a BLIF file with the SIS script shown in Figure 6.14. BLIF representation is very similar to the EXLIF file format used by FORTE. EXLIF is an extension of the LIF format in general. The EXLIF holds, in addition to combinational circuit truth tables, constructs to model sequential elements, like transparent latches and master slave flip-flops. It has constructs for describing structural hierarchy, tri-state drivers, various kinds of assertions, etc. Some simple modifications allowed BLIF to be converted to EXLIF and in turn read by FORTE also shown in Figure 6.14. These manual edits could be worked into a Perl script very easily.

```
//sis commands
read_kiss <filename>
state_minimize
state_assign <nova> or <jedi>
source script.rugged
write_blif <filename>
```

BLIF to EXLIF Manual Edits:

- Remove start_kiss, end_kiss, and kiss code embedded in file

- Remove external don't care section (.exdc)

- Add to the .latch definitions a clk signal and the type of flop it is (rising, falling)

- Remove the .latch_order and .code portions

Figure 6.14: SIS Commands and EXLIF Requirements for FORTE Flow

Once the models are converted to EXLIF files, FORTE can begin to process them for refinement. The algorithm which FORTE uses to prove refinement between to EXLIF files is in Algorithm 6.

The running time for such an algorithm is approximately $O(m * n)$ where n is the number of states and m is the number of transitions. This algorithm (6) and corresponding implementation code was not created as part of this work but supplied by Intel.

Algorithm 6: FORTE Refinement Check for EXLIF Files

    Input : Two EXLIF Models, A and R with State Space Σ_A and Σ_R

    Output: Answer to the Refinement Question (R, A)

1 Let $q_A \in \Sigma_A$ and $q_R \in \Sigma_R$;

2   Given a set of states, the set S, S^C is $(\Sigma_A \cup \Sigma_R) \setminus S$;

3 Let $\vec{x}$ be a vector of inputs common to both A and R;

4 Let $\vec{y}_A(q_A, \vec{x})$ be a vector of outputs for A given the state q_A and the inputs $\vec{x}$;

5 Let $\vec{y}_R(q_R, \vec{x})$ be a vector of outputs for R given the state q_R and the inputs $\vec{x}$;

6 Let E_n be a set of sets of states reachable in n input sequences;

7 Let σ be a set of sets $\{(q_A, q_R) \mid q_A \in \Sigma_A, q_R \in \Sigma_R\}$;

8 Let$(Tr)(q_A, \vec{x}, q_A') = $ true if there is a transition from q_A to q_A' under input $\vec{x}$;

9 Let$(pre)(\sigma) = \{(q_A, q_R) \mid \exists \vec{x} : Tr(q_A, \vec{x}, q_A') \cap Tr(q_R, \vec{x}, q_R') \cap (q_A', q_R') \in \sigma\}$;

    /*Start of Algorithm   */

10 $E_0 = \emptyset$;

11 $E_1(q_A, q_R) = \forall \vec{x}, \vec{y}_A(q_A, \vec{x}) \odot \vec{y}_R(q_R, \vec{x})$;

12 $k = 0$;

13 **repeat**

14 | $k = k + 1$;

15 | $E_{k+1}(q_A, q_R) = E_k(q_A, q_R) \setminus pre E_k^C(q_A, q_R)$;

16 **until** $E_{k+1} = E_k$;

17 **if** $\forall\, q_R \in \Sigma_R,\, \exists\, q_A$ *such that* $(q_A, q_R) \in E_k$ **then**

18 | **Return** *YES*;

19 **Return** *NO*;

MOCHA Accommodations

Since the CFA backend produces a Reactive Module, MOCHA can be used to do refinement checking as well. However, this process requires some manual preparation of the file produced by the backend. [AHM$^+$98] describes refinement as a *trace inclusion* problem. To check that $X \preceq^{Ref} Y$ requires:

1. For every initial state, s of X, the projection of s to the variables of Y is an initial state of Y. Basically they need the same initial state.

2. For every reachable state of s of X, if X has a transition from s to t then Y has a matching transition.

The search can be done symbolically or enumerated with MOCHA. In the case that the test fails, it generates a counterexample of a trace on X which is not a trace of Y. This may be computationally complex. Therefore some restrictions are placed on the modules, to verify $X \preceq^{Ref} Y$.

1. The module Y has no private variables - This requirement has to do with observability.

2. Every interface variable of Y is an interface variable of X - This requirement is a syntactic issue issue which allows the tool to function without the user explicitly providing a list detailing variable correspondence.

3. Every external variable of Y is an external variable of X - This requirement is also a syntactic issue.

Recalling the requirements for refinement, the 2nd and 3rd conditions are already met. However, a module created with the CFA Backend will have private variables representing states. The solution for this is to create a *Witness Module*, W. This is a module whose interface variables are the private variables of Y. Also, W should not contain any of the external variables of X. In turn a module, Y', will be created with the original $Y's$ private variables declared as interface variables. Once this process has been performed then $X||W \preceq^{Ref} Y'$ as shown in [AH99]. The procedure is naturally:

1. Create a module Y' from Y by changing private variables to interface.

2. Define a *Witness Module*, W, whose interface variables are the private variables of Y but exclude the observable (external) variables of X.

3. Check $X||W \preceq^{Ref} Y'$ with MOCHA

Since this process is not automatic it represents a potential bottleneck in the flow. The creation of a *Witness Module* requires creativity on the part of the user since the variable reassignment may be

216

nontrivial in order to maintain correct functionality. In addition the parallel composition is also manual. There is much information required to fully understand the MOCHA tool which is not described here. The reader is referred to the references provided for more information.

Composite Design Flow

In conclusion, in order to demonstrate a *proof of concept* for this surface refinement methodology, this work assembled the previously described components into a complete flow as shown in Figure 6.15.

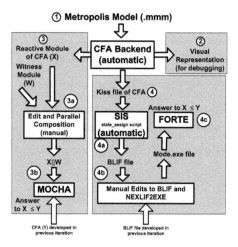

Figure 6.15: Surface Refinement Flows for METROPOLIS

As one can see from Figure 6.15, the process begins with a METROPOLIS model. Using the METROPOLIS compilation engine one can simply run it through the CFA backend automatically. This will return a reactive module file, a KISS file, and a visual representation. The reactive module will be fed to MOCHA but first it must be augmented with a witness module, W, manually to do refinement checking on it. This was described previously. The visual representation is simply for viewing and debugging. The main trunk of the flow requires that you submit the KISS file to SIS. The script shown previously in Figure 6.14 is run to assign state encoding and logic to the symbolic states in the KISS file. This information can then be written out in BLIF format. Then the slight manual edits as described previously must be done to the BLIF file to convert it to EXLIF for FORTE. Finally one runs NEXLIF2EXE (provided by FORTE) to convert the EXLIF to an executable format for FORTE. Both the FORTE and MOCHA trunks of this flow assume the presence of another file previously created will represent the more abstract model to be compared. These files will be inserted in the flow at the appropriate locations as shown. Aspects of this methodology as applied to METROPOLIS have been shown in [DRSV04]. The asymptotic analysis of various aspects of this flow (as detailed previously) is collected in Table 6.6. Further results of this design flow will be shown in Chapter 9 on an industrial case study.

Description	Analysis	Comments
CFA Overall Size	O(GNT)	$GNT = \|Grouping\_Node\_Types\|$
CFA Creation Time	O(NV)	$NV = \|Nodes\_Traversed\_by\_Visitor\_Functions\|$
CFA Visual Rep. Creation Time	O(Q)	$Q = \|CFA\_Control\_Locations\|$
KISS File Creation Time	O(2*(GL*IV + GL*OV))	$\|Group\_Locations\|$ and $\|Input\|$, $\|Output\|$ Vars.
Reactive Module Creation Time	O(L)	$L = $ KISS Lines
Forte Running Time	O(m * n)	$N = \|States\|$ and $M = \|Transitions\|$

Table 6.6: Asymptotic Analysis of Surface Refinement Flows

6.6   Compositional Component Based Service Refinement

Finally, compositional component based refinement will discuss the how changes internally made to a component can be related. Whereas the previous approaches examined relationships between components (event based), and changes to observable at the periphery (interface based), this approach allows changes to be related at the "lowest" level. Very frequently designers want to change protocols offered by services, size of storage elements, memory access or dequeuing polices, or add more service functionality (modes, operands, etc). This approach allows for the designer to specify each individual service as a small component. A change to this component can be modeled as an individual component as well. Refinement verification is then performed against these two small relatively simple components. Once this has been performed, large systems can be composed from these smaller systems. The two verified components can then be swapped in and out of the two designs without performing additional refinement verification. The refinement problem is greatly simplified in this way since the refinement effort to check the small individual components is much more manageable than verifying the larger, composed system. Also the modular nature of this approach allows for easy system modification. The bulk of this section is based directly on work from [KL03a]. Therefore, in this section, new contributions unique to this work will be denoted with ♠ to avoid confusion between new and established work.

6.6.1 Proposed Methodology

The methodology for this approach uses the work of [KL03a] and their tool SynCo [KL03b]. The approach overall is termed *Depth Refinement*. The contribution of the methodology contained here is a set of extensions for system level design (METROPOLIS, METRO II and SystemC for example). These extensions include how to represent events, refinement domains (subsystems), relations between subsystems, gluing relations between subsystems, visible events, and visible properties. These are defined similarly to the definitions in Section 6.2.3. The reader needs to begin this discussion by reviewing the LTS foundations mentioned in this chapter's background section.

Definition 6.6.1 Event ♠ - *Events, E_E^N, in LTS S are a set of transition labels such that: $E_E^N \subseteq E$.*

Events are essentially a subset of the labels on transitions. All events are labels but not all labels are events. When LTSs are created from an environment such as METROPOLIS they should be created such that events are captured when they cause a change in the state of the system or are involved in a property of interest.

Definition 6.6.2 Refinement Subsystem ♠ - *Let SR be a refined component, $< Q_R, Q_{0R}, E_R, T_R, l_R >$. A refinement subsystem is a collection of states in the refined component, Q_{Rn}^{RS} defined:*

1. *$Q_{Rn}^{RS} \subseteq Q_R$,*

2. *$Q_R = Q_{R1}^{RS} \cup ... \cup Q_{Rn}^{RS}$, and*

3. *for all pairs of Q_{R1}^{RS} and Q_{R2}^{RS}, $Q_{R1}^{RS} \cap Q_{R2}^{RS} = 0$.*

The first requirement is that the states in the refinement subsystem are a subset of the states in the refined component. Basically this means refinement subsystems can't introduce new states. The second rule is that the union of all the refinement subsystems equals the state space of the refined component. This means that when all subsystems are considered, the state space is the refined component. Finally the third requirement is that refined subsystems do not include states from other refined subsystems.

Definition 6.6.3 Subsystem Refinement Relation ♠ - *Let SRR be a relation between SR (refined component) and SA (abstract component). The states $q_{Rn}^{RS} \in Q_{Rn}^{RS}$ and $q_{An} \in Q_A$ are related when written $q_{Rn}^{RS} \vee q_{An}$.*

This is a simple way of stating which states in the refined model correspond to those in the abstract model.

Definition 6.6.4 Subsystem Gluing Relation ♠ - *Let GI be a gluing invariant between SR and SA. The states $q_{Rn}^{RS} \in Q_{Rn}^{RS}$ and $q_{An} \in Q_A$, $\forall q_{Rn}^{RS} \vee q_{An}$ are glued, written $q_{Rn}^{RS} \mu q_{An}$, $l_R(q_{Rn}^{RS}) \Rightarrow l_A(q_A)$.*

This is the same style of gluing relation which was discussed previously in Section 6.2.3.

Definition 6.6.5 Visible Event ♠ - *For a system S, a visible event set,E_{EV}^N, is defined as:*

1. $E_{EV}^N \subseteq E_E^N$,

2. *for all pairs Q_{R1}^{RS} and Q_{R2}^{RS} where Q_{R1}^{RS} and $Q_{R2}^{RS} \subseteq Q$ and $q_{R1}^{RS} \in Q_{R1}^{RS}$ and $q_{R2}^{RS} \in Q_{R2}^{RS}$, $e_i \in E_{EV}^N$ iff $\exists t_i \in q_{R1}^{RS} \times e_i \times q_{R2}^{RS}$.*

The first requirement is that the visible events are a subset of the events. The second requirement is that for each pair of refinement subsystems, where the subsystems are part of the same component, the events are visible if they are labels between the two different refinement subsystems (i.e. not labels inside the refinement subsystem).

SystemC, METROPOLIS, or METRO II models can be used to automatically extract E_E^N. Refinement subsystems and subsystem refinement relations are then defined by the designer. The subsystem gluing relation now is produced automatically as a result (whereas GI was defined manually before). Refinement properties can then be defined over E_{EV}^N (which also fall out of the proposed definitions). These properties can be used for verification purposes (such as refinement).

Definition 6.6.6 Visible Property ♠ - *For a system S with set E_{EV}^N, a visible property is the set of variables, Var = $\{X_1, ..., X_n\}$ with the respective domain, $\mathbb{D}$ assigned to a path of states along a set of transitions assigned visible events.*

For each of the LTS based service components one can correlate them to existing SystemC, METROPOLIS, or METRO II code through E_E^N since the code uses its own notion of events to do synchronization. We can use each METROPOLIS/METRO II/SystemC *notify(e_n)* call (or METROPOLIS request() or await()) as an E_E^i in LTS. State variable sets will be defined for each SystemC *module*, METROPOLIS *media or process*, of METRO II *component* ($l : Q \rightarrow SP$). This process is shown in the pseudo-algorithm 7.

Algorithm 7: LTS USE IN SYSTEM LEVEL ARCHITECTURE SERVICE REFINEMENT ♠

Require: $\mathbb{M}$ is a set of SystemC *modules*, $\{m_1,...,m_N\}$ **or**

Require: $\mathbb{C}$ is a set of METRO II *components*, $\{c_1,...,c_N\}$ **or**

Require: $\mathbb{P}$ is a set of METROPOLIS *media*, $\{p_1,...,p_N\}$

Ensure : $\mathbb{X} = \mathbb{M} \cup \mathbb{P} \cup \mathbb{C}$

1 $\mathbb{S}$ is a set of LTS where $s : x_n \rightarrow s_n$.

2 Q for s_n is defined by $l : Q \rightarrow SP$ /*SP is defined manually   */

3 Synchronization ((α **when** p) is defined manually for $\mathbb{S}$

4 $\mathbb{S}_i^c$ (Context-in Component) is produced automatically

5 Q_{Rn}^{RS} is defined manually

6 $q_{Rn}^{RS} \vee q_{An}$ is defined manually

7 Subsystem Gluing Relation is produced automatically

8 E_{EV}^N is produced automatically

9 LTL/CTL/Refinement properties verified automatically over visible events in LTS

In addition to identifying refinement opportunities and definitions in order to formalize depth refinement, naturally refinement itself for LTS systems must be formally defined. Since this work will describe services as LTS, *compositional component based weak refinement* from [KL03a] will be used. This specifies the following rules for refinement, where η is the refinement relation :

1. **Strict transition refinement** - $(q_R \; \eta \; q_A \wedge q_R \xrightarrow{e} q_R' \in T_R) \Rightarrow \exists q_A'(q_A \xrightarrow{e} q_A' \in T_A \wedge q_R' \; \eta \; q_A')$

2. **Stuttering transition refinement** - $(q_R \; \eta \; q_A \wedge q_R \xrightarrow{\tau} q_R' \in T_R) \Rightarrow (q_R' \; \eta \; q_A)$

3. **Lack of old or new deadlocks** - $(q_R \; \eta_f \; q_A \wedge q_R \nrightarrow R) \Rightarrow ((q_A \nrightarrow A) \vee ((q_A \xrightarrow{e} q_A' \in TA) \Rightarrow (q_R \in D)))$

4. **Lack of τ-divergence** - $(q_R \; \eta \; q_A) \Rightarrow \neg \; (q_R \xrightarrow{\tau} q_R' \xrightarrow{\tau} q_R'' \xrightarrow{\tau} ... \xrightarrow{\tau} ...)$

5. **External non-determinism preservation** - $(q_A \xrightarrow{e} q_A' \in T_A \wedge q_R \; \eta \; q_A)$
   $\Rightarrow \exists q_R', q_A'', q_A'' \; (q_R' \; \eta \; q_A \wedge q_R' \xrightarrow{e} q_R'' \in T_R \wedge q_A \xrightarrow{e} q_A'' \in T_A \wedge q_R'' \; \eta \; q_A'')$

We note $q \nrightarrow$ when $\forall q' \; (q' \in Q \wedge e \in E \Rightarrow (q \xrightarrow{e} q') \notin T)$

The first rule essentially states that if there is a transition in the refined LTS from one state to another, then there must be the same transition in the abstract LTS. There are also syntactic restrictions that the transitions have the same label. The second rule states if there is a new (τ) transition in the refined

LTS, then its beginning state and ending state must correspond to the same state in the abstract LTS (this correspondence must be defined in the gluing relation). The third rule states if there is a deadlock in the refined LTS, then there is either a deadlock in the abstract LTS or the refinement LTS introduced a new deadlock. This allows that individual components can deadlock in the refinement as long as the composition of components still makes progress. The fourth rule is that there are no new transitions in the refinement that go on forever (τ loops for example). The fifth and final rule is if there is a transition in the abstract LTS and the corresponding (glued) refined LTS state does not have any transition then two conditions must be true: 1) there must be another refined state, qR', that corresponds (is glued) to the abstract state, qA, 2) qR' must take a transition to another refined state, qR'', and in the abstract LTS there must exist a state, qA'', which is glued to to the refined state, qR''. Illustrations of rules 1, 2, 4 and 5 are shown in Figures 6.16, 6.17, 6.18, and 6.19. In the Figures, qR refers to a state in the refined model whereas qA is a state in the abstract. Each state is grouped into abstract or refined groups. Arcs between the two groups indicate gluing relations.

Figure 6.16: Strict Transition Refinement

Figure 6.17: Stuttering Transition Refinement

Figure 6.18: Lack of τ-Divergence

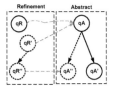

Figure 6.19: External Non-Determinism Preservation

The design flow for *depth* refinement verification now consists of the following three steps. These are based upon a design flow using SynCo. The results of such a design flow on specific communication structures of the FLEET architecture are shown in Chapter 9.

1. The first step is to create a .fts file for each component. This file defines LTS transitions and states for each service. This creation has the potential for automation but is not currently. An automation scheme would involve capturing the states where the combination of each event's presence is unique (enabled or disabled) and when events are involved in a service interface or when they are communicated between services using, *wait()* or *notify()* (in SystemC for example). This process would be similar to the CFA creation process outlined for depth refinement. A sample set of .fts files for a consumer LTS (part of a larger producer/consumer example) are shown in Figures 6.20 and 6.21. The abstract consumer is either waiting to consume or consuming. The refined consumer allows for a "clean up" procedure (a purging of sorts) after waiting but before it begins to consume again.

```
Transition System
//Two state values
type SIGNAL = {consume, wait}
local con : SIGNAL

//Can only be in one state
Invariant
(con = consume) \/ (con = wait)

//Initial state
Initially (con = wait)

//Transition to consume (''get'' event)
Transition get :
enable (con = wait) ;
assign con := consume

//Transition to wait (''stallC'' event)
Transition stallC :
enable (con = consume) ;
assign con := wait
```

Figure 6.20: .fts for Abstract Consumer LTS

```
Transition System
//Three state values (added clean)
type SIGNAL = {consume, wait, clean}
local conR : SIGNAL

Invariant
(conR = consume) \/ (conR = wait)
\/ (conR = clean)

//Initial state
Initially (conR = wait)

//Transition to consume (''get'' event)
Transition get :
enable (conR = clean) ;
assign conR := consume

//Transition to wait (''stallC'' event)
Transition stallC :
enable (conR = consume) ;
assign conR := wait

//Transition to cleanup (''cl'' event)
Transition cl :
enable (conR = wait) ;
assign conR := clean
```

Figure 6.21: .fts for Refined Consumer LTS

2. The second step is to create a .inv file for each set of components (the abstract and refined versions). This file defines the gluing invariants between abstract and refined states. In Figure 6.22 the two "consume" states are glued. The abstract consumer's "wait" state is glued to the refined consumer's "wait" and "clean" states.

```
((con = consume) <--> (conR = consume))
/\((con = wait) <--> ((conR = wait) \/ (conR = clean)))
```

Figure 6.22: .inv for Consumer LTSs

3. The third step is to create a .sync file for the whole system. This file defines synchronization and interactions between LTS components. There should be a .sync file for the refined and abstract systems. One file for the abstract LTSs' interaction and one for the refined LTSs' interaction. When composing modules together, the total number of states in the system is less than the product of number of states in each component. This is one of the strengths of synchronization and its partial specification. In Figure 6.23 the .sync file for the abstract producer and consumer example system is shown. Each event of the LTSs is enabled depending on the state of the collection of LTSs.

These three sets of files are provided to SynCo for both the abstract and refined systems. SynCo will then check the validity of the refinement rules outline previously. This design flow can be automated partially. As mentioned, the .fts file creation can be automated starting from a METROPOLIS or SystemC description. The .inv file creation can be automatic but must start from some designer specification. The state correspondence must either be explicitly described by the designer in a separate file or implicitly via a state naming conventions. The .sync file must be manually created. In large systems this can be done hierarchically to make the process more manageable. This type of automation is potential future work.

6.7 Conclusions

This chapter has introduced three approaches to architecture service refinement and its verification. These are: event based (vertical, horizontal, diagonal), interface based (surface), and compositional component based (depth) refinements. Each is a potential tool in a system level architecture service development design flow. Each has their own unique strengths and weaknesses. For example it has been shown that an event based approach is scalable and allows two distinct system level design exploration scenarios. However, event properties may be difficult to specify and capture. Interface behavior capture allows for IP integration and relies on a very nice formalism which is currently verified by existing

```
//Buffer Events (reads and writes)
//''write1'' event is enabled when the LTSs are in the following states
(write1) when
((prod = produce) /\ (buf = empty)   /\ (con != consume)),

(write3) when
((prod = produce) /\ (buf = notempty)  /\ (con != consume)),

(read1) when
((prod != produce) /\ (buf = notempty)  /\ (con = consume)),

(read3) when
((prod != produce) /\ (buf = full)   /\ (con = consume)),

//Producer Events
make when
(prod = wait),

stall when
(prod = produce),

//Consumer Events
get when
(con = wait),

stallC when
(con = consume)
```

Figure 6.23: .sync for Producer/Consumer LTSs

(free) tools. However, it can be time consuming and requires certain syntactic conditions and manual steps which may require more effort and knowledge on the part of the designer. Finally compositional component verification also employs a clean formalism and allows specific changes to be made in the granularity of individual service offerings. However, it requires that a manual correspondence between states be made in a gluing relation and also requires that the overall behavior of the system (synchronization) be specified manually. In implementing a design flow, one should use each of these techniques in particular situations to maximize their strengths while minimizing their weaknesses. In Chapters 8 and 9 specific examples of each of these techniques will illustrate their potential uses.

Chapter 7

Architecture Service Characterization

"The first rule of any technology used in a business is that automation applied to an efficient operation will magnify the efficiency. The second is that automation applied to an inefficient operation will magnify the inefficiency." – Bill Gates, Microsoft Co-Founder

In [Don04], the relative importance of ESL design tasks was explored for a variety of product scenarios. The design tasks identified are "early software development", "functional verification", "performance analysis", and "design space exploration". Crossing the "methodology gap" introduced in Chapter 2 requires a suitable ESL technology be in place for each of these design tasks. Furthermore, ESL tools must address the design tasks in a way that matches the designer's specific product scenario. An ESL solution for application specific standard products (ASSPs) will differ from the solution for structured ASICs. Figure 7.1 formulates, as an example, an ESL roadmap to support the transition of RTL designers to ESL. This illustration is based on Figure 2.3 which illustrated the productivity progress required for ESL adoption. The exact number and sequence of steps varies according to the priorities of a given market segment. Each technology that is in place is a step up from RTL to ESL and a complete set of steps is required for a smooth transition. If one or more of the steps are missing, the risk of migration will deter designers that are not close to their "maximum tolerable design gap". Designers that reach their maximum tolerance before the ESL steps are available are in a pathological scenario because their product is no longer cost effective to develop. Naturally, the steps must also occur in a timely manner or system complexity will overtake productivity again.

Chapter 5 and this work in general deal with the design space exploration step. However this chapter focuses on one specific step in the transition from RTL to ESL, ESL performance characterization (and analysis). ESL performance characterization allows designers to predict whether a given system architecture can meet a requested level of performance. For this process to be truly useful, performance

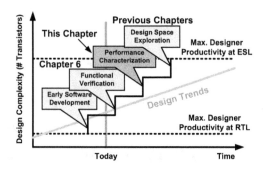

Figure 7.1: Transcending RTL Effort with ESL Design Technologies

characterization must be integrated with architecture service modeling. The contribution of this work lies in that integration.

This chapter will provide a methodology for characterizing architecture services for performance analysis. This methodology produces results that are highly accurate while maintaining modularity.

Today, the two most common performance metrics are computational throughput and power consumption. In some markets, computational throughput is the dominant metric whilst, in others, power consumption takes pole position. In this work, we will focus on computational throughput. As with all ESL design tasks, performance characterization relies on simulation or analysis of abstract system models to derive system performance in a given situation. Abstraction allows the system to be described early and at a reasonable cost but it also casts a shadow of doubt over the accuracy of performance characterization data. Since the data guide the selection of one system architecture over another, the veracity of data recovered from ESL performance characterization techniques must be weighed carefully by the designer. Accuracy is paramount for ESL acceptance and legitimacy.

Fear of inaccuracy in ESL performance characterization is a major impediment to the transition from RTL to ESL. However it is not the only impediment. Beyond the fundamental abstraction accuracy tradeoff, current ESL methods and tools lack a coherent set of performance modeling guidelines. These guidelines are important because they allow a single system model to be reused over multiple ESL design tasks: the same basic model must be instrumented for performance characterization without significant compromise to its usefulness in early software development. Clear guidelines and coding standards also allow analytical data to flow out from a model into multiple ESL vendor tools. Close coupling of an instrumentation interface to a single ESL vendor is generally perceived as a bad thing unless the ESL vendor's tools precisely complement the designer's target market segment. Modularity is paramount for ESL acceptance and legitimacy.

As seen in the naïve flow in Figure 5.1, typically, average and worst case cost estimates for system features are often used today in ESL performance models. The estimated cost model for Xilinx Virtex II Pro architecture services was described in Chapter 5 as well. As will be shown, the inaccuracy of these measures quickly accumulates in the data recovered from ESL performance analysis tools and restricts the designer to measure only the relative performance of two systems. In this chapter, a novel technique is described that combines very accurate performance characterizations of a target platform with abstract architecture service models in an ESL design environment. This process is an enhancement present in the flow proposed in Figure 1.11 and Figure 7.2 highlights and expands this. It is proposed that characterization data recovered from an actual target device can be gathered easily and can be annotated into ESL architecture service models to enhance the accuracy of performance estimates. In the prototype of this approach, two specific modeling environments (METROPOLIS and METRO II) are selected. One set of target characterizations can be exchanged for the estimated data to aid in the selection of a specific instance of the target architecture. All of this effort specifically looks to address accuracy as required by the overarching goals of this work while maintaining modularity.

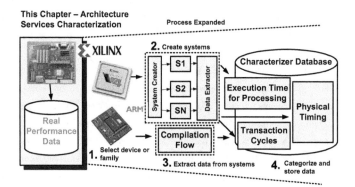

Figure 7.2: Characterization of Architecture Services in the Proposed Design Flow

7.1 Chapter Organization

The remainder of this chapter is organized as follows: first in Section 7.2, the technique's general process is discussed along with its set of requirements and assumptions. The next section (7.3) offers more details on the pre-characterization process by discussing the automatic generation of the group of target systems. An important part of this discussion is how to automate the extraction of reusable performance data from the target system's physical design flow. Section 7.4 provides an example of the characterization data that can be obtained with this automated method. Storage, organization, and categorization of this data are discussed next in Sections 7.5 and 7.6. A Motion-JPEG encoder example of pre-characterizing processor systems on Xilinx Virtex Platform FPGAs is contained in Chapter 8 (along with other design examples such as UMTS). This chapter concludes by summarizing the results and discussing potential future work.

7.2 Platform Characterization

Platform characterization is the process of systematically measuring a set of properties of a physical platform or its components. Ultimately, this measurement data will be annotated to ESL architecture service models of the platform's components. Subsequent simulations of the now annotated model yield performance estimates that are correlated with the actual target platform. As such, they are more accurate than equivalent models annotated with "ballpark" performance metrics. In short, platform characterization extracts a relevant group of performance measures from the physical implementation of a platform.

For characterization to be applicable to a system design, an appropriate, measurable implementation of the target platform must already exist. Clearly, ASIC designs are less amenable to this approach because a suitable implementation of the implementation target is not available early enough in the design process. Fortunately, programmable platforms based on FPGA technology and ASSP based systems are common targets for new design starts [Dat08]. Both technologies are amenable to the proposed approach. In the case of an ASSP, its physical properties and system architecture are fixed at fabrication. For FPGAs, the physical properties of the platform are fixed at fabrication, but the system architecture is flexible. Clearly one cannot apply this technique to the designer's final platform: the system architecture for the designer's product has not yet been determined. Instead, it is necessary to pre-characterize a broad set of potential system architectures that the designer may use. Systems destined for this kind of target require the designer to choose the characterization data that is most representative of the system architecture they intend to create. Additionally, the designer can explore the effect of different system architectures through their characterizations.

Systematically characterizing a target platform and integrating the data into the platform model is a three-way tradeoff between the following factors:

- Characterization effort (code size, maintenance, run time);

- Portability of characterization data (system agnostic design); and

- Characterization accuracy (correspondence to real systems).

The more accurate a characterization, the more effort it will take to extract and the less portable it will be to other system models. Alternatively, a highly portable characterization may not be accurate enough to base a design decision on. This process must offer more accuracy than standard transaction level approaches [Don04], require less effort than an RTL approach, and have more portability than an ASIC (static architecture) based target. Table 7.1 relates this approach (platform characterization) to RTL and transaction level modeling (TLM) with regards to the three main tradeoffs. Each column is an approach and each row is a design description. "Low", "Medium", and "High" are just used to show the relative ranking of each approach in the context of the others.

Tradeoff	TLM	RTL	Platform Characterization
Effort	Medium	High	Medium-Low
Portability	Medium	Low	High
Accuracy	Medium-Low	High	Medium-High

Table 7.1: Performance Characterization Tradeoffs

7.2.1 Characterization Requirements

In order to develop a robust environment for platform characterization processes there are several requirements:

- **Direct correlation between the platform metrics characterized and architecture service models** - the designer must make sure that the service models developed can be easily paired with characterization data and that the models have the essential behaviors which reflect the aspects captured by characterization.

- **IP standardization and reuse** - in order to make characterization scalable, designs must use similar components. If each design is totally unique and customized there will not be existing information regarding its characterization. For example, Xilinx FPGAs accomplish this by employing the IBM CoreConnect bus and the CoreConnect interfaces on the IPs in their Xilinx Embedded Development Kit (EDK). These are industry standard IP used in embedded system design. This requirement will be expanded in the next section when extraction is discussed.

- **Tool flow for measuring target performance** - whichever platform one chooses to model, the actual target device must have a method and tool flow to gather measures of the characterization metrics.

- **System Level Design environment with support for quantity measurements** - the framework that the characterization data is integrated with must support the measurement of quantities (execution time, power, etc). In addition it must allow for the data used to calculate these quantities come from an external source.

This chapter will not focus on the details of each requirement. The discussion in this chapter will describe characterization in the context of both METROPOLIS [BHL+03] and METRO II [DDM+07] design environments and Xilinx Virtex II Pro FPGAs and ARM9/7 general purpose microprocessors. Other design environments and platform types that meet the requirements above may also be characterized with this approach.

The next section will discuss how to begin the process of gathering data for use in the characterization process.

7.3   Extraction of Platform Characterization Data

Extraction of data from the target platform's physical tool flow is at the heart of this characterization methodology. Extraction is a multiphase process concerned with:

- **Selecting a programmable platform or device family** - by selecting a family of products in a programmable product line, one increases the opportunity that the extraction will be portable to other systems. Ideally, the selection of a platform family is done without regard to application domain but, in practice, this will influence the designer's decision. An example of a programmable platform family is the Xilinx Virtex-4 family of platform FPGAs. If a static device is chosen, a tool flow (instruction set simulator and virtual hardware simulation) must be available. Two such examples are [kei] and [BA97].

- **Selecting programmable platform components or device instance** - the properties of the components will vary depending on the granularity and type of the programmable platform. For example an FPGA could consist of IP blocks, embedded processing elements, or custom made logic blocks. The device instance for a static target will detail the microarchitectural specifics (i.e. ISA restrictions, arithmetic precision).

- **Selecting systems for pre-characterization** - from the selected components, assemble a template system architecture. From this template architecture create many other permutations of this template. In many cases the permutation of the template architecture is automatic. For a static architecture, these permutations are software routines that will be characterized.

  - Permutations can be made incrementally using various heuristics regarding the desired number and types of components. For example, one might want to constrain the number and type of embedded processors instantiated or the number of bus masters/slaves. The entire permutation space does NOT need to be generated. This is true for the selection of software routines as well.

- **Independently synthesize and implement each system permutation** - the ability to quickly synthesize the created architecture instances is what differentiates programmable platforms from static, ASIC like architectures. Each of the systems should be pushed through the entire synthesis and physical implementation flow (place, route, etc). Static architectures typically go through a compilation and simulation process.

- **Extracting desired information from the synthesis process and its associated analysis** - the conclusion of the synthesis process will give information about each system. Such information includes (but is not limited to) various clock cycle values, longest path signal analysis, critical path information, signal dependency information, and resource utilization. Standard scripting languages like Perl [O'R07] can be used to automatically extract the appropriate information from the platform tool reports.

Figure 7.3 illustrates the pre-characterization process for programmable platforms. This is a sequence of six steps as shown. These steps roughly correspond to the steps just outlined (Section 7.3).

Figure 7.4 illustrates the pre-characterization process for general purpose processing elements. This process involves the use of embedded profilers or instruction set simulators, in particular, Keil ARM Development Tool [kei], as well as GNU compiler tools, virtual hardware, and a custom designed code

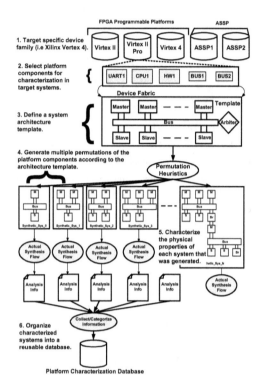

Figure 7.3: A Design Flow for Pre-characterizing Programmable Platforms

annotator. Initially the code is cross compiled for the particular architecture target we are interested in. This executable is then fed to a virtual hardware simulator (in this case, simplescalar [BA97]). These results along with the original source code and corresponding binaries are fed to a code annotator which produces annotated code detailing the actual running time of individual segments of the original code for the given architecture target. This annotated code is then mined for the execution times which are stored later for annotation of the architecture services.

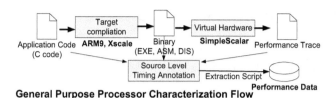

General Purpose Processor Characterization Flow

Figure 7.4: A Design Flow for Pre-Characterization of General Purpose Processors

7.3.1 Data Extraction Requirements

The issues that need to be observed during the extraction of characterized data are Modularity, Flexibility, and Scalability. These are important aspects of steps 5 and 6 in Figure 7.3:

- **Modularity** - After the initial selection of components and the architecture template, the rest of the extraction can be performed by many independent extraction processes. These processes can be distributed over multiple workstations. This reduces the time to generate N permutations and characterize them to a constant time M where M is the duration of the longest permutation.

- **Flexibility** - Ultimately the extracted characterization data must be correlated to designs during simulation. Therefore the closer the permutated templates are to the actual designs the better. In most cases they will be identical but it is possible that some architecture service model designs will have parameters that differ from the characterized system. In the event that the differences do not affect the performance under test, the characterization data already obtained can be used.

- **Scalability** - The extraction process is independent of the storage mechanism for the data so it in no way limits the amount of characterization data that can be extracted. Constraints can be added or relaxed on the permutations of the initial template. Theoretically, all permutations of the target's component library are candidates for characterization. Even though the characterizations can happen at the platform vendor well in advance of the designer using the data, the set of permutations will be constrained. This is necessary to maintain a reasonable total runtime for the overall extraction process initially. This method does support incremental addition of permutations later if the need arises however.

7.4   Example Platform Characterization

To exemplify this process (Sections 7.2 and 7.3), a set of typical FPGA embedded system topologies was pre-characterized [DDSV06a]. Each topology was generated from a template to create a microprocessor hardware specification (MHS) file for the Xilinx embedded system tool flow. Architectures were generated with permutations of the IPs listed in Table 7.2. The table also shows the range in the number of IP instances that can be present in each system permutation along with the potential quantities of each. In addition to varying the number of these devices, also permuted were design strategies and IP parameters. For example, the system's address decoding strategy was influenced by specifying tight (T) and loose (L) ranges in the peripheral memory map. A loose range in the memory map means the base and high addresses assigned to the peripheral's address decoder are wider than the actual number of registers in the peripheral. For a tight range, the converse is true. Also permuted was the arbitration policy (registered or combinatorial) for systems that contained an On-Chip Peripheral Bus (OPB). These axes of exploration were used to investigate the relationship between peripherals and the overall system timing behavior. These design factors are not usually considered in system characterization. This is due to the fact that they are not traditionally considered in influencing the system size. System size is a heuristic often used since it has the ability to influence system performance (e.g. system clock speed). These often overlooked features effects on performance will be of particular interest.

Component	MicroBlaze	PPC	Combo
PowerPC (P)	-	1-2	1-2
MicroBlaze (M)	1-4	-	1-4
BRAM (B)	1-4	1-4	1-2 (per bus)
UART (U)	1-2	1-2	1-2 (per bus)
Loose vs. Tight Addressing	Yes	Yes	Yes
Registered or Combinational Arbitration	Yes	N/A	Yes
Total Systems	**128**	**32**	**256**

Table 7.2: Example CoreConnect Based System Permutations for Characterization

The columns of Table 7.2 show three permutation "classes" that were used. The implementation target was always a Xilinx XC2VP30 (Virtex II Pro) device. The first class (column MicroBlaze), refers to designs where MicroBlaze and OPB were the main processor and bus IPs respectively. The second class (column PowerPC) represents PowerPC and Processor Local Bus (PLB) systems. The third class

(Combo) contain both MicroBlaze and PowerPC. The number of systems generated is significant (but not unnecessarily exhaustive) and demonstrates the potential of this method. Note each system permutation can be characterized independently and hence, each job can be farmed out to a network of workstations. For reference, the total runtime to characterize the largest "Combo" system with Xilinx Platform Studio 6.2i on a 3GHz Xeon Windows machine with 1 GB of memory was 15 minutes. The physical design tools were run with the "high effort" option and a User Constraint File (UCF) that attempts to maximize the system clock frequency. An observation of the characterization data shows that as resource usage increases (measured by FPGA slice count; a slice contains two 4-input function generators, carry logic, arithmetic logic gates, muxes, and two storage elements) the overall system clock frequency decreases. Figure 7.5 shows a graph of sample Combo systems, their size, and reported performance. Nested loops of each IP were used to generate the system permutations, giving the systems generated predictable increases in area and complexity. The major, periodic increase in area is as anticipated and indicates that a MicroBlaze processor was added to the system topology and all other peripheral IPs were reset to their lowest number. Note that the graph's performance trace is neither linear nor monotonic. Often area is constant while frequency changes drastically. This phenomenon prevents area-based frequency estimations. The relationship between the system's area utilization and performance is complex, showing that building a static model is difficult, if at all possible, and confirming the hypothesis that actual characterization can provide more accurate results.

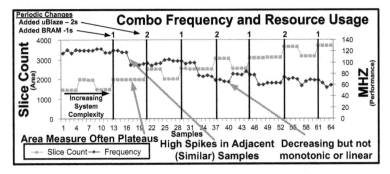

Figure 7.5: Combo Systems Resource Usage and Performance

Table 7.3 highlights an interesting portion of the data collected in the PowerPC class. Each row is a PPC system instance: the leftmost columns shows the specific IP configuration for the system

((P)owerPC, (B)RAM, and (U)ART) and the remaining columns show area usage (slice count), max frequency, and the % change (Δ) between the previous system configuration (representing potentially a small change to the system). This work contends that a difference of 10% is noteworthy and 15% is equivalent to a device speed-grade. Note that there are large frequency swings (14%+) even when there are small (<1%) changes in area. This is not intuitive, but seems to correspond to changes in addressing policy (T vs. L) and indicates that data gathered in pre-characterization is easy to obtain, not intuitive, and more accurate than analytical cost models. The data shown here is not what would be estimated in an area based approach. As a result systems using area based techniques would not be nearly as accurate.

P	B	U	Addr.	Area	f(MHz)	MHz Δ	Area Δ
1	2	1	T	1611	119	16.17%	39.7%
1	2	1	L	1613	102	-14.07%	0.12%
1	3	0	T	1334	117	14.56%	-17.29%
1	3	0	L	1337	95	-18.57%	0.22%
1	3	1	T	1787	120	26.04%	33.65%

Table 7.3: Non-linear Performance Observed in PPC Systems

Figure 7.6 illustrates Table 7.3 and shows area and separate performance traces for PPC systems in two addressing range styles (one tight and one loose). One set of data points correspond to area measurements and the other reflect frequency measurements. The graph demonstrates that whilst area is essentially equivalent (the area curves overlap visually), there are clear points in each performance trace with deviations greater than 10%.

7.5   Organization of Platform Characterization Data

The organization of the raw, extracted data is the process of categorizing the information in such a way that system simulation remains efficient, data remains portable, and flexible data interactions can be explored. This is a very important part of the characterization process and if a poor job is done in this stage, many of the benefits of the previous efforts will be lost. This is an aspect of step 6 in Figure 7.3 as well as the final step in Figure 7.4. More concisely the goals are thus:

- **Maintain system efficiency** - if the simulation performance of the system using estimated data (a naïve method) is P_E and the performance of the system using characterized data (the proposed method) is P_C, the following relation must hold, $P_C \geq P_E$. Performance in this case is a measure of

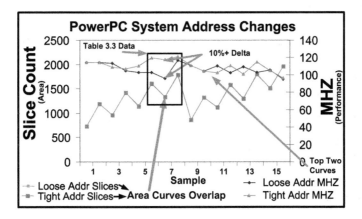

Figure 7.6: PowerPC System Performance Analysis

simulation effort or cycles consumed which directly affect the execution time of the simulation or runtime memory requirement (higher performance results in lower execution time or lower runtime memory requirement).

- **Portable Data** - in order to reuse data, it must be stored in such a way that it is maximally portable amongst various models. This requires three things: 1) A standard interface for accessing the stored data 2) A common data format for the stored data and 3) The ability for the data set to grow over time.

- **Flexible Data Interaction** - data interaction refers to the ability to allow many ways in which data can interact in order to give information regarding the performance of the simulation. For example if data regarding transactions per instruction can be combined with information regarding cycles per transaction one can determine the cycles per instruction. Another example is that if Transaction$_1$ can use signals S_1 or S_2 and it is known that S_1 resides along a longer path than S_2, Transaction$_1$ can utilize S_2 for greater performance. It is best to place no restriction on data interaction in so much as it does not conflict with any of the other characterization goals.

7.5.1 Data Categorization

With the goals defined for "characterization data organization" the second aspect that must be determined is how data is categorized. Data can be categorized in many ways depending on what is being modeled. For the sake of this discussion, it will be in the context of what is required typically for programmable architecture service models of embedded systems. To this end there are three categories:

- **Physical Timing** - this information details the physical time for signals to propagate through a system. Typically this information is gathered via techniques such as static timing analysis or other combinational or sequential logic design techniques to determine clock cycle or other signal delays.

- **Transaction Timing** - this information is a unit of measure which details the latency or stages of a transaction. A transaction is an interaction between computational units in a point to point manner or through other communication mechanisms (buses, switches, etc). This could be a cycle count in reference to a particular global clock which determines the overall speed of the system. Or it could alternatively be an asynchronous measure.

- **Computation Timing** - this information is regarding the computation time taken by a specific computation unit. This could be both HW and/or SW based routines. For example it could be a cycle count given by the time a HW unit (Adder, Shifter, etc) takes to complete its operation. Alternately it could be the cycle time taken by a particular software routine (Discrete Cosine Transform perhaps) running on a particular embedded processor.

These three areas interact to give estimated performance for a system under simulation. The following example (Table 7.4) shows how all three areas can be used along with their ability to flexibility interact to provide performance analysis:

Instruction	Timing Categorization	Performance Implication
read(0x64, 10B)	Transaction - 1 cycle/Byte	10 cycles
execute(FFT)	Computation - FFT 5 Cycles	5 cycles
write(0x78, 20B)	Transaction - 2 cycles/Byte	40 cycles
Total Cycles	Physical - 1cycle/10ns	**550ns**

Table 7.4: Sample Simulation Using Characterization Data

The leftmost column provides three different instructions. The center column gives the characterization of each instruction and what category it falls under. The rightmost column gives the resultant performance implication given the instruction and its characterization. The final row illustrates the execution time of this sequence of instructions given the physical time of one execution cycle.

7.5.2 Data Storage Structure

Finally, it must be decided what actual structure will hold the now categorized, characterized data. The primary concerns are related to the goals initially mentioned in this chapter regarding portability and efficiency. This should be a structure that can grow to accommodate more entries. Ultimately what structure will be used is determined by which system level design environments are intended to be used. However the following issues should be considered:

- What is the overhead associated with accessing (reading) the data?

- What is the overhead associated with storing (writing) the data? Both this and the reading overhead are affected by code size and complexity and ultimately affect simulation speed.

- Can data be reorganized incrementally? This is of use if new data is added or the categorization mechanism changes.

- Can data be quickly sorted? Searched? This can increase the speed of access and allows exotic relationships between data elements and their use.

More specifics on data structures for characterization data will be touched on in the next section when specific example executions are discussed. For now this work leaves the reader with an illustration of an abstract structure in Figure 7.7. The left hand side of the illustration shows the data categorization and in which stage of the design flow that data is generated. Shown are the three types of data categories as well as where that data is collected. Notice that each element is connected to the others, illustrating that they should be flexible in their interaction. Also there should be an input interface (to enter data), as well as an output interface to retrieve the data. The right hand side shows a sample entry in the data storage structure where each system categorized has its own index and may have independent or shared entries in the storage structure. With each index there is associated physical timing, computation timing, and transaction timing data. This data can be shared or be unique to a particular index. Additionally each index is not required to have an explicit entry for each category and can utilize a globally stored default value.

240

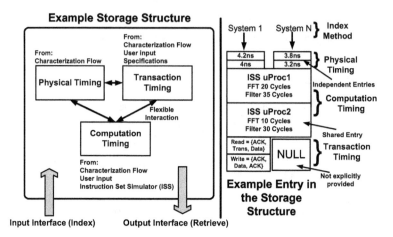

Figure 7.7: Characterized Data Organization Proposal

7.6 Integration of Platform Characterization and Architectural Services

Once the data has been extracted and organized it now must be integrated into a system level design environment for simulation. The following discussion will highlight the key issues associated with this integration and provide an example of each in the METROPOLIS and METRO II environments.

- **Separation of architecture models and their scheduling** - this requirement allows for the data structure containing the extracted data to be independently added and modified apart from the actual system.

    - In METROPOLIS, architecture models are a combination of two netlists (as described in Chapter 4). The first netlist is called the scheduled netlist and contains the topology and components that make the architecture instance (CPUs, BUS, etc). The other netlist is the scheduling netlist and contains schedulers for each of the components in the scheduled netlist. When multiple requests for a resource are made in the scheduled netlist, it is the other netlist which resolves the conflict (according to any number of algorithms). The schedulers themselves are called quantity managers since the effect of scheduling is access to update a quantity (time, power, etc) of the simulation. See [DDSZ04] for more information on METROPOLIS architecture modeling.

– In METRO II the three phase execution semantics ensures this separation. In phase 1 the architecture service models will execute. Once all processes which constitute services have been suspended, the phase will terminate and the simulation manager will switch to the next phase. In the third phases those services will be scheduled. The manager uses information from components (phase 1) and schedulers (phase 3) in a completely independent fashion.

• **Ability to differentiate concurrent and sequential requests for resources** - the simulation must be able to determine if requests for architecture resources occur simultaneously and are allowed to be concurrent or if they should be sequential, in which case the ordering should be known. This is important since each request will be accessing characterization data and accumulating simulation performance information which may be order dependent.

– In METROPOLIS there is a *resolve()* phase during simulation. This is the portion of simulation where scheduling occurs. This scheduling selects from multiple requests for shared resources. This is done by quantity managers in METROPOLIS.

– In METRO II both the notion of rounds (three phases of simulation) along with the tags on events allow the simulation infrastructure to order event requests for services. Events having tags annotated with logical time are used to create a partial order.

• **Simulation step to annotate data** - during simulation there should be a distinct and discernible time (simulation step) where data is annotated with characterized data

– METROPOLIS is an event based framework which generates events in response to functional stimulus. These events are easily isolated and augmented with information from characterization during scheduling (with the *request()* interface). This annotated data is stored in the event's "value" set. Events are defined as demonstrated by the tagged-signal model of [LSV98].

– In METRO II the second phase of simulation is explicitly the annotation phase.

The overall message of this integration discussion is that once the data is ready to be integrated into the design environment 1) it must be able to be added non-destructively 2) it must be able to augment existing simulation information detailing performance 3) the simulation must be able to correctly recognize concurrent and sequential interactions/requests for the characterized data.

7.6.1 Sample Annotation Semantics

This final section will demonstrate an example execution of a system integrated with the characterized data. This will be shown first on the METROPOLIS design environment. In this case, the structure holding the data is a *hash table*-like structure indexed by information regarding the topology of the system. Figure 7.8 illustrates these steps in METROPOLIS. Each step in the figure corresponds to a step below.

1. An active METROPOLIS architecture thread generates an event, *e*. This event represents a request for a service. This event will have been generated by a functional model mapped to this architecture needing a service (CPU, BUS, etc). This event can represent a transaction or computation request. In the case of the figure, the event is generated by a *thread.serviceRead()* interface call.

2. The service will make a request to its scheduler, with the *request(e)* method. This passes the request from the scheduled netlist to the scheduling netlist where *e* joins a list of pending events. While this event is awaiting scheduling, the task that generated it remains blocked (unable to proceed). In the figure this is the *serviceScheduler.request(e)* call.

3. Once all events that can be generated for this simulation step have been generated, the simulation proceeds to a *resolve()* phase where scheduling decisions (algorithms vary depending on the service they schedule) are made which remove select events from the pending lists. The figure illustrates this in the scheduling netlist's *serviceScheduler.resolve(e)* object call.

4. *serviceScheduler.Annotate(e)* selects events by indexing the characterized database according to event information. In practice more than just the event is passed. In addition a "request class" is passed also to provide information for indexing the database. This allows access to simulation quantities (like simulation global time) which can now be influenced by annotated events. Note that this requires no more impact on simulation performance as compared to estimated data (a requirement of our methodology; $P_C \geq P_E$).

5. Report back to the task that it can now continue (unblock the thread). This is the communication between the scheduler and the thread, *unBlock(e)* in the figure. This process is actually communicating to the process through a statemedia using the *setMustDo()* function.

6. The process can occur recursively when transactions like read() use CPU, BUS, and MEM services. These calls would generate their own sets of events. The figure illustrates a potential *nextSer-*

vice.serviceRead() call by the existing service which would initiate a similar sequence once again further down the netlist.

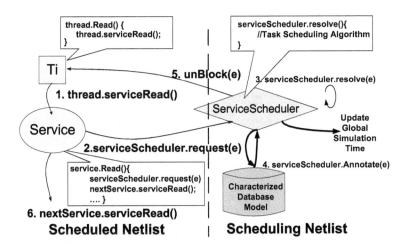

Figure 7.8: METROPOLIS Sample Annotation Semantics Using Characterized Data

The flow for using characterized data in the METRO II environment is shown in Figure 7.9. Initially an event (*use_service_event*) is proposed which requests an architectural service. After all processes in the simulation have proposed events, the simulation will switch to the second phase. The process that proposed the event is now suspended until the event is enabled. During phase 2, the annotator's table which keeps information regarding the cost of services is updated with the information from the characterized database. This can be done dynamically in the event that the database is updated between rounds. The annotator will manipulate the event (typically by changing the tag) in such a way that reflects the appropriate cost. In the 3rd phase this information can be used for scheduling by changing the logical time of the system simulation.

7.7 Conclusions

Before the gap between designer productivity and design complexity becomes an impassible chasm, architects must complete a transition from RTL to ESL design methods. However, a complete

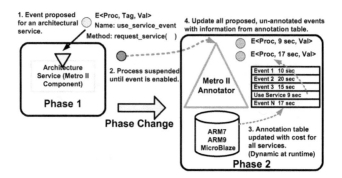

Figure 7.9: METRO II Sample Annotation Semantics Using Characterized Data

path from RTL to ESL has not yet been established. The reasons for the ESL methodology gap include the difficulty of isolating a set of design technologies that solve ESL design problems for the diverse range of system types. Designers desirous of ESL performance analysis tools are also wary of the accuracy of the data they can recover from existing tools and models.

In this chapter, an ESL performance analysis technology for programmable platforms and general purpose processing elements was presented. This approach united characterizations of actual platforms with abstract, designer created, architecture service model simulations. The result is an integrated approach to ESL performance modeling that increases the **accuracy** of performance estimates. The use of METROPOLIS quantity managers and METRO II annotators also eases design space exploration by separating the architectural models of a system from the specific timing model used during system simulation.

Future efforts with system level pre-characterization will begin with a deeper exploration of the tradeoff between accuracy and a given system model's level of abstraction. Additionally, formal techniques can be applied to analyze the bounds of our approach which is currently simulation based.

With both the modeling and characterization of programmable architecture models described, the subsequent chapters will explore how all the techniques outlined in previous chapters were applied to both multimedia and distributed design case studies.

Chapter 8

Multimedia System Design

"The code is pretty much overrated. It's more important to know such things as structure, features and functionality." – Larry Ellison, co-founder of Oracle Corporation

We live in a multimedia world. From the flashing advertisements of Times Square in New York City, to the explosion of personal communication devices in India, at no point in the history of the world have people been able to communicate with the frequency and regularity they do today. Not only is the volume of communication at an all time high but the complexity has also dramatically increased as well. Where once simple text messages were the norm, it is now not uncommon for individuals to send images and videos to one another wirelessly. The same devices that send these images are often able to also capture them, store them, and manipulate them. Work that was once the domain of supercomputers is now available in the palm of your hand. Voice dialing, GPS navigation, and video-on-demand are no longer features relegated to only the high end product offerings but now are requirements of the low end. It is extremely important for any discussion of embedded systems to discuss this vibrant systems domain.

The multimedia domain can be broadly characterized by the presence of large amounts of data streamed between different stages in the system. Data streaming applications such as audio, video, and image codecs as well as wireless communication, all characterized as multimedia systems, are predominant in many consumer electronics devices.

This chapter will present several case studies from the multimedia domain that illustrate the applicability of the design flow.

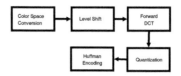

Figure 8.1: JPEG encoder block diagram

8.1 Chapter Organization

The outline of this chapter is as follows: In Section 8.2, the deployment of a JPEG encoder application onto the Intel MXP5800 platform will be covered. The key aspect of this case study concerns the choice of the CMD that allows for future automation and accurate performance estimation. The second case study, covered in Section 8.3, deals with this automation. After showing how characterization can allow for accurate performance models, we leverage the previously developed CMD to automatically allocate and schedule the system using a mathematical programming approach. The third case study, in Section 8.4, considers exposing varying amounts of concurrency from the h.264 application in order to explore different points in the design space for the Xilinx platform. Finally, the fourth case study in Section 8.5 considers design effort and METRO II framework overhead in addition to simulation accuracy.

8.2 JPEG on Intel MXP5800

8.2.1 JPEG Encoder Application

The JPEG encoder [Wal91] application is required in many types of embedded multimedia systems, from digital cameras to high-end scanners. The application compresses raw image data and emits a compressed bitstream.

The JPEG encoder application is chosen here since it is representative of a wide class of multimedia applications. In particular, the DCT, quantization, and Huffman blocks in this application are utilized in several other image/video compression applications, including h.264 [WSBL03].

A block diagram for the JPEG encoder is shown in Figure 8.1. The input for the application is an image described with raw RGB data. Each pixel is characterized with three bytes of information: one each for the red, green, and blue components.

The first step is color space conversion, where the raw data is first converted into YCbCr format. YCbCr format consists of luminance (Y), blue chrominance (Cb), and red chrominance (Cr) information, with each of the three components being represented by a single unsigned byte. Since the human eye is more sensitive to the luminance components of an image, the chrominance components can be compressed further. This makes the YCbCr colorspace better suited for image compression than RGB.

Different variants of the JPEG algorithm can be used depending on the ratio of the chrominance to luminance information used in the compression. If all three components of the colorspace appear in the same ratio, then the mode is known as 4:4:4. If every other chrominance component is used in the horizontal dimension, the mode is known as 4:2:2. Sampling every other chrominance component in both the vertical and horizontal dimensions is referred to as 4:2:0 mode. In this work, we only utilize the baseline 4:4:4 mode.

The next step in the algorithm involves level shifting each of the component values such that they can be stored as signed bytes. The pixels are then bundled together into 8x8 blocks from top left to bottom right in scan order (row-major order). The blocks are processed independently for the following steps in the algorithm.

The subsequent step of the algorithm is a forward integer DCT transform. In this step, the 8x8 YCbCr spatial data is transformed into frequency data. Besides errors introduced through rounding, this step of the algorithm is non-lossy. Different algorithms can be used to carry out the forward DCT transform. In this work, we utilize the Chen Wang [Wan84] fast DCT algorithm.

The next step in the algorithm is quantization. Each component in each 8x8 block is divided by a user-supplied coefficient from a quantization table. Two separate tables are used, each with 64 coefficients: one for the luminance components and the other for the chrominance components. Quantization is the main information-losing step in the JPEG algorithm. Larger coefficients lead to lower image quality and higher compression. Standard quantization tables are provided with the JPEG standard, and they are used here.

After the division has taken place, the next step is to rearrange the component values within each 8x8 block from scan order into zig-zag order. This ordering tends to group the higher frequency components together, preferably leading to long sequences of zeros.

The first part of the Huffman encoding step is run-length compression which takes long strings of zeros and represents them in a concise intermediate form. The second stage is the actual Huffman table lookup, which translates the intermediate form into compact bit sequences. Like the quantization tables, the Huffman tables are statically specified by the user. Huffman encoding transforms the bytestream into a bitstream.

The final JPEG image file consists of header data along with the compressed bitstream. The header data includes the quantization and Huffman tables for both the chrominance and luminance components. The JPEG file interchange format (JFIF) is the standard for representing JPEG-encoded images.

8.2.2 The Intel MXP5800 Platform

The Intel MXP5800 digital media processor [int] is a heterogeneous, programmable processor optimized for document and image processing applications. It implements a data-driven, shared register architecture with a 16-bit data path and a core frequency of 266 MHz. The MXP5800 provides specialized hardware to accelerate frequently repeated image processing functions along with a large number of customized programmable processing elements.

The basic MXP5800 architecture, shown in Figure 8.2, consists of eight Image Signal Processors at the top level (ISP_1 to ISP_8) connected with programmable Quad Ports (8 per ISP). Quad Ports are used for data I/O and are each essentially FIFOs of size two. They provide blocking read and write semantics which ensure that all communication is data driven. Quad Ports are statically configured by the system developer during mapping according to the data flow topology of the application. In addition to Quad Port connections, all of the ISPs are connected to DMA units and some are connected to other expansion ports. Each ISP consists of five programmable Processing Elements (PEs), instruction/data memory, 16 16-bit General Purpose Registers (GPRs) for passing data between PEs, and up to two hardware accelerators for key image processing functions. Two of the PEs are used for Data I/O: The Input PE (IPE) which is used to read data from the Quad Ports, and the Output PE (OPE) for writing data to a Quad Port. Of the remaining 3 PEs per ISP, one is for general purpose use (GPE) while two PEs have Multiply/Accumulate (MACPE) capabilities in addition to the general purpose functionality.

Each general purpose register in an ISP has a set of 8 data valid (DV) flags - one per PE. If all the DV flags for a register are cleared, a PE may atomically write data to the register and set the DV flags for all of the destination PEs. Each of the destination PEs can clear its own flag when it reads the data. In this way, the global registers serve as single-place blocking-read, blocking-write buffers for possibly multiple writers and readers.

A Memory Control Handler (MCH) provides the interface to the SRAM data memory block within each ISP. The MCH has support for a number of different read/write modes which support variable offsets and stride lengths. Access to the MCH is provided using global registers, just as for the other PEs.

Each ISP is optimized for a particular function and the hardware accelerators in the ISP reflect that optimization. ISP_2, ISP_5 and ISP_6 each have variable-tap and single-tap triangular filters. ISP_4 and ISP_8 contain Huffman encode/decode engines that are useful for many compression/decompression applications. ISP_3 contains G4 encode/decode blocks. ISP_7 contains 8x8 DCT/iDCT hardware. Finally,

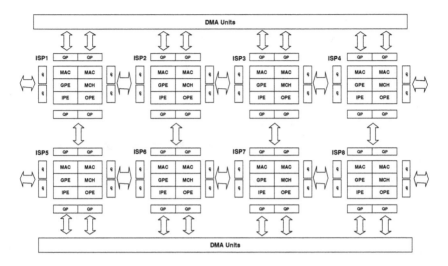

Figure 8.2: Block Diagram of MXP5800

ISP_1 has an additional 16 KB of data SRAM instead of a hardware accelerator.

The major characteristic of this architecture platform is the extremely high degree of parallelism and heterogeneity. Harnessing the diverse capabilities of the PEs to realize high application performance is the main design challenge.

8.2.3 Prior Work: Models of Computation

In this section, some of the models of computation that have been suggested for data-streaming systems will be described in further detail. The expressiveness and suitability for analysis of these MoCs will be evaluated. All of the MoCs are specializations of Process Networks, where concurrently executing processes communicate with each other using explicit messages.

Kahn Process Networks

Kahn Process Networks [Kah74] is a well-studied MoC where concurrent processes communicate with each other through one-way point-to-point FIFOs. Read actions from these FIFOs block until at least one data item (or token) becomes available. The FIFOs have unbounded size, so write actions are non-blocking. Reads from the FIFOs are destructive, which means that a token can only be read once.

More formally [LSV98], each channel in a KPN is a signal which carries a finite (possibly empty) or infinite sequence of tokens. The set of all possible signals is denoted S while the n-tuple of signals is denoted as S^n. The relation $\sqsubseteq$ is defined as the binary prefix relation on signals. For instance, $s_1 \sqsubseteq s_2$ means that the sequence of tokens contained in the signal s_1 is a prefix of the sequence of tokens contained in s_2. This definition generalizes to an element-wise prefix order, which can be defined on S^n. This element-wise prefix order is a CPO [DP02].

Any process P in the KPN with m inputs and n outputs is a mapping from its input signals to its output signals, $P : S^m \rightarrow S^n$. The semantics of KPN places a restriction on the type of mapping that a process can represent. A process must be monotonic in its mapping from input signals to output signals under the element-wise prefix order, $s_1 \sqsubseteq s_2 \Rightarrow P(s_1) \sqsubseteq P(s_2)$. This means that supplying additional inputs to a process results in additional outputs being produced, tokens which have already been produced cannot be retracted.

Under this restriction, and given that the processes are monotonic functions on a CPO, the least fixpoint theorem tells us that a least fixpoint exists for a network of these processes. According to the denotational semantics of KPNs, this least fixpoint represents the behavior of the KPN [Kah74]. This is the behavior that we want to find with simulation. However, the procedure for finding this least fixpoint is not given under monotonicity conditions alone. To find this procedure, we must apply a stronger condition on processes.

The stronger condition that is required is that of continuity, which requires that the result of this function to an infinite input is the limit of its results to the finite approximations of this input. Under this stronger condition, a procedure for finding the least fixpoint behavior exists. This procedure involves an initial condition of empty channels. Then, the processes are allowed to act on the empty channels until no further change takes place. This is the procedure that we need for finding the behavior of the KPN since it corresponds directly to simulation. To guarantee continuity, the sufficient (but not necessary) condition imposed on Kahn processes involves blocking read semantics [KM77]. Since continuous functions are compositional, is suffices to ensure that each process in a KPN is continuous to guarantee that the entire network has a deterministic behavior.

This is the appealing characteristic of the KPN model of computation – that execution is deterministic and independent of process interleaving. Also, this model of computation allows natural description of applications since it places relatively few requirements on the designer other than blocking reads.

Implementing KPN specifications on resource-constrained architectures has a key challenge:

that of realizing a theoretically infinite-sized communication channel with a finite amount of architectural memory. Indeed, a KPN implemented in this manner no longer satisfies the original definition of non-blocking writes, since a lack of storage space in the communication channel may force further write actions to be blocked. This additional constraint of blocking writes may possibly introduce deadlock into the execution of the system. This undesirable occurrence is referred to as *artificial* deadlock [GB03]. It is undecidable in general to determine if a KPN can execute in bounded memory, therefore deadlocks cannot be avoided statically.

The resolution of artificial deadlock requires dynamically supplying extra storage to some communication channel which is involved in the deadlock. This is the basis of Parks' algorithm [Par95]. However, choosing the channel and the amount of memory to allocate such that the deadlock is resolved with a minimum of extra memory is undecidable in general. A "bad" strategy will allocate memory to channels in such a way that the deadlock is not truly resolved, just postponed. In this case, the system will eventually run out of memory and the system will need to be reset.

Dataflow Process Networks

Dataflow process networks are a special case of Kahn Process Networks where the execution of processes can be divided into a series of atomic firings [LP95]. This MoC in general suffers from the same undecidability as Kahn Process Networks [Buc93]. However, certain variants are more suitable for analysis.

Homogeneous and static dataflow [LM87] are two such examples. In homogeneous dataflow, each process consumes and produces the same number of data tokens on each firing, for all channels. In static dataflow, the number of tokens produced and consumed must be constant for each firing on all channels, but can vary between channels. Cyclo-static dataflow [BELP95] is a slight generalization that permits the firing characteristics of each process to vary in a cycle. The main advantage of a cyclo-static dataflow model is reduced buffer size requirements. Heterochronous dataflow [GLL99] permits firing characteristics to vary according to the current state in a finite state machine. Boolean dataflow [Buc93] and integer dataflow [Buc94] allow special actors where the number of tokens produced and consumed are based on Boolean and integer-valued control tokens respectively.

8.2.4   Chosen model of computation

In order to apply the design flow, the first step is to choose a common modeling domain (CMD) as described in Chapter 5, Section 5.6. In this section, one component of that decision is described:

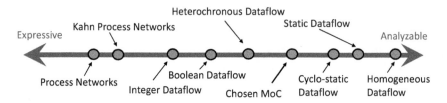

Figure 8.3: Classification of MoCs

choosing a model of computation. The granularity of the computation services that comprise the CMD is decided later, in Sections 8.2.5 and 8.3.4.

A rough classification of these MoCs along an axis of expressiveness vs. analyzability is shown in Figure 8.3. The approximate location of the MoC that we have chosen is also shown in the diagram. Like other dataflow models, processes in the chosen MoC consume and produce tokens according to firing rules. Multiple firing rules can be specified for each process. Each process cycles between its firing rules in a fixed pattern. Therefore, this MoC is most similar to cyclo-static dataflow, but contains two extensions that allow for more concise specification.

First, only one writer is permitted per channel, but multiple reader processes are allowed. For all channels, each reader process can read each data token exactly once. Tokens are removed from the FIFO only after all reader processes have read them once. Note that this extension allows for more succinct specification, but does not change the expressiveness of the model.

Second, we allow limited forms of data-dependent communication. If a data-dependent number of tokens is to be exchanged on a channel, the sender is required to first indicate how many such tokens will be sent in a "header" token. In this way, the property of effectiveness [GB03] is guaranteed.

To enable support for executing multiple processes on a single processing element, this MoC has support for cooperative multitasking. Specifically, a process may only be suspended between firings. Scheduling, buffer sizing, and mapping are decidable problems for this MoC. Processes may be scheduled statically, allowing for lower overhead implementation.

An example of a process within this MoC is shown in Figure 8.4. The process reads from an input channel "input1" and writes to an output channel "output1". A variable number of tokens are first read from the input channel. Computation is then carried out on the input data. Finally, all of the processed tokens are then written to the output channel. This sequence of actions occurs within an infinite loop.

```
void main()
{
  int n;
  int data[5];

  while(1)
  {
    read(input1, n);
    assert(n >= 0 && n < 5);

    for(int i = 0; i < n; i++)
      read(input1, data[i]);

    // computation

    write(output1, n);

    for(int i = 0; i < n; i++)
      write(output1, data[i]);
  }
}
```

Figure 8.4: An example process within the chosen MoC

Like many specialized dataflow models, our dataflow model induces stronger constraints on the application as opposed to the architecture. In fact, a variety of dataflow models can be supported by multiprocessor architectures that allow efficient blocking read and blocking write operations.

8.2.5 Manual Design Space Exploration: JPEG on MXP5800

Having chosen the appropriate MoC for the applications and architectures, they can be modeled at different abstraction levels and then mapped. In this case study, we deploy the JPEG encoder application onto the Intel MXP5800 architecture and demonstrate that the chosen MoC is able to accurately represent the system at a specific abstraction level. After the MoC is shown to be suitable for these types of systems we can consider automating the mapping in a subsequent case study in Section 8.3.4.

JPEG Application Modeling

Starting from both a sequential C++ implementation [ijg] and the concurrent assembly language implementation provided in the Intel MXP5800 development kit, we assembled an architecture-

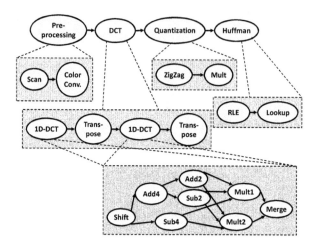

Figure 8.5: METROPOLIS model of JPEG encoder

Implementation	Language	Concurrency	Lines of Code
IJG	C++	Sequential	18,000
MXP5800 Library	ASM	Concurrent	915
METROPOLIS	MMM	Concurrent	2,695

Table 8.1: JPEG encoder models

independent model of the JPEG encoder in METROPOLIS expressed in our dataflow model. The model carries out a full implementation of the 4:4:4 JPEG encoder baseline algorithm and is described hierarchically at multiple abstraction levels. A total of 20 FIFO channels and 18 separate processes are used in the finest granularity representation of the application model. An overview of the METROPOLIS model is given in Figure 8.5. Characteristics of the original C++, assembly, and Metamodel designs are provided in Table 8.1. Note that the IJG model implements a superset of the functionality implemented by the METROPOLIS and ASM models.

To describe the application model in further detail, we will concentrate on the breakdown of the discrete cosine transform step in the algorithm. At the top level, the required two-dimensional DCT can be broken down into four basic steps: two one-dimensional DCT operations, each followed by a transpose operation. This is shown in Figure 8.6. Each of the 1D-DCT block reads in 8 spatial data

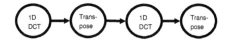

Figure 8.6: 2D-DCT block diagram

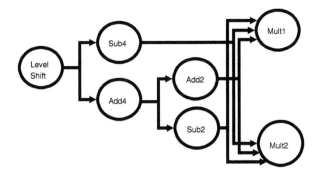

Figure 8.7: Dataflow model for 1D-DCT

values (corresponding to either a row or a column) and outputs 8 frequency values. The algorithm used to carry out the 1D-DCT is based on the implementation given by Chen-Wang [Wan84].

Each one dimensional DCT operation is broken down into concurrent steps, as shown in Figure 8.7. The diagram shows that all the channels require multiple readers. This observation, along with the fact that the MXP5800 architecture supports multiple readers through data valid bits, is the reason why our MoC explicitly supports this type of communication, as described in Section 8.2.4.

Architecture Modeling

The MXP5800 architecture platform can be modeled in METROPOLIS by using processes, media, and quantity managers from the METROPOLIS Metamodel. A single ISP is modeled as shown in Figure 8.8. The rectangles in the diagram represent tasks, the ovals represent media, while the diamonds are the quantity managers. The ISP contains the Huffman hardware accelerator, and is sufficient for implementing the JPEG encoder application. If we want to use the DCT hardware accelerator, another ISP is needed. In this case, the two ISPs will be connected through Quad Ports. Modeling various ISPs with different hardware accelerators is very similar, we use the diagram in Figure 8.8 as an example.

Each PE is modeled as a medium, which supports multiple tasks running on it. Each task is

256

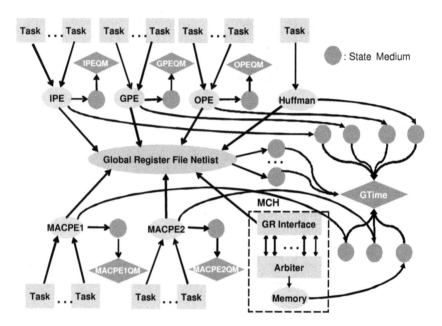

Figure 8.8: MXP5800 ISP Modeling in METROPOLIS

modeled as a process. These processes represent the possible functionality executed on the PE. After mapping, the behavior of each task will be restricted by the corresponding process from the application; there is a one-to-one correspondence between architecture processes and application processes.

The scheduling of multiple tasks on a single PE is carried out by the quantity manager connected to this PE. The quantity managers support static scheduling, which is required by the MoC.

The communication between programming elements occurs through either global registers or local SRAM. The global register file is modeled as a netlist that contains 16 global registers. Each global register is modeled as a medium which implements blocking write and allows multiple simultaneous reads. These global registers can be accessed by all PEs, hardware accelerators, and the MCH. To reduce the modeling complexity, a single interface medium in the netlist represents the entire global register file and is used to communicate with the PEs, accelerators, and MCH.

The SRAM is controlled by a memory command handler (MCH). The MCH contains a global register interface (GR interface), arbiter and memory. The GR interface is used to communicate with the global register file and is modeled as a process that waits for the appropriate data valid bits to be set in the global register file. The arbiter obtains memory access requests from the GR interface through multiple FIFO channels, then uses a round-robin access scheme to select one of them to access the local memory, which is modeled as a medium.

To model running time, a global time quantity manager is used. All PEs and global registers and the local memory are connected to it. Both computation and communication costs can be modeled by sending requests to this global time quantity manager and obtaining time annotations.

Design Space Exploration and Results

Given the application and architectural models in METROPOLIS, the design space can be explored by attempting different mappings between the application model and the architectural model. Each mapping scenario is specified in METROPOLIS with two types of information. The first is a specific set of synchronization constraints between the events in both models corresponding to the services that constitute the MoC. Along with these events – which represent the read, write, and execution services defined in our MoC – the parameters such as register location or local memory address can also be configured. The second is the set of schedules for the PEs that determine the execution order between the tasks. Both of these are relatively compact, meaning that new mapping scenarios can be created quickly and without modifying either the application or the architectural models.

The application is a total of 2,695 lines as shown previously in Table 8.1. The architectural

Process	Hardware	Balanced	OPE Emph.	OPE Heavy
Level Shift	IPE	IPE	IPE	IPE
Add4-R		OPE	OPE	OPE
Sub4-R		OPE	OPE	OPE
Add2-R		IPE	OPE	OPE
Sub2-R		IPE	OPE	OPE
Mult1-R		MACPE1	MACPE1	MACPE1
Mult2-R	DCT HW	MACPE2	MACPE2	MACPE2
Add4-C	Accelerator	IPE	IPE	OPE
Sub4-C		OPE	OPE	OPE
Add2-C		IPE	IPE	OPE
Sub2-C		OPE	OPE	OPE
Mult1-C		MACPE1	MACPE1	MACPE1
Mult2-C		MACPE2	MACPE2	MACPE2

Table 8.2: Mapping assignments

model is 2,804 lines, while the mapping code is 633 lines. Each additional mapping scenario can be described with approximately 100 lines of additional code, and without modifying any of the other code.

To show the fidelity of our modeling methodology and mapping framework, we initially abstracted two mapping scenarios from the implementations provided in Intel MXP5800 algorithm library and carried out simulation in the METROPOLIS environment. We also tried an additional two scenarios which did not have corresponding assembly language implementations. For all of the scenarios, only the mapping of the fine granularity 1D-DCT processes was varied.

The first scenario makes use of the DCT hardware accelerator and clearly has the highest performance. The other three scenarios use various software implementations of the row-wise and column-wise 1D-DCT operations. For these three scenarios, the transpose process is mapped to the MCH, which natively supports this type of operation. Register mappings are taken from the Intel library implementations and consist of 1, 2, or 4 global registers per FIFO channel. The second scenario uses a balanced partitioning of the processes among the available PEs, while the third and fourth scenarios put progressively more load on the OPE. The details for all four scenarios are provided in Table 8.2. The rows indicate the 13 processes in the refined two-dimensional DCT model. The columns indicate the mappings for the four different scenarios.

Cycles for different scenarios

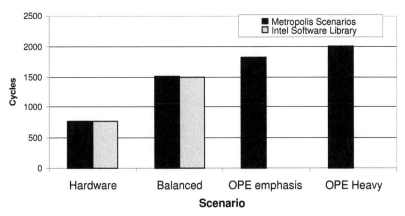

Figure 8.9: Performance Comparisons

For each scenario, the number of clock cycles required to encode an 8x8 sub-block of a test image was recorded through simulation in METROPOLIS. For the first two scenarios, implementations from the MXP5800 library are available and were compared by running the code on a development board. The results are shown in Figure 8.9. The cycle counts reported with the METROPOLIS simulation are approximately 1% higher than the actual implementation since we did not support the entire instruction set for the processing elements. The latter two scenarios provide reasonable relative performance, but assembly implementations were not available for comparison.

As long as the granularity of the each dataflow process is small (such as for most DSP-like systems), we expect that this model will provide very accurate estimates of performance. Regardless of the computational granularity, the schedules and deadlock analysis capabilities of this MoC will still remain valid.

8.2.6 Conclusions

The main steps of the design flow exercised in this case study are choosing the model of computation, modeling the application and architecture in a common framework using this MoC, and evaluating

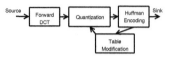

Figure 8.10: Motion JPEG Encoder

different mappings.

If the MoC chosen captures the important characteristics of the system, as shown, then accurate performance estimates can be obtained at a fraction of the cost and much faster than with other verification methods and tools. The main tradeoff when choosing a MoC, as described in Section 8.2.4, is between expressiveness and analysis capabilities. After determining that a particular MoC captures the main characteristics of a particular class of systems, automated design space exploration techniques can be developed.

8.3 Motion-JPEG on Xilinx Platforms

The application and architecture for this case will be described in Sections 8.3.1 and 8.3.2, while Sections 8.3.3 and 8.3.4 will look at two goals: accurate characterization and automated design space exploration.

8.3.1 Motion JPEG Application

The Motion JPEG encoder application [Wal91] carries out video encoding without inter-frame compression. Motion JPEG encoding is commonly implemented in consumer and security cameras as well as high-resolution video editing.

The Motion JPEG encoder application is quite similar to the JPEG encoder, except that quantization and/or Huffman tables are changed adaptively between frames. The table modifications are carried out based on discrepancies between the actual and desired compression rates for the encoded video stream. Unlike the JPEG encoder, there is no standard file format for a Motion JPEG encoded stream. A block diagram of this application is shown in Figure 8.10.

In this work, only the quantization tables are modified between frames, based on a linear scaling of the quantization coefficients provided in the standard tables. The modification is based on the size of the previously compressed frame as compared to the desired size.

8.3.2 Xilinx Virtex II Pro Platform

The architectural platform used in this work is the Xilinx [Cora] Virtex II FPGA. It is a platform FPGA that may be customized at the lookup table (LUT) level to implement a variety of functionality. For the purposes of this work, we will only utilize a limited subset of the flexibility offered by the platform. Specifically, hard and soft processor cores, bus and FIFO interfaces between cores, and one type of customized IP block is utilized in this work. This case study uses a 2VP30 part on Xilinx XUP board with a maximum frequency of 100 MHz.

The MicroBlaze [Corb] is 32-bit RISC Harvard processor which can be instantiated on the FPGA fabric. Caches and hardware support for operations such as shifting, multiplication, and division can be enabled or disabled based on designer choice. The uBlaze can interface with three types of interconnect. The Local Memory Bus (LMB) is used for fast access to data and instruction memory. The Fast Simplex Link (FSL) is used to interface with other MicroBlaze processors or hardware accelerators instantiated on the fabric. Finally, the On-chip Peripheral Bus (OPB) is used to interface with other peripherals, such as timers, network controllers, and UARTs. METROPOLIS models of these components were described in Chapter 5, Section 5.7.2.

FSL [Cor05] FIFO links are used in this work to carry out data transfers between multiple MicroBlaze cores, and between MicroBlazes and synthesized hardware accelerators. FSL depth can range from 1 to 8,192 entries, each of which may be 4 bytes in width. Reads and writes to the FSL FIFOs from the MicroBlaze take only a single cycle. Both blocking and non-blocking read/write access to FSLs is provided.

Hardware accelerators can be directly synthesized onto the fabric. In this work, we make use of a DCT-specific processing element with FSL interfaces created using the XPilot [CCF$^+$05] [CFH$^+$06] synthesis system.

8.3.3 Characterization Aided Fidelity Example

To demonstrate how the programmable platform performance characterization method described in Chapter 7 can be used to make correct decisions during design space exploration, the following multimedia example is provided. This example deals with evaluating various architecture topologies and illustrates the importance of accuracy in characterization and exemplifies the fidelity achieved with the proposed design flow's method. In an exploration like this one, the designer is interested in choosing the design with the best performance. Therefore it is not as important that the exact performance be known, but rather that the ordering of the performances amongst the candidates is correct (hence the emphasis on fidelity). Without the methods covered in this work, estimated values would be used to inform the designer of the predicted performance. These values may come from datasheets, previous simulations, or even best guesses (techniques described in Chapter 5, Section 5.7.3). None of these are preferable to actual characterization as will be shown.

We've investigated four MJPEG architectural service models here. A single functional model was created in METROPOLIS which isolated various levels of task concurrency between the DCT, Quantization, and Huffman processes present in the application. These processes from the functional model were then mapped to the architectural model. The topologies are shown in Figure 8.11. Each of the topologies represents a different level of concurrency and task grouping. A key is provided to show what functionality is mapped to which aspect of the architectural model. The diagrams show the architecture topologies after the mapping process. This was a one-to-one mapping where each computational unit was assigned a particular aspect of MJPEG functionality. The computation elements were MicroBlaze soft processor models realized by METROPOLIS media and the communication links were Fast Simplex Link (FSL) queues also realized as METROPOLIS media. In order to facilitate the mapping, METROPOLIS mapping processes were provided one-to-one with the MicroBlaze service models. In addition to the METROPOLIS simulation, actual Xilinx Virtex II Pro systems running on the Xilinx ML310 development platforms were created. The goal was to compare how closely the simulations reflected the actual implementations and to demonstrate that the simulations were only truly useful when using our characterization approach.

The results of a 32x32 pixel image MJPEG encoding simulation are shown in Table 8.3. The table contains the results of METROPOLIS simulation and the results of the actual implementation. The first column denotes which architectural model was examined. This column corresponds to Figure 8.11. The second column shows the results of simulation in which estimations based on design area and assembly code execution were used. The third column shows the simulation results using the characterization method described previously. Provided with the results is the percent deviation from the real cycle values. Notice that the estimated results have an average difference of 35.5% with a max of 52% while the characterized results have an average difference of 8.3%. This is a significant indication of the importance of the proposed method. In addition, the fifth column shows the rank ordering for the real, characterized, and estimated cycle results respectively. Ideally all three values would be the same. Draw your attention to the rankings for Arch 2 and Arch 3. Notice that the estimated ranking does not match that of the real ordering. Even though the accuracy discrepancy is significant, it is equally (if not more) significant that the overall fidelity of the estimated systems is different. Finally the maximum frequency according to the synthesis reports, the execution time (cycles * period), and area (slice) values of the implementation are

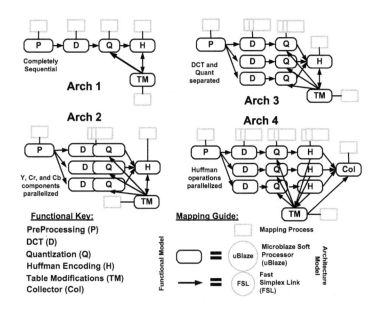

Figure 8.11: MJPEG Architecture Topologies in METROPOLIS

shown. It is important to notice several trends may not have been taken into account using an estimated method. One is that the largest area design (Arch 4) requires the fewest cycles. However, it also has the lowest clock frequency. This confirms that while one might be tempted to evaluate only the cycle counts, it is important to understand the physical constraints of the system only available with characterized information.

System	Est. Cycles	Char. Cycles	Real Cycles	Ranking (Real, Char, Est)	Max Mhz	Execution Time (Secs)	Area (Slices)
Arch 1	145282 (52%)	228356 (25%)	304585	4, 4, 4	101.5	0.0030	4306
Arch 2	103812 (33%)	145659 (6%)	154217	3, 3, 2	72.3	0.0021	4927
Arch 3	103935 (29%)	145414 (1.2%)	147036	2, 2, 3	56.7	0.0026	7035
Arch 4	103320 (28%)	144432 (< +1%)	143335	1, 1, 1	46.3	0.0031	9278

Table 8.3: MJPEG Encoding Simulation Performance Analysis

When discussing the efficiency of this method in terms of simulation time, the two points of

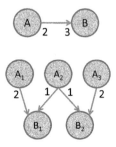

Figure 8.12: Transforming an SDF graph into a data precedence DAG

interest are the simulation times for the simulations using estimated data versus those using characterized data. In this case, due to the unique METROPOLIS execution semantics described earlier, the simulation times are the same for each method. The increased fidelity therefore comes at no extra "cost" to the design.

8.3.4   Automated Design Space Exploration: Motion-JPEG on Xilinx

The MoC chosen for this class of systems in Section 8.2.4 is amenable to analysis. In this section, the analysis capabilities are leveraged to develop an automated approach for allocation and scheduling. The Motion JPEG application and the Xilinx Virtex II FPGA platform are used in this work.

Mathematical programming techniques such as Mixed Integer Linear Programming (MILP) provide the ability to easily customize the allocation and scheduling problem to application or platform-specific peculiarities. The representation of the core problem in a MILP form has a large impact on the solution time required. In this work, we investigate a variety of such representations and propose a taxonomy for them. A promising representation is chosen with extensive computational characterization. We demonstrate that our approach can produce solutions that are competitive with manual designs.

Problem Statement

Statically schedulable dataflow descriptions can be automatically transformed into acyclic data precedence graphs. These graphs are then used in the automated mapping procedure for multiprocessors [PBL99]. An example of such a transformation for SDF systems is given in Figure 8.12. The graph on the top is a SDF graph where process A writes 2 tokens to the channel each time it is fired while process B reads 3 tokens from the channel each time it is fired. The associated data precedence DAG is shown in the lower part of the figure and captures the relationships between the process firings that are required to return the system to its initial state of having no tokens within the channel. The simplest such DAG requires process A to fire three times while B fires twice.

The core problem to be solved is to map the weighted directed acyclic graph (DAG) representing the application onto a set of architectural nodes. In the application DAG, nodes represent process firings while edges represent data precedence relationships. The architecture is represented with a weighted directed graph where nodes are processing elements (PEs). Edges that represent communication channels may be added explicitly to the architecture graph if connectivity between PEs is restricted, otherwise, a fully-connected graph is assumed. The execution time for each task on each processor is fixed and given. The amount of communication between each application task is given by weights on the application graph edges. Communication cost only depends on the edge weight, not on the allocation of the source and sink processes. The objective is to allocate and schedule the tasks onto the PEs such that the completion time – or makespan – is minimized.

This work has four main contributions. First, a classification of existing MILP representations for this problem into a taxonomy. Second, based on the taxonomy, a core MILP formulation and useful customizations. Third, computational characterization and a comparison of our approach with a competing approach. Finally, a representative case study that illustrates the ability or our approach to produce competitive designs in terms of both throughput and area.

Prior Work: Allocation and Scheduling

The scheduling problem we consider is a *generalization* of $R|prec|C_{max}$ [GLLK79] and is strongly NP-hard. R refers to the usage of multiple heterogeneous PEs with unrelated processing times, *prec* indicates that the application description includes precedence constraints, and C_{max} indicates the objective is to minimize the makespan or the maximum completion time. Unrelated processing times means that different processors may execute different tasks with arbitrary worst-case execution times.

For the special case of $R||C_{max}$ (no precedence constraints), there exist polynomial-time approximation algorithms that can guarantee a solution within a factor of 2 of the optimal [ST93]. No poly-time approximation algorithm exists that can provide a solution for $R||C_{max}$ within 1.5 times the optimal, unless $P = NP$ [LST87]. If precedence constraints are added, there are no known good approximation results; an overview of related work for $R|prec|C_{max}$ is provided in [KMPS05]. A comprehensive listing of known lower and upper approximation bounds for a variety of scheduling and allocation problems can be found in [CKH$^+$00], while an overview of heuristics is given in [KA99].

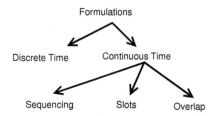

Figure 8.13: Taxonomy of MILP Approaches

In [JSRK05], a MILP-based approach for mapping onto a multiprocessor FPGA platform is described. However, their approach is more specialized since it does not handle task precedence constraints nor determine scheduling.

MILP Taxonomy

Solution time for MILP instances is strongly affected by the representation used for the core allocation and scheduling problem. We observe that the effective encoding of task precedence relationships is key not only for approximation algorithms as mentioned in Section 8.3.4, but also for MILP representations. Along these lines, we propose a taxonomy of known MILP representations in Figure 8.13.

Discrete time approaches introduce a variable for each instant of time on each PE. The resultant scheduling constraint requires that each such time instant be allocated to at most one task. The advantage of this method is that the formulation can easily be constrained to use only integer or binary variables. A rich variety of SAT solvers [ES03] can be utilized to solve these problems. However, this formulation has a significant drawback: the number of time variables introduced can quickly become very large, especially if diverse task execution times are present.

Continuous time approaches represent time with real-valued variables in the formulation. Vastly different execution times can easily be handled by these approaches, but the choice of variables and constraints used to specify a correct scheduling on each PE becomes critical in determining performance [GG02].

Sequencing: Variables are used to indicate sequencing to schedule tasks on PEs. These sequencing variables indicate whether a task is executed after another task on the same PE [Ben96, CRS03]. This choice of variables can be viewed as a straightforward extension of the well-known formulations used in uniprocessor scheduling [NW88]. Typically, a large number of constraints or variables is required to enforce the scheduling requirements on each PE. Many of these constraints can be attributed to the linearization of bilinear terms [Lib05].

Slots: This method uses explicit slots on each PE [MPC$^+$97, DLMM04] to which at most one task can be allocated. The start and finish time for each slot is not fixed a priori. With slots, the scheduling constraints between tasks on each PE become simpler to represent. However, since the exact number of slots on each PE is unknown, a conservative amount need to be used. As a result, this approach may suffer from variable blow-up if the typical number of tasks allocated to each PE is large.

Overlap: Variables are used to indicate temporal overlap (independent of PE assignment) in the execution of tasks [PPer, SBC00, Tom03]. Constraints that prevent overlap on the tasks allocated to each PE are used to enforce scheduling. Since the scheduling constraints can be expressed succinctly, this type of formulation scales well with respect to variables and constraints than the formulations in the other categories.

In this paper, we focus on continuous-time MILP formulations that use overlap variables, since this category seems the most promising for generating problems with fewer constraints and variables.

Core Formulation

The core formulation is based closely on the formulation presented in [Tom03].

Let $\mathbf{F}$ represent the set of tasks in the application DAG while $\mathbf{E} \subset \mathbf{F} \times \mathbf{F}$ represents the set of communication edges. The set $\mathbf{A}$ indicates the set of architectural PEs. The parameter $t \in \mathbb{R}^{\mathbf{F} \times \mathbf{A}}$ specifies the execution time of each task on each PE.

The variable $d \in \mathbb{B}^{\mathbf{F} \times \mathbf{A}}$ indicates if a task is mapped to a PE. Variables $s \in \mathbb{R}^{\mathbf{F}}$ and $f \in \mathbb{R}^{\mathbf{F}}$ indicates the start and finish times respectively for each task. $o \in \mathbb{B}^{\mathbf{F} \times \mathbf{F}}$ is a variable which is used to determine overlap in execution times between a pair of tasks.

$$min \qquad \max_{i \in \mathbf{F}} f_i \tag{8.1}$$

$$s.t. \qquad \sum_{x \in \mathbf{A}} d_{ix} = 1 \qquad \forall i \in \mathbf{F} \tag{8.2}$$

$$f_i \leq s_j \qquad \forall (i,j) \in \mathbf{E} \tag{8.3}$$

$$f_i = s_i + \sum_{x \in \mathbf{A}} (t_{ix} d_{ix}) \qquad \forall i \in \mathbf{F} \tag{8.4}$$

$$f_j - s_i \leq M o_{ij} \qquad \forall i,j \in \mathbf{F}, i \neq j \tag{8.5}$$

$$o_{ij} + o_{ji} + d_{ix} + d_{jx} \leq 3 \qquad \forall i,j \in \mathbf{F}, x \in \mathbf{A}, i \neq j \tag{8.6}$$

The objective function in 8.1 minimizes the maximum finish time over all tasks. This has the

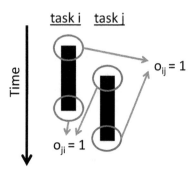

Figure 8.14: Overlap between two tasks i and j

effect of minimizing the makespan. Constraint 8.2 ensures that each task is mapped onto exactly one PE. Constraint 8.3 requires that the precedence relationships between edges in the application DAG hold. Constraint 8.4 relates the start times and finish times of each task based on the execution time of the task on the appropriate PE. Constraint 8.5 ensures that if a task j finishes after task i begins, the corresponding variable o_{ij} is set to 1. In this constraint, M represents a large constant, which can be no less than the maximum finish time. Finally, Constraint 8.6 is a particularly elegant means of ensuring that no two tasks mapped onto the same PE may overlap in time.

In this constraint, the sum $o_{ij} + o_{ji}$ has a value of 2 iff the executions of the two tasks i and j overlap. An example is shown in Figure 8.14 where o_{ij} is set to 1 since task j finishes after task i begins. o_{ji} is set to 1 since task i finishes after task j begins. Both of these ensure that tasks i and j overlap. Therefore, they may not be allocated to the same PE.

Note that Constraint 8.6 only needs to be defined over $i, j \in \mathbf{F}$ such that neither i nor j are in each other's transitive fan-out (TFO). For all other cases, the sum $o_{ij} + o_{ji}$ must be 1.

Customizing the Formulation

The core formulation does not support communication cost, restricted architectural topologies, partially specified task allocation on the platform, and real-time requirements on portions of the application. Of these, the first three are crucial for the case study we target in Section 8.3.4.

The additional set $\mathbf{C} \subseteq \mathbf{A} \times \mathbf{A}$ represents the directed edges between PEs. The parameter $c \in \mathbb{R}^{\mathbf{E}}$ denotes the communication cost for each edge in the application DAG. The parameter $n \in \mathbb{B}^{\mathbf{F} \times \mathbf{F}}$ indicates whether two tasks are required to be mapped onto the same PE. Parameter $e \in \mathbb{B}^{\mathbf{F} \times \mathbf{A}}$ indicates if a particular task must be mapped onto a particular PE. The variables $r \in \mathbb{R}^{\mathbf{F}}$ and $w \in \mathbb{R}^{\mathbf{F}}$ represent the reading and writing time required for each task.

$$f_i = s_i + \sum_{x \in \mathbf{A}}(t_{ix}d_{ix}) + r_i + w_i \quad \forall i \in \mathbf{F} \tag{8.7}$$

$$r_i \geq \sum_{(j,i) \in \mathbf{E}} c_{ji}(d_{ix} - d_{jx}) \quad \forall i \in \mathbf{F}, x \in \mathbf{A} \tag{8.8}$$

$$w_i \geq \sum_{(i,j) \in \mathbf{E}} c_{ij}(d_{ix} - d_{jx}) \quad \forall i \in \mathbf{F}, x \in \mathbf{A} \tag{8.9}$$

$$d_{iy} + d_{jz} \leq 1 \quad \forall (i,j) \in \mathbf{E}, (y,z) \notin \mathbf{C} \tag{8.10}$$

$$d_{ix} \geq e_{ix} \quad \forall i \in \mathbf{F}, x \in \mathbf{A} \tag{8.11}$$

$$n_{ij} - 1 \leq d_{ix} - d_{jx} \leq 1 - n_{ij} \quad \forall i, j \in \mathbf{F}, x \in \mathbf{A} \tag{8.12}$$

Constraint 8.7 replaces Constraint 8.4 from the core formulation and considers the reading and writing time for each task. Constraint 8.8 charges time for reading iff the predecessor task is assigned to a different PE. Likewise, Constraint 8.9 charges the corresponding write time. Constraint 8.10 ensures that the mapping conforms to the restricted architectural topology. Constraint 8.11 is a forcing constraint that allows some allocations to be fixed. Finally, Constraint 8.12 restricts certain pairs of tasks to be allocated to the same PE. This is useful when considering applications derived from dataflow specifications, where multiple invokations or firings of the same actor may be constrained to occur on a single PE.

Characterizing the Formulation

In this section, we compare our formulation against the sequence-based formulation and identify characteristics of problem instances that affect the runtime of the MILP formulation.

The experimental setup involves coding the sequence-based formulation from [CRS03] and our core formulation in AMPL [FGK93] and evaluating them with a set of 45 test cases. The test cases were generated with the TGFF [DRW98] tool with three random seeds. Five problem sizes, ranging from 10 to 50 tasks, were generated from each seed. Each task graph was allocated to different numbers of PEs to keep the average task/PE ratio the same. The test cases were solved using CPLEX 9.1.2 on 2.8GHz Linux machines with 2GB of memory under a time limit of 1000 seconds.

Comparison: Sequencing vs. Overlap

For the 45 test cases, on average, our overlap-based formulation has 30% more variables than the sequence-based formulation. However, our formulation also has 63% fewer constraints, which substantially reduces overall problem size.

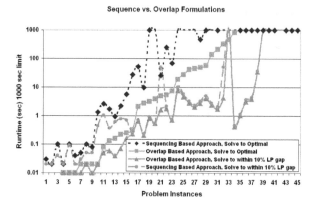

Figure 8.15: Sequence vs. Overlap Runtime

For solving problems to optimality in a balanced branch-and-bound exploration of the solution space, our approach is an *order of magnitude* faster than the sequence-based approach, as shown in Figure 8.15. LP relaxations of the problems are usually quite tight, often within 10-15% of the optimal value. This means that good lower bounds can be obtained in polynomial time for these problem instances. For instances that could not be solved to optimality within the time limit, feasible solutions within 14% of optimal were obtained on average.

If a solution within 10% of the optimal is sufficient, we can bias the branch-and-bound exploration to find feasible solutions. These results are also plotted in Figure 8.15 and show that solution time can be decreased by 1-2 orders of magnitude with biasing and a 10% optimality gap. For very few cases, feasibility biasing may increase solution time.

Factors Influencing Solution Time

Solution time is typically analyzed with respect to the number of tasks, the number of constraints or the number of PEs for a given problem instance. None of these factors is a good indicator of solution time for this formulation. For a problem with same number of tasks, as the number of PEs available decreases, we discover a counterintuitive trend: the number of constraints (and variables) drops, but the runtime increases.

The rising solution time for test cases with fewer PEs and constraints can be explained with three observations. First, when there are relatively fewer PEs, more unrelated tasks (tasks not in each other's TFOs) have to be sequentialized onto each PE. A formulation that relies on binary variables and big M constants to enforce non-overlapping of tasks (Constraint 8.5) has a weaker LP lower bound with more tasks/PE. Secondly, when many unrelated tasks have similar processing times, many feasible solutions have similar makespans, this prevents effective pruning of the branch and bound tree based on known feasible solution upper bounds. Thirdly, the number of feasible permutations of task ordering explodes with more tasks/PE. If we have k unrelated tasks allocated on the same processor, many of the $k!$ permutations must be considered in the branch and bound tree. The inverse application graph has edges between unrelated nodes (those without precedence constraints). The total number of permutations increases as a function of the maximum clique (fully connected component) of the inverse application graph. Making more PEs available disperses unrelated tasks - fragmenting the cliques and improving the LP lower bound.

Case Study

We now turn our attention to demonstrating the applicability of our customized MILP formulation on a case study. The chosen application is the Motion JPEG encoder described in Section 8.3.1. The architectural platform we consider contains soft-core processors and processing elements on a Xilinx Virtex II Pro FPGA fabric as described in Section 8.3.2. For various manual and automated mappings, we compare the performance in terms of system throughput and area utilization. For applications derived from dataflow specifications, the makespan of the data precedence DAG of an unrolled dataflow graph is equivalent to throughput [PBL99].

Manual Design Space Exploration   The goals in manual design space exploration are to utilize various numbers of uBlaze processors and DCT-specific PEs to maximize the throughput of the Motion-JPEG application. A nominal frame size of 96x72 is assumed for all implementations. We start from a baseline topology where the entire application is mapped onto a single uBlaze processor. As additional PEs are utilized, portions of the application are migrated to these PEs to improve throughput.

The PEs in the various designs are connected with FSL queues that are accessed in blocking-read, blocking-write mode. Data is fed to and retrieved from the device with a 100 Mbps Ethernet connection to a host PC. An Ethernet MAC device is instantiated on the fabric to handle this communication. One of the uBlaze PEs in each design is designated as the I/O processor and connects to the Ethernet device. In addition, this uBlaze is connected to peripherals to allow for debugging and performance mea-

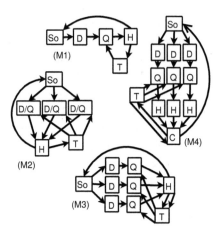

Figure 8.16: Topologies of Manual Designs

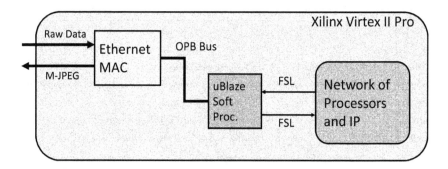

Figure 8.17: Experimental Setup

Design	uBlaze	DCT	fps	Area
Base	1	0	26.5	21%
M1	5	0	51.1	39%
M1D	4	1	72.0	53%
M2	6	0	85.1	47%
M3	9	0	85.3	62%
M3D	6	3	85.6	94%
M4	12	0	148.8	83%

Table 8.4: Manual Designs

surement. In all designs except the baseline topology, a single uBlaze is reserved for quantization table updates. The experimental setup is shown in Figure 8.17.

The different manual designs obtained are shown in Figure 8.16. The blocks used include the data source (So), DCT (D), quantization (Q), Huffman encoder (H), and table update (T). A combiner (C) is necessary when the Huffman block is split into 3 parts. Salient characteristics of each implementation – the number of uBlaze and DCT PEs used, the frames processed per second, and the area (% slices occupied on the FPGA) – are summarized in Table 8.4. Designs $M1D$ and $M3D$ are obtained from $M1$ and $M3$ respectively by substituting uBlazes with DCT PEs where possible.

The manual designs exploit the task-level and data-level parallelism in the application. The designs first attempt to use task-level parallelism between the different stages and the exploit the natural data-level parallelism between the three components in the color space.

Automated Design Space Exploration Automated Design Space Exploration uses the MILP formulation developed in Section 8.3.4 to determine the scheduling and allocation for tasks in the case study. The aim is to show that the cost model in the formulation accurately captures the design space and can be used to implement competitive designs.

The first step is to create a representation for the application which identifies the maximal amount of available concurrency. Since we would like to compare against the manual implementations, we create a task representation which extracts no more concurrency than is utilized in the manual designs. This corresponds to design (M4) from Figure 8.16. We also reserve a separate uBlaze for the table update portion of the application, just as in the manual designs. Note that both of these restriction can be relaxed to obtain higher quality automated designs.

Task	Cycles
Source	200
DCT	4,760
Quant	2,572
Huffman	3,442
Combiner	2,542
DCT Acc.	328

Table 8.5: Profiling Information

The next step is to characterize the application so that the task execution times (the t parameters in the formulation) can be obtained. For both the uBlaze processor and the DCT accelerator, the cycle times are obtained from the timer peripheral. The cycle times for the tasks are shown in Table 8.5.

If these parameters are used in our MILP formulation, any legal solution produces a static estimate for throughput. The accuracy of this estimate is important in determining the effectiveness of an automated approach. We compare the estimates for the base design and the 6 manual designs from Section 8.2.5 against the actual implementation results obtained from the development board.

The makespan estimated from the formulation is, on average, within 5% of the execution time measured on the development board. Most of the predictions overestimate the makespan, since the formulation does not consider simultaneity between reads and writes on each FIFO.

With this accurate model of the design space, we can automatically evaluate a number of different solutions from the MILP formulation. Based on the characteristics of the manual designs, we picked three promising MILP solutions and implemented them on the development board. These three automated solutions (A1, A2, and A3) use 3, 5, and 8 uBlazes and were obtained with a 100s time limit on the solver. The MILP solution also confirms that design M4 is optimal given the chosen granularity of the task graph.

Both the manual and automated solutions can be plotted in terms of frame rate vs. FPGA slices consumed - which is roughly proportional to the number of uBlazes and DCT PEs used. Figure 8.18 shows this tradeoff and indicates that the automated designs do indeed result in competitive implementations that lie on the Pareto curve.

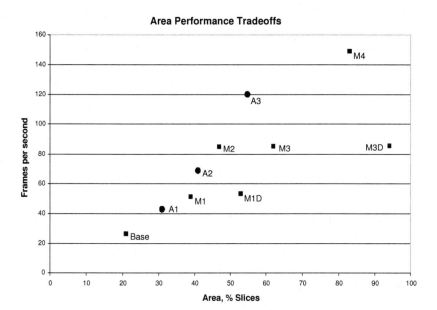

Figure 8.18: Manual vs. Automated Designs

Conclusions and Future Work

In this work, we considered a number of MILP approaches for solving the task allocation and scheduling problem. Based on their treatment of precedence constraints, a taxonomy of known MILP representations was proposed, and the most promising core formulation was selected. Extensions were then added to customize the formulation for our needs. With extensive computational testing, we showed that our overlap-based formulation has better solution time than a competing sequence-based formulation, and demonstrated that tight lower bounds could be obtained in polynomial time. We also identified key metrics for determining the difficulty of problem instances - demonstrating that for a given application, larger platforms actually decrease solution time. Our formulation was applied to a case study that considers the deployment of an MJPEG application on a Xilinx FPGA platform. We showed that our formulation can accurately predict the performance of the system and quickly produce solutions that are competitive with manual designs.

In the future, we plan to further extend the customizations used here to enable the targeting of more complex platforms. For instance, by handling the allocation of distributed memory. Also, we would like to consider more specialized solution techniques. In particular, the addition of constraints corresponding to critical paths during the branch-and-bound process seems promising.

8.4 H.264 Deblocking Filter

It is clear from the design flow that the more architecture topologies that can be created from service models, the larger the potential design space available for exploration. Therefore the more modular the individual services, the more unique interactions that can be explored and hence more topologies can be created. This section will demonstrate the advantages of this modularity and show how unique design points can be analyzed with a high level of accuracy.

The first stage of this process is the functional modeling of an application. Functional model exploration is twofold. The first stage is *behavior capture*. This is the process of examining the various ways to express the behavior of an application. An important area of exploration is the examination of the various levels of concurrency which can be present in an application. This process is covered in [KWD+06] using an algebraic representation. This process will not be discussed in depth here and the reader is directed to read the provided reference for more information. For this work it is sufficient to understand that certain aspects of an application can occur in parallel. These parallel aspects can then also be sequentialized. Given all the operations in a design, each can be classified as sequential or parallel in relation to each other. The design space then becomes the manipulation of these relationships and the partitioning of the sets in which these relationships are considered. Sequential operations can be executed on one service while parallelism requires a service for each parallel operation.

The second stage is to take one of the candidate functional representations and assigned aspects of the functionality to architectural services which compose a architecture instance. This is *mapping* in METROPOLIS. This requires a methodology to partition the functional model (this is the first stage [KWD$^+$06]) as well as a set of architecture components (as shown in Chapter 5). The METROPOLIS framework then evaluates potential performance by mapping the functional model onto an architectural model for simulation. The architecture models for this flow are based on the Xilinx Virtex II Pro FPGA platform [Cora] created in METROPOLIS. These models were described in Chapter 5. Specifically this work will be examining architectures based on MicroBlaze soft-microprocessor cores and Fast Simplex Links (FSLs). An FSL is a FIFO-like communication channel, which connects two MicroBlazes in a point-to-point manner. These components were selected because they can easily correspond to dataflow applications like the one to be presented. Because of the way in which these models were created, this section will demonstrate that any functional model that could be created using the algebraic methods, can be presented with a corresponding architecture model for a one-to-one mapping.

What will follow is a discussion of the application details, how mapping is performed, and an analysis of the results obtained.

8.4.1 Application Details

This work chose to explore the H.264 deblocking filter algorithm since it is responsible for a significant percentage *(approx. 33%)* of the total computational complexity of H.264 [HJKH03]. The deblocking filter function is applied to a block (4×4 pixels) border of an image for the luminance and chrominance components separately, except for the block borders at the boundary of the image. Note that the deblocking filter function is performed on a macroblock basis after the completion of the image construction function.

The filtering is applied to a set of eight samples across a block border as shown in Figure 8.19. When block border $V0$ is selected, eight pixels denoted as a_i and b_i with $i = 0, \cdots, 3$ are filtered. The other fifteen rows along $V0$ are also filtered. Likewise when block border $H1$ is selected, eight pixels denoted as c_i and d_i with $i = 0, \cdots, 3$ are filtered as well as the other fifteen pixel set along $H1$. Vertical block borders are selected first from left to right on the macroblock (in the order of $V0$, $V1$, $V2$, and $V3$ in Figure 8.19) followed by horizontal block borders from top to bottom of the macroblock (in the order of *H*0, *H*1, *H*2, and *H*3 in Figure 8.19).

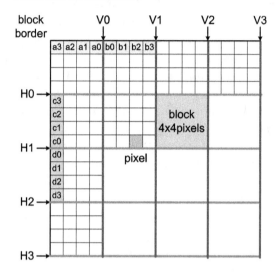

Figure 8.19: Macroblock and Block Border Illustration for H.264 Deblocking Filter

The filter function can be roughly divided into two parts. The first function is a derivation function of boundary filtering strength and the second function is a filtering function for samples across the block border.

Figure 8.20 is the pseudo code for the deblocking filter derived from H.264 reference software [WSBL03], [JM9]. *DeblockMB* checks whether neighbor macroblocks (16×16 pixels) are available for a target macroblock. *GetStrength* outputs a boundary strength ($str_{i,j,k}$) for the filter, and *EdgeLoop* does filtering for the eight samples depending on the boundary strength. The boundary strength is in the range of 0 and 4 (integer number) and the number is determined depending on slice type, reference pictures, the number of reference pictures, and the transform coefficient level of every block according to the encoding profile. This exploration is carried out for the worst case among five boundary strengths. It was observed that the total cycle count is the worst (largest) when the boundary filtering strength is one. *GetStrength* and *EdgeLoop* function transactions will be the system level units of granularity for this exploration.

$D : DeblockMB()$;

```
for i=0,1 do
```
$\quad I_i :$
```
    for j=0,···,3 do
```
$\qquad J_j :$
```
        for k=0,···,15 do
```
$\qquad\quad P_k : str_{i,j,k} = GetStrength(i,j,k);$
```
        end for
        for k=0,···,15 do
```
$\qquad\quad Q_k : EdgeLoop(i,j,k,str_{i,j,k});$
```
        end for
    end for
end for
```

Figure 8.20: Deblocking Filter Pseudo Code

8.4.2 Mapping Details

Once the functional model topology has been created, one must transform this into a METROPO-LIS functional model. This case study ultimately is interested in investigating potential clock cycle counts when the functionality is mapped and simulated with an METROPOLIS architecture model. Therefore it is important that functional model actions have consequences in the architecture model. METROPOLIS' higher abstraction level allows functional model statements to be classified into three primitive functions: *read*, *write*, and *execute* as previously described. The MicroBlaze elements have corresponding functions. Mapping amounts to correlating these functions to each other so that the appearance of a call in the functional model triggers its corresponding call in the architecture model. The total number of clock cycles required is found by accumulating cycles for *read*, *write*, and *execute* functions triggered in the architecture services. Figures 8.21 and 8.22 show how *GetStrength* and *EdgeLoop* in the functional model are composed of these primitive functions, where an argument *type* in *execute* is the the type of execution operation being carried out and arguments of *read* and *write* are the amount of transfered data in bytes. The arguments to *read*, *write*, and *execute* are translated by the METROPOLIS characterizer databases. This process translates into a cycle count for each operation (Chapter 7 details this process). This work will refer to a process with *GetStrength* and *EdgeLoop* functionality as a "filter process" henceforth.

```
GetStrength(){
    execute(type1);
    mem_read(2wd);
    execute(type2);
    mem_read(8wd);
    ...
    execute(type3);
    write(strength);
}
```

```
EdgeLoop(){
    execute(type4);
    mem_read(8wd);
    ...
    read(strength);
    mem_read(8wd);
    execute(type5);
    mem_write(8wd);
}
```

Figure 8.21: Decomposition of GetStrength Function

Figure 8.22: Decomposition of EdgeLoop Function

The mapping in this exercise is carried out in such a way that a filter process and a communication channel in the functional model are mapped onto a MicroBlaze and an FSL in one-to-one manner respectively as shown in Figure 8.23. The left hand side of this illustration is the functional model and the right hand side is the architectural model. The functional model is partitioned into sequential and parallel operations. Shaded areas indicate how many services are required (some services may be supporting more operations depending on how many circles are in each shaded area). P indicates *GetStrength* and Q indicates *EdgeLoop* activities. These shaded areas are each given a process identification number (PID). Arcs between each side indicate how the mapping was performed. Only two examples are shown here. These are topologies H and C. All the topologies will be shown in Figure 8.24.

A source process in the functional model is defined as follows: A "source process" (SRC) is a storage element with stream data and baseband data. A source process communicates with "filter processes" in such a way that a filter process sends 32-bit wide data (a read/write flag, a target address, and a target data length in this order) and afterwords a source process sends or receives data in a burst transfer manner. The source process has in/out ports connected to all filter processes as shown in Figure 8.23 and receives requests from the filter processes in a first-come-first-served basis with non-blocking reads.

The source process is also mapped onto a MicroBlaze. The length of a FIFO connected between the source process and filter processes is large enough so that processes are not blocked on write operations. For this case study, the source process FIFOs have a depth of 16. The length of a FIFO between filter processes changes in this case study, and is represented by N in Figure 8.23.

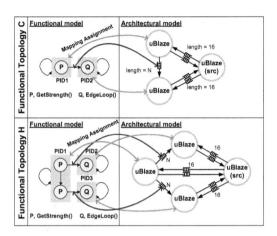

Figure 8.23: Mapping a Functional Model onto an Architectural Model for H.264

Provided that the number of MicroBlazes is three and under, 14 functional topology candidates are obtained and shown in Figure 8.24. As in Figure 8.23 a gray zone represents what will be executed on each MicroBlaze (a "partition"; called a "mapping" in this work) and resource ID is denoted by PID once again in the figure. For example, (C) in Figure 8.24 implies that resource 1 (PID1) has computational block *P* (GetStrength) and resource 2 (PID2) has computational block *Q* (EdgeLoop). Another example is topology (F) which illustrates two processes as well. PID2 contains Q functionality. PID1 has a collection of P and Q functionality. These candidates will form the basis for the design space exploration to follow.

8.4.3   Design Space Exploration Results

The results detailing execution cycle counts for the functional topology candidates explored in Figure 8.24 are discussed in this section first. Figure 8.25 shows the total execution cycle count breakdown (computation cycles, communication cycles with a source process, and waiting cycles) when the length of a FIFO between filter processes (N denoted in Figure 8.23) is size one. The waiting cycles accumulate in two following cases: when a filter process waits for other filter processes to finish their transaction with a source process and when a filter process waits for data to come to a FIFO from other filter process.

The vertical axis in Figure 8.25 is the number of clock cycles required and horizontal axis

282

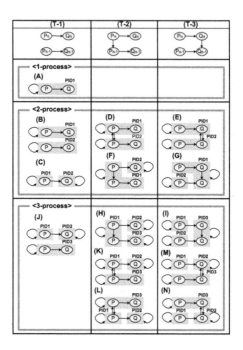

Figure 8.24: H.264 Functional Topology Mapping Candidates

shows topologies (A through N) as shown in Figure 8.24. B through G have two bars, where the first bar corresponds to process 1 denoted by $PID1$ in Figure 8.24 and the second is process 2 denoted by $PID2$. H to N have three bars, where the first bar corresponds to process 1 denoted by $PID1$ and the second and third bars are results of process 2 and process 3 denoted by $PID2$ and $PID3$ each.

The simulation results demonstrate that workload balance has a strong effect on execution time for a multiprocessor system. Case H is the best case in terms of workload balance and as a result, the total amount of cycles is the smallest. Compare case J with case L. L has more communication channels than case J. Nonetheless process 3 in L spends less time waiting than process 3 in J, which implies that the memory traffic of L is lighter than that of J due to synchronization between process 1 and process 3. Compare K with L. Their topologies are the same, but the process execution order differs. As a result, the completion times are different. Similar conclusions can be drawn for topologies M and N.

There are several broad conclusions that can be drawn from these results. First, apparently small changes in the functional topology can actually have dramatic effects on execution time. Secondly, the breakdown of overall execution time is important to examine for these types of applications in order to better understand how communication bottlenecks play a role in each topology. Finally, METROPOLIS was able to perform efficient functional design space exploration with ease and with only minor changes to the functional and mapping models. Fourteen topologies were able to be explored with very few changes to the architectural model. In fact, only the top level architectural netlist needs to be changed since the the structures are very regular and modular. This modification process could be automated and even more topologies explored if the 4 MicroBlaze restriction was relaxed.

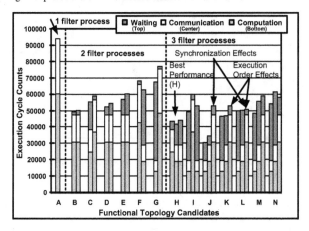

Figure 8.25: METROPOLIS H.264 Simulation Results for All Candidate Topologies

Optimal FIFO Size

In the second set of experiments, the effect of FIFO length between filter processes was examined. Figure 8.26 shows execution cycle counts of three topologies: C, F and I when the length of a FIFO between filter processes changes (N in Figure 8.23). The results show the optimal length in terms of achieving minimum cycle counts. Changing the length of a FIFO does not have an effect on the total cycle count, but rather on the cycle counts of individual processes.

Let process P be a producer process and process Q be a consumer process, and FIFO F connect P to Q. As it turns out, when process P takes more time than process Q for its computation and communication time (with a source process), the length of F does not matter. Meanwhile, when process P takes

284

less time than process Q, the length of F has an effect on waiting time of P, not on Q. However, the total cycle counts do not change. Therefore, this illustrates that FIFO length exploration is less important in terms of the total amount of execution clock cycle counts.

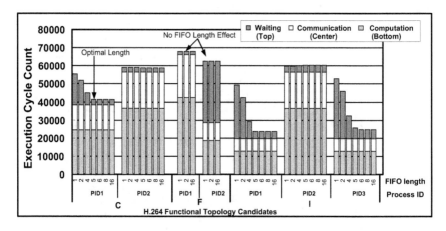

Figure 8.26: METROPOLIS H.264 Simulation Results for Various FIFO Sizes

Table 8.6 breaks down the performance of all the topologies further. Total execution time in clock cycle counts (second column), the optimal FIFO length (third column), and topology decomposition (fourth, fifth, and sixth columns) are shown. Optimal FIFO length is the smallest length which maintains the lowest clock cycle count. Resource cost is given as program binary code size below the table. The 4th, 5th and 6th columns then can be interpreted as how much memory is consumed for program memory on each process. PQ is the result given by combining P and Q computational blocks on the same architectural resource. In the case where FIFO length does not make any difference for the counts, the optimal length is set to 1.

This simulation demonstrates that users can make a decision regarding the optimal functional model based on parameters related to performance and costs such as total execution cycle counts (work-load balance), communication overhead, memory traffic, FIFO length, shared memory size, the number of processors, program code size, context switching overhead, register, cache, dedicated hardware logic size, and so forth. Again, METROPOLIS provides a easy-to-use framework for this type of functional exploration thanks in no small part to the modular and flexible architecture construction.

Table 8.6: H.264 Performance and Cost Results for All Topologies

Topology	Counts	Length	Proc1	Proc2	Proc3
A	94021	1	PQ	-	-
B	50188	1	PQ	PQ	-
C	58839	5	P	Q	-
D	54505	1	PQ	PQ	-
E	60124	1	PQ	PQ	-
F	67981	1	PQ	Q	-
G	76182	6	PQ	P	-
H	43932	1	P	Q	Q
I	60215	5	P	Q	P
J	52031	3	P	Q	PQ
K	52971	1	P	Q	PQ
L	50780	1	P	Q	PQ
M	58941	6	P	Q	PQ
N	61190	6	P	Q	PQ

Binary Data Size PQ: 47.9KB; P: 47.0KB; Q: 45.9KB

Simulation Accuracy

All of the previous results are meaningless unless METROPOLIS simulation accurately corre-lates to the actual implementation. Figure 8.27 illustrates how closely METROPOLIS' simulation com-pares to experimental results. Six of the more interesting topologies were selected. Each design was implemented on a Xilinx ML310 design board and the execution time was measured. Shown in the figure are the percentage differences between simulation and implementation. The maximum difference be-tween implementation and simulation is 7.3%. This is a high correlation while maintaining a high level of abstraction in the METROPOLIS models. In addition, it confirms that H has the lowest cycle count of any design and demonstrates that making an absolute design decision based on METROPOLIS simulation would have been the correct choice.

8.5 UMTS METRO II Design Exploration

This section details a UMTS data link layer case study done in METRO II. After describing the functional and architectural models, we enumerate the 48 different points in the design space that were explored. In the results subsection, we detail the estimated execution times and processing element

286

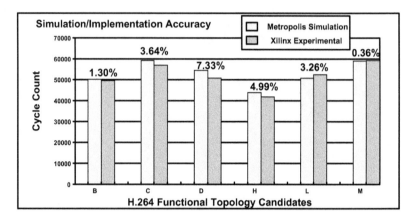

Figure 8.27: METROPOLIS H.264 Accuracy Versus FPGA Implementation

utilization, the design effort, the simulation cost breakdown, and an analysis of how events are processed during each simulation phase.

8.5.1 Functional Modeling

The functional model covered in more detail in Section 8.5.3 is described as a process network or actor-oriented model, where concurrently executing processes communicate with each other through point-to-point channels. The communication semantics can be adjusted on a per-channel basis. For instance, data transfer between a pair of processes might take place with rendezvous semantics or blocking read, non-blocking write semantics. In METRO II, there are a number of ways in which this coordination between processes may be specified. Two such mechanisms are detailed in Figure 8.28.

In the explicit synchronization style (top half of the figure), imperative code is written to prevent process P_2 from reading data out of an empty FIFO $F1$ and also preventing it from writing data to a full FIFO $F2$. The code makes use of interface methods provided by the FIFOs as well as events exposed from within the FIFOs through view ports (not shown in the figure). This imperative specification is quite similar to what might be specified in SystemC. These interface methods will themselves propose *begin* and *end* events to the system for scheduling along with any other methods used to query the state of the FIFOs. These additional events have the ability to affect simulation as will be discussed in Section 8.5.6.

The second specification style uses constraints to enforce the same behavior. Now, instead of

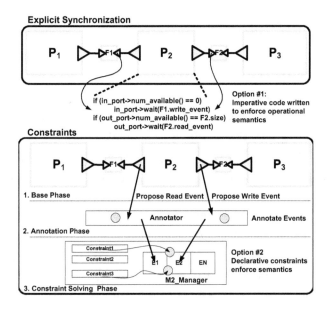

Figure 8.28: Two options for expressing coordination

imperative code being specified in the phase 1 components, the corresponding constraints are passed to the phase 3 constraint solver. This separates computation from coordination, and different coordination models can be used for the same process P_2. This separation is similar to the concept of a "director" within the Ptolemy [LXL01] environment, but within METRO II, the coordination may be specified declaratively.

The first style has more events and may also cause the phase 1 processes to become suspended due to internal blocking (as opposed to blocking due to event disabling in phase 3). The benefits of the second style in this regard are described in Section 8.5.6.

8.5.2 Architectural Modeling

METRO II supports the development of architecture service models to complement the functional modeling effort. These models should be modular (support various configurations and parameterizations), flexible (offer a variety of mapping solutions), and accurate (have a firm grounding in the real-world counterparts they represent). The goal is to create a model which provides meaningful simulation data for design space exploration when mapped to a functional design. This section will detail the development of such an architecture platform which will be used in the UMTS case study covered in Section 8.5.4.

General Structure

Before detailing the specifics, it is important to make several key points. Architecture models are ultimately viewed as providing *services*. These services support functional model requirements and provide meaningful *costs* when these services are requested. To this end they minimally need to:

1. Contain components with provided ports to provide services to the functional model for mapping. These components (in this case called *architectural tasks*) each have their own thread of execution. These threads will generate the events associated with ports. These ports will be *required ports* and events associated with their interface functions will be mapped to the functional model via M2_MAP (FUNC_COMP, FUNC_EVENT, TASK, REQUEST_JOB).

2. Register the events associated with the architectural task required port function calls to the necessary annotators and schedulers. This event association will provide the costs of the architectural services and ultimately the cost of the overall simulation.

The architecture models to be discussed in this work are composed of the following three portions:

1. Tasks - active components which serve as the mapping target for each component in the functional model.

2. Operating System - explicit, imperative mechanism for scheduling and assigning tasks to processing elements.

3. Processing Element - workhorse of the architectural model. Used to model the core cost of a service.

Figure 8.29 illustrates these three portions along with METRO II mappers. This diagram will be explained further in the following subsections.

Architectural Tasks

Tasks are lightweight, active components in the architecture model. The thread for each task constantly proposes begin events for its provided services. Mapping creates a rendezvous constraint

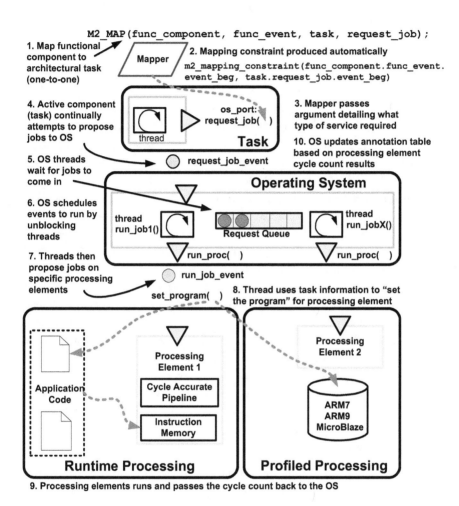

Figure 8.29: Topology of Architecture Service Environment

between the event generated by the task thread (in this case, *request_job*) and the functional event. Therefore, there is a 1:1 mapping between these tasks and functional components. Due to the rendezvous constraint, the task remains blocked until the corresponding event from the functional model is proposed. Steps 1-4 in Figure 8.29 illustrate the task's role in architecture model execution.

Operating System

The operating system is used to assign tasks to processing elements (in a many-to-one relationship). In addition, it also carries out phase 1 scheduling - reducing the work to be done in phase 3. This is done by pruning the events that are proposed in the first phase. An investigation of scheduling policies will be seen in Section 8.5.6. An OS is an active component with N threads (where N is the number of processing elements it controls). It maintains a queue of requested jobs which processing elements query to decide if they will execute or not. The queue contains events proposed for processing, which processing element they wish to use, the round they were proposed in, and the statically assigned priority for the event. Scheduling controls how events are added to and removed from this queue. Access to this queue is coordinated such that there are a limited number of outstanding requests for a given processing element. Steps 5-7 in Figure 8.29 illustrate the OS's role in the architecture model execution.

The OS is also used to access the annotation tables for events. The annotation tables are used by the annotator in phase 2 (where event tags can be written). These tables relate event costs to architectural services. The OS updates the appropriate entry in the table after a request is completed and the true cost known. In addition to the processing cost, it may also add cost related to OS overhead (e.g. context switching). It is in this way that the tables are updated dynamically at runtime and do not require that the tables be statically created with the netlist. Tasks themselves need know nothing about this process and only need to indicate which service they require. Again this separates the computation behavior from its performance cost.

Processing Elements

The third piece of the architecture platform consists of the actual processing elements. Once the OS decides to run a task request, it calls the corresponding *run_proc()* function call on one of its N required ports. The interface supported by all processing elements is the same (to provide modularity and flexibility) but there are different ways in which the cost may be calculated. Steps 8-9 in Figure 8.29 illustrate two different types of processing elements that may be used and the interface to inform them which processing routine they should compute a cost for. The type of processing element may be changed

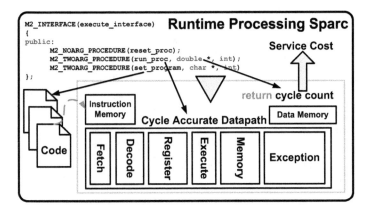

Figure 8.30: Sparc Runtime Processing Element

easily to provide the necessary balance between speed of simulation and required pre-simulation effort.

Runtime Processing

The first architectural modeling style is *runtime processing*. In this style, the processing elements are cycle accurate, microarchitectural models which execute code dynamically. An example of runtime processing is a cycle accurate model of a Leon 3 SPARC shown in Figure 8.30. There are three key functions in the interface.

The first step in using this type of processing element is for the OS to provide information gathered from the task as to which operation is requested. This will be done via the *set_program()* function where the instruction memory of the processing element is loaded with pre-compiled code for the operation. The second step is then to execute that code at runtime with *run_proc()* which will return the cycle requirements for that code. The operating system will use that information to update the annotation table. Finally, the OS will call *reset_proc()* to ready the processor for the next request.

While this style will result in a slower simulation time as compared to the following approach, it simply requires that the code for the function be available. It requires no other modeling work by the user and is as accurate as the level of detail in the microarchitectural model.

Profiled Processing

The second style is *profiled processing* where precomputed performance metrics are stored for lookup. Again the OS will indicate to the processing element which services are requested. In turn, the processing element will lookup the costs for the given operations. These can be trivial table lookups or more complex (but still static) calculations based on the current state of the processing element. Ways to characterize processing elements for this approach have been shown in [MSVSL08], [DDSV06b], and [DSP08]. An advantage of this approach is that the lookup is extremely fast as compared to the runtime processing approach. The drawback is that the modeling of these elements is often more limited in its usage and requires that characterization be carried out prior to simulation. This precharacterization however only needs to be done once per computation routine. It does requires a more complex set of transformations as compared to simple compilation (runtime processing approach). Chapter 7 detailed the characterization flows applicable to this work.

8.5.3   Functional Model

We focus on the User Equipment Domain of the UMTS protocol [Pro04], which is of interest to mobile devices and is subject to stringent implementation constraints. The protocol stack of UMTS for the User Equipment Domain has been standardized by the 3rd Generation Partnership Project (3GPP) up to the Network layer, including the Physical (PHY) and Data Link (DLL) layers. Our model includes the implementation of the DLL layer and the functionality of the PHY layer.

The DLL layer contains the Radio Link Control (RLC) and the Medium Access Control (MAC) sublayers and performs general packet forming. The RLC communicates with the MAC through different *logical* channels to distinguish between user, signaling, and control data. Depending on the required quality of service, the MAC layer maps the logical channels into a set of *transport* channels, which are then passed to the Physical Layer. The Physical Layer handles lower level coding and modulation in order to reduce the bit error rate of the transmitted data.

For the purposes of this case study, the UMTS application was largely separated into both receiver and transmitter portions and then further by RLC and MAC functionality. Simulation consists of processing 100 packets, each packet being 70 bytes. The functional model is shown in Figure 8.31. The semantics is dataflow, with blocking read and blocking write semantics for the FIFOs. Both timed and untimed models were created to determine the advantages of functional/architectural separation - the timed model mixes both, while the untimed model relies on mapping with an architectural model to obtain performance metrics.

The pure functional model allows processes to communicate in zero time provided data is present on the input and space available at the output. The timed model, on the other hand, introduces a scheduler and timer. The scheduler is modeled as a finite state machine which controls the execution of the system. The activation of each process is controlled by the typical firing conditions of process networks, i.e., the availability of data at the input queues, and the availability of space at the output queues. These conditions are notified to the processes every time data is written to or read from the attached FIFOs. When a firing condition is satisfied, the process triggers the scheduler by sending a "Ready_to_Run" signal through the dedicated bi-directional scheduling channel and then waits for permission to start computation, which will be granted by the scheduler when the resources are available and when no higher priority process is ready to run (the timed model scheduler has a notion of priority and preemption; see Section 5.7.4). The process runs to completion, and stops before the results are written to the output FIFO. Computation is carried out in logically zero time. The scheduler will again trigger the process to post its outputs at the correct time, which will not only account for the process execution latency, but also for the time spent in running higher priority processes that had become active and preempted its execution. In this manner, a process is never physically suspended as a result of preemption, thus reducing the overhead due to context switches. Instead, the scheduler verifies if any preemption has occurred, and, if so, updates the completion time by delaying it by the appropriate amount.

Preemptive scheduling

Several policies can be implemented by the scheduler in the timed functional model. An example of a fixed-priority preemptive scheduler is shown in Algorithm 8. The scheduler manages a priority list whose items are process descriptors which include the time, T_{init}, at which the process initiates its computation (negative if not yet scheduled) and a variable, τ, indicating the time left for the process to finish its computation.

The procedure may be triggered either by a new process entering the enabled state or by a timeout from the timer that signals that a process has terminated execution and needs to post its data (lines 1 and 5).

In the case of a new process, the algorithm first compares its priority with the one of the currently running process. If the new process has higher priority, then we update (decrease) the time to completion τ of the current process with the time it has already been executed, which is the difference between the current time and the time it was last given the resource (line 3). In all cases, the new process is added to the list of processes (line 4).

Algorithm 8: Fixed Priority Preemptive Scheduler

---

1 **if** (*new process new_P*) **then**

2   **if** (*current_P.priority* $\leq$ *new_P.priority*) **then**

3   current_P.τ $-$ = current time - current_P.T_{init};

4   Add_item(new_P);

5 **else if** *(timeout current_P)* **then**

6   notify current_P post data;

7   list.pop( );

8 current_P = list.top( );

9 **if** (*current_P.T_{init}* ≤ 0) **then**

10   trigger current_P execution (notify event);

11 current_P.T_{init} = current time;

12 reset_timer(current_P, current_P.τ);

If a timeout occurs, it signals that the current process has reached the end of its computation. It is therefore removed from the list, and granted permission to post its data to the output (line 6).

Resources are then given to the process at the top of the list (line 8). If the process starts execution for the first time, then its body is actually invoked (line 10). Time will not advance during its execution, since all timing is accounted for by the scheduler. To do so, the process descriptor is updated to record the starting time (line 11), and the timer is reset with the remaining time to completion for the process (line 12).

Other scheduling policies, such as round robin or EDF, can be implemented as well. However, we do not implement them in the timed functional model, but rather implement the round robin, priority based, and first-come, first-serve (FCFS) algorithms in the architectural OS component. This will prove to offer these alternate policies at a low simulation runtime cost than in the functional scheduler. Mapping between functionality and architecture uses the untimed model allowing for costs and scheduling to come purely from the architecture services.

8.5.4   Architecture Model

The architecture model assigns one task for each of 11 UMTS components (TR Buffer and PHY were not mapped as they represent the environment). The OS employs three different scheduling

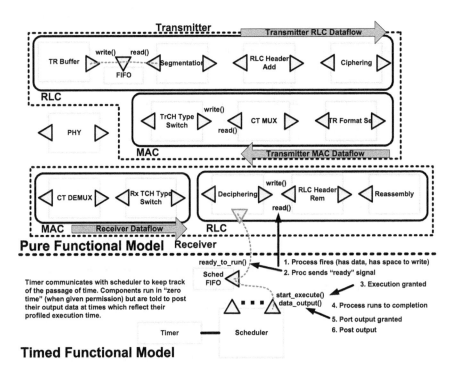

Figure 8.31: UMTS METRO II Functional Model

policies for each processing element. The first is a round robin scheduler where each processing element is simply selected sequentially. This is a cyclic process beginning with processing element 0 and moving through the number of PEs. If a PE does not have a request pending then the next PE in the list is allowed to proceed. The second algorithm is a priority based scheduling algorithm where higher priorities are assigned to tasks with higher processing requirements. These requirements are determined during the pre-profiling stage. Preemption is not employed as in the timed functional model. Priority scheduling here examines all the requests for processing in a given round and selects the one with the highest priority. The selected priority is noted and in the next round it can not be chosen again if there are still events pending from previous rounds. The final algorithm is a first-come, first-serve (FCFS) algorithm which marks requests with the round that they enter the system and ensures that they are handed in order of appearance. In the event of a tie, this falls back to a round robin scheme.

The runtime processing elements were fed C code reflecting the kernels of each UMTS component. The runtime processing element available for this case study is a cycle accurate datapath model of the Leon 3 Sparc processor. The pre-profiled processors use the same code but carry out offline characterization as detailed in [MSVSL08] and [DDSV06b]. The processors profiled were the ARM7, ARM9, and Xilinx's MicroBlaze. All of the processing elements are common in embedded and SoC applications and are widely documented.

8.5.5 Mapped System

Table 8.7 describes the 48 mappings that were investigated. These reflect systems with a high level of concurrency (11 PEs) all the way down to 1 PE. The mappings reflect system partitions broken down by Rx and Tx as well as RLC and MAC functionality. The partitions are enumerated at the bottom of the table. Each mapping is categorized into one of 9 separate *classes*. Classes are assigned to make analysis more manageable. The classes are based on the number of processing elements and the mix of pre-profiled and runtime processing elements. Mappings are further categorized as purely runtime processing based (RTP) elements, purely profiled processing (PP) elements, or a mix (MIX) of both.

8.5.6 Results

In this section, results relating to design effort, estimated processing time, framework simulation time, and event processing are analyzed. Five different models were utilized in the experiments: a timed SystemC UMTS model [DSP08], a timed METRO II UMTS model, an untimed METRO II UMTS functional model, a SystemC runtime processing model, and a METRO II architectural model. In specific

#	Type	Partition	#	Type	Partition	#	Type	Partition
1	1: RTP	11 Sp	17	6: PP	2 μB (2), 2 A9 (3)	33	7: MIX	A7 (4), Sp (5), μB (6), A9 (7)
2	2: PP	11 μB	18	6: PP	2 A9 (2), 2 μB (3)	34	7: MIX	A7 (4), Sp (5), A9 (6), μB (7)
3	2: PP	11 A7	19	6: PP	2 A7 (2), 2 A9 (3)	35	7: MIX	A7 (4), μB (5), Sp (6), A9 (7)
4	2: PP	11 A9	20	6: PP	2 A9 (2), 2 A7 (3)	36	7: MIX	A7 (4), μB (5), A9 (6), Sp (7)
5	3: RTP	4 Sp (1)	21	7: MIX	Sp (4), μB (5), A7 (6), A9 (7)	37	7: MIX	A7 (4), A9 (5), μB (6), Sp (7)
6	4: PP	4 μB (1)	22	7: MIX	Sp (4), μB (5), A9 (6), A7 (7)	38	7: MIX	A7 (4), A9 (5), Sp (6), μB (7)
7	4: PP	4 A7 (1)	23	7: MIX	Sp (4), A7 (5), μB (6), A9 (7)	39	7: MIX	A9 (4), Sp (5), μB (6), A7 (7)
8	4: PP	4 A9 (1)	24	7: MIX	Sp (4), A7 (5), A9 (6), μB(7)	40	7: MIX	A9 (4), Sp (5), A7 (6), μB (7)
9	5: MIX	2 Sp (2), 2 μB (3)	25	7: MIX	Sp (4), A9 (5), A7 (6), μB (7)	41	7: MIX	A9 (4), μB (5), Sp (6), A7 (7)
10	5: MIX	2 μB (2), 2 Sp (3)	26	7: MIX	Sp (4), A9 (5), μB (6), A7 (7)	42	7: MIX	A9 (4), μB (5), A7 (6), Sp (7)
11	5: MIX	2 Sp (2), 2 A7 (3)	27	7: MIX	μB (4), Sp (5), A7 (6), A9 (7)	43	7: MIX	A9 (4), A7 (5), μB (6), Sp (7)
12	5: MIX	2 A7 (2), 2 Sp (3)	28	7: MIX	μB (4), Sp (5), A9 (6), A7 (7)	44	7: MIX	A9 (4), A7 (5), Sp (6), μB (7)
13	5: MIX	2 Sp (2), 2 A9 (3)	29	7: MIX	μB (4), A7 (5), Sp (6), A9 (7)	45	8: RTP	1 Sp
14	5: MIX	2 A9 (2), 2 Sp (3)	30	7: MIX	μB (4), A7 (5), A9 (6), Sp (7)	46	9: PP	1 μB
15	6: PP	2 μB (2), 2 A7 (3)	31	7: MIX	μB (4), A9 (5), A7 (6), Sp (7)	47	9: PP	1 A7
16	6: PP	2 A7 (2), 2 μB (3)	32	7: MIX	μB (4), A9 (5), Sp (6), A7 (7)	48	9: PP	1 A9

(1 = Rx MAC, Tx MAC, Rx RLC, Tx RLC), (2 = Rx MAC, Rx RLC), (3 = Tx MAC, Tx RLC)

(4 = Rx MAC), (5)(Rx RLC), (6)(Tx MAC), (7 = Tx RLC) (Sp = Sparc, μB = Microblaze, A7 = ARM7, A9 = ARM9)

Table 8.7: Mapping Scenarios for UMTS Case Study

configurations, constraints were used as opposed to explicit synchronization in the functional model. The selection of constraints, functional model configuration, architectural model parameters, and mapping assignment is all achieved through small changes to the top level netlist. All results are gathered on a 1.8 GHz Pentium M laptop running Windows XP with 1GB of RAM.

Estimated Processing Time

Figure 8.32 shows the UMTS estimated execution times in cycles for the various mappings along with the average processing element utilization. The x-axis (mapping #) is ordered such that execution times increase moving toward the right. The data is collected for each of the three scheduling algorithms implemented by the OS architecture service model. RR is round robin, PR is priority, and FCFS is first come, first serve.

Examining the round robin scheduling first, as expected, the lowest and highest execution times are obtained with mapping #1 (11 SPARCs) and mapping #46 (1 μBlaze) respectively. Mapping #1 is 2167% faster than mapping #46. This shows a large range in potential performances across mappings. It is interesting to note that there are 23 different mappings which offer better performance than the 11 μBlaze or 11 ARM7 cores (mappings 2 and 3). This illustrates that communication becomes a bottleneck

298

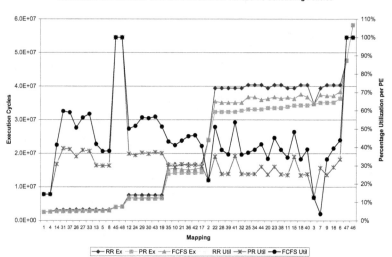

Figure 8.32: UMTS Estimated Execution Time vs. Utilization

for some designs and despite having maximum concurrency those designs cannot keep pace with smaller, more heavily loaded mappings. Among all 4 processor systems, mapping #14 has the lowest execution time (two ARM9s used for receiver, two SPARCs used for transmitter). Mapping #31 has a similar execution time with 4 different processors (Rx MAC on μBlaze, Rx RLC on ARM9, Tx MAC on ARM7, and Tx RLC on the Sparc). Many of the execution times are similar and the graph shows that there are essentially 4 areas of interest.

For round robin, the lowest utilization values occur in the 11 processor setups (average of 15%). The highest is 100% for all single processor setups. The max utilization before 100% is 39%. This gap points to a deficiency in the round-robin scheduler used in the OS. It should be a goal of the other scheduling algorithms to improve this if possible. One should also notice that for similar execution times, utilization can vary as much as 28% (mappings #41 and #32 for example).

The priority based scheduling keeps the same relative ordering amongst the execution times but reduces them on average by 13%. The highest reduction is an 18% reduction (mapping #22 for example) and the smallest reduction is 9% (mapping #8 for example). The utilization numbers are actually reduced (made worse) as well by an average of 2%. The largest reduction was 7% (in mapping #6 for example) and the smallest was 1% (in mapping #31 for example). As expected there was no change to the utilization or execution times for mappings involving either 11 processing elements (fully concurrent) or those with 1 element (no scheduling options). The utilization drop results from high priority, data dependent jobs running before low priority, data independent jobs.

FCFS scheduling also does not change the relative ordering of execution times but is not as successful at reducing them. The average reduction is only 7%. The maximum reduction is 11% (in mapping #24 for example) and the minimum reduction is 4% (in mapping #5 for example). However, utilization is improved (increased) by 27%. The max improvement was 45% (in mapping #31 for example) and the minimum improvement was 20% (in mapping #5 for example). FCFS improves utilization due to the fact that many jobs which would be low priority often request processing in the same round as high priority jobs. While technically they are both "first", priority would negate this fact. FCFS's round robin tie breaking scheme helps smaller jobs in this case.

The analysis of execution and utilization for UMTS shows that high utilization is difficult to obtain due to the data dependencies in the application. Also, some of the partitions explored do not balance computation well amongst the different processing elements in the architecture. Many of the more coarse mappings only make this problem worse. A solution is to further refine the functional model to extract more concurrency. From an execution time standpoint, scheduling can improve the overall execution time but not as much as is needed to make a large majority of these mappings desirable for an actual implementation. In fact only 11 (23%) of the mappings would actually be feasible for an actual UMTS system.

Design Effort

The untimed METRO II UMTS functional model contains 12 processes while the architectural model may contain up to 26 processes depending on the configuration. The specification of the entire design is spread across 85 files and 8,300 lines of code. This is clearly a large design. The changing of a mapping is trivial however as it requires us to change a few macros and recompile 2 of the 85 files (2.3%) which takes less than 20 seconds. The 48 mappings could be done in less than 16 minutes.

The conversion of the SystemC timed functional model to an untimed METRO II functional model removes 1081 lines of code (related to scheduling and timing - both of which are in the architecture model). By having an architecture model we strip out much of the overhead associated with making the SystemC model synchronize and run in the first place.

Also METRO II constraints for the read/write semantics of a FIFO only require 60 lines of code which is 1.4% of the total code cost. In fact the average difference of the entire conversion to METRO II was only 1% per file. More than half of these lines (58%) have to do with registering the constraints with the solvers.

The conversion of a SystemC runtime processing model (for the SPARC processing element) to METRO II only requires 92 additional lines for a direct conversion. This was a mere 3.4% increase (2773 lines to 2681). This includes adding support for loading new code at runtime, returning the cost of operation to the highest element in the hierachical netlist, and exposing events for mapping. Since METRO II was created to easily import existing code, this result is encouraging.

Framework Simulation Time

Figure 8.33 illustrates the percentage of the actual simulation runtime spent in each of METRO II's simulation phases for the 9 classes of mappings. The SystemC entry indicates the time spent in the SystemC simulation infrastructure upon which METRO II is built.

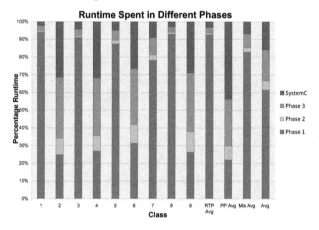

Figure 8.33: METRO II Phase Runtime Analysis

As shown, on average 61% of the time is spent in Phase 1 (lowest section on the bar graph), 5% in Phase 2 (second section), and 17% in Phase 3 (third section). For models with only runtime processing elements (RTP) the averages are 93%, 0.9%, and 3% respectively. This indicates that in runtime processing, the METRO II activities of annotation and scheduling are negligible in the runtime picture. For pure profiled (PP) mappings they are 21%, 7% and 26%. In this case one can see that

MetroII now accounts for a greater percentage of runtime than would normally occur (phase 1 alone is representative of other simulation environments). For mixed classes the numbers are 82%, 2.6% and 7.6%. Again the runtime processing elements dominate. It should be noted that while PPs have higher averages, the average runtime to process 7000 bytes of data was 54 seconds. Phase 1 runtime (and SystemC overhead) are the main contributors to overall runtime.

If we consider the SystemC timed functional model, METRO II timed functional model, and the METRO II untimed functional model mapped to an architecture, the METRO II timed functional model had an average increase of 7.4% in runtime for the 9 classes while the mapped version had a 54.8% reduction. This reduction is due to the fact that METRO II phases 2 and 3 have significantly less overhead than the timer-and-scheduler based system required by the SystemC timed functional model.

In a comparison of the METRO II timed model running without constraints and one running with them, the average decrease for the 9 classes of mappings was 25%. There was little discernable difference between the runtimes for each class here. This is not surprising since contraints have no direct role in the scheduling of processes - only in how the phases switch (simulation scheduling).

Class	Event/Ph.	Comp. %	Comm. %	Coord. %	Avg Wait
1	0.091	0.083	0.083	0.833	3839.240
2	0.091	0.083	0.083	0.833	3839.240
3	0.169	0.125	0.042	0.833	6276.190
4	0.169	0.125	0.042	0.833	6276.190
5	0.131	0.170	0.114	0.716	5117.003
6	0.169	0.170	0.114	0.716	6276.190
7	0.150	0.101	0.088	0.811	5691.130
8	0.176	0.319	0.043	0.638	6718.550
9	0.176	0.319	0.043	0.638	6718.550
Avg	0.147	0.166	0.072	0.761	5639.143
RTP Avg (1,3,8)	0.145	0.176	0.056	0.768	5611.327
PP Avg (2,4,6,9)	0.121	0.174	0.070	0.755	4622.034
MIX Avg (5,7)	0.140	0.136	0.101	0.763	5404.067

Table 8.8: METRO II Phase Event Analysis

Framework Event Analysis

Table 8.8 shows the average number of events which change state per phase and the average number of phases an event waits in the system (from the round that it is first proposed to the round in which it is finally enabled).

On average, only 0.14 events are annotated or scheduled per round. Because of the architectural model integration with the UMTS functional model there are a limited number of synchronization points (which satisfy a rendezvous constraint and hence an event state change). As shown in Figure 8.33, Phases 2 and 3 do not account for a large portion of the runtime so while the event state change activity is low, it does not translate to increased runtime. Runtime is not increased directly by changing an event's state but rather by the total number of events that must be examined in phases 2 and 3.

Events in classes 1 and 2 on average wait 42% less than the worse case. These classes are precisely those which provide maximum concurrency (11 processing elements). The worst is in classes 8 and 9 (single processing elements). As one would expect, when the scheduling overhead is lower and more processing elements are available, events wait much less for resource availability.

Finally it should be noted that runtime processing vs. pre-profiled processing does not have an impact on this aspect of the simulation. Comparing classes 1 with 2 or 3 with 4 confirms this. Therefore, this contrasts heavily with the runtime of the simulation (in which PE type is a key factor). The runtime processing which takes place in the microarchitectural model is treated as a black box by METRO II such that the internal events are unseen and do not trigger phase changes. This indicates that SystemC components can be imported quite easily into METRO II without affecting the three-phase execution semantics. In fact, the import of external code into any one of the three phases is easily accomplished. Due to the orthogonalization of intent in the three phases, we attempt to minimize import that would impact multiple phases at once, which would be a more complex undertaking.

The third, fourth, and fifth columns of Table 8.8 indicate the categorization of proposed events in phase 1. Computational events are those that request processing element services directly. Communication events are those that transfer data between FIFOs and coordination events maintain correct simulation semantics and operation. As broken down by class, the table indicates that events in the system are heavily related to coordination. Classes 8 and 9 have the lowest percentage of coordination events (64%) since these are 1 PE systems.

Figure 8.34 illustrates how each functional UMTS component proposes events. For each of the 11 components there are four scenarios. The first scenario is when the component is involved in the timed METRO II functional model. These are on the far left of the figure. The second scenario is when the timed

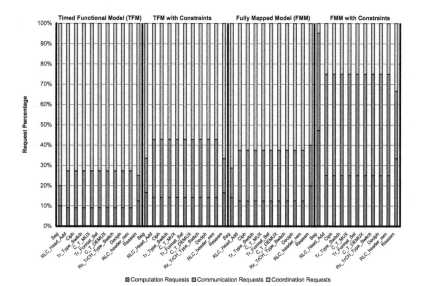

Figure 8.34: UMTS Functional Model Request Analysis

functional model uses constraints to coordinate its reading/writing operations to the FIFO communication channel. The third scenario is where the pure functional model is mapped to the architecture model and uses explicit synchronization. The final scenario is when the pure functional model is mapped to the architecture and uses constraints. As shown, the percentage of coordination-based requests again dominate for all scenarios except the mapping model with constraints. If the timed functional model without constraints is taken as the baseline, the raw data (not the percentage breakdown shown in the Figure) indicates that coordination is reduced by 50% when the timed model uses constraints, by 37% on the mapped implemenation, and by 87% on the mapped version with constraints. This is due to the fact that declarative constraints remove many of the events related to coordination.

8.5.7 Conclusions

In this work, we have shown how the METRO II event-based design framework may be used to carry out design space exploration and architectural modeling. The results show that our framework is able capture the functional modeling, architectural modeling, and mapping for a UMTS case study with little overhead as compared with a baseline SystemC model. We show that the design effort involved in carrying out 48 separate mappings with a variety of architectural models is minimal when using our methodology. Within the framework, we detail the runtime spent in the three different phases of execution and provide and idea of how events move through the system.

In the future, we would like to take our learnings from this case study and apply them to make METRO II more efficient. For instance, given the low number of events processed per round, we would like to investigate optimizations for identifying the subset of events that are relevant for annotation and/or scheduling, and hiding the remainder from the three-phase execution semantics. We would also like to add more support for a wider variety of declarative constraints than are supported today. For instance, implications can be useful to trigger events based on other events, such as those exposed from blackbox IP through the use of view ports. We are currently working on carrying out a similar analysis for a video decoder application within the framework.

8.6   Conclusions

The multimedia domain features increasingly complex applications deployed on highly heterogeneous parallel platforms. In this work, we have considered four case studies from the multimedia domain. The conclusion from the first case study is that the choice of the CMD is essential for enabling accurate performance modeling and automated design space exploration. The second case study demonstrates that automated design space exploration techniques can produce design competitive with manual implementations. The third case study illustrates how concurrency in the application can be formally represented, enabling the creation and evaluation of many points in the design space. Finally, the fourth case study looks in detail at the demands such case studies make on design frameworks like METRO II.

Future work in the area of multimedia applications involves an investigation of h.264 decoding in METRO II. This will involve the use of both preprofiled and runtime processing elements similar to the case study in Section 8.5. Processing components will include ARM7, ARM9, and MicroBlaze elements all configured for h.264 core kernel routines. This exploration will serve to continue to improve the rapid design space exploration capabilities of METRO II as well as validate the findings of the UMTS case study regarding simulation overhead and the use of constraint based synchronization.

Chapter 9

Distributed System Design

"The future is here. It's just not widely distributed yet." – William Gibson, cyberpunk author

Increasingly electronic systems are being deployed in new and interesting ways. Advanced "social" robotic systems, smart cars, and large area sensor networks are all examples of designs which have a large set of distributed electronic components. Today it is not uncommon for electronic components to have to communicate with other components over large distances or in hostile environments. These designs are often in safety critical systems and represent a large financial investment. Embedded system design is no longer relegated to the design of an individual integrated circuit but rather to an entire system of components often with hard real time execution and communication requirements.

Distributed systems are characterized by processing elements connected with standardized interconnect. Such systems are usually control-oriented. Sensor networks, automotive systems, computer networks, and highly concurrent single-chip processors all fall within this category.

Distributed systems are of interest because they represent a unique challenge with regard to both their timing and execution requirements as well as the need for standardization and scalability. Systems in this domain are often in need of "guarantees" regarding their worst case performance as well. In the case of highly modular designs such as automotive systems, components must be interchangeable and designs are centered around the development of an entire family of systems with the need for in field diagnostic capabilities and a high level of fault tolerance.

This chapter will describe a set of case studies which illustrate how design for distributed embedded systems can leverage the techniques outlined in previous chapters. In particular refinement verification and mathematical programming approaches are highlighted.

9.1 Chapter Organization

This chapter begins with Section 9.2 which provides a discussion of a highly concurrent, asynchronous architecture. We illustrate how compositional component based service refinement can be used to manipulate the communication structures present in the design. In Section 9.3 we discuss an SPI-5 packet processing design and show how we proceed from early specifications to a final design along with illustrating an interface based service refinement approach. Section 9.4 describes design space exploration for a stability control application. In Section 9.5 a mathematical programming approach is outlined for automatically assigning task and message periods for distributed automotive systems. Finally Section 9.6 provides conclusions and future work.

9.2 FLEET Architecture Modeling Exploration

In addition to targeting programmable architectures, highly concurrent system architectures are excellent candidates for the design flows discussed in Chapter 5. The reason being that the individually executing processes are very separated from the asynchronous switch fabric by which they communicate. This forces a natural separation of the computation and communication models. In addition, each computation engine operates using its own scheduling mechanism. These systems can therefore by highly modular allowing design aspects to be considered in isolation. The specific highly concurrent system architecture this thesis chose to explore is the FLEET architecture. FLEET is developed at U.C. Berkeley and Sun Microsystems. Another reason for examining this architecture is that it does not have a strict specification on the amount of concurrent communication or computation and many aspects of its design are unspecified in general giving the model a lot of flexibility in terms of its implementation. FLEET is a class of architectures, not one specific instance. It is this underspecification which will allow a microarchitectural exploration of design changes in this chapter. In particular we investigate the role that refinement can play in the design process.

The simplest description of FLEET is as a collection of SHIPs. SHIPs can specify almost any functionality and at the time of this work's writing, the set of SHIPs is quite unspecified (Adder SHIPs and MemoryAccess SHIPs are some examples). SHIPs have an input and output interface. FLEET executes only one instruction, the *MOV*. A MOV specifies a source and destination. The "source" specifies a SHIP output address and the "destination" specifies a SHIP input address. The data transfers move throughout an asynchronous fabric categorized as the instruction horn, source funnel, and destination horn. This fabric makes no guarantees which MOV instruction will reach a particular shared destination first. However, it does guarantee that instructions issued to a shared source receive data from the SHIP in strict program order (this is called the "source sequence guarantee"). MOV instructions are held in *code bags* and fetched by a dedicated Fetch SHIP. The coordination of data through the switch fabric is controlled by special units called InBoxes and OutBoxes. For more information please see [Sut06] and [CLJ+01].

Figure 9.1 illustrates a high level view of a SHIP model. In the center of the figure a SHIP is shown. The SHIP type is denoted with an "ID". This ship will perform some type of computation (add, multiply, etc). In order to perform this operation, it will read data from its inputs (on its right side). This read can be a destructive or non-destructive read. The results of computation are presented on its outputs (left side). Boolean values indicate when the input can be consumed and when the output has been produced. MOV instructions propagating through the switch fabric ("instruction horn") will remove the output data from the SHIP. Again, this can be a destructive or non-destructive read operation. This removed data enters the "source funnel". This SHIP is denoted as the "source" since it is the "source" of the MOV instruction. MOVs from the same source are executed in strict program order. The removed data then enters the "destination horn" where it will reach the input of another SHIP. Data sent from two different sources to the same destination make no guarantees on the order on which they arrives. In order to remedy this situation, explicit coordination SHIPs are used.

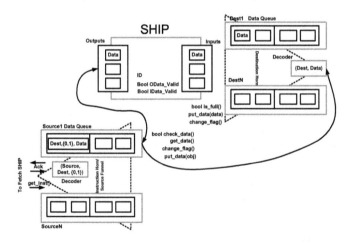

Figure 9.1: FLEET SHIP Architecture

The switch fabric can be thought of at the system level as a collection of source and destination queues (one for each SHIP). In order to replicate its asynchronous nature, a series of handshakes occur in which local variables are manipulated using "set" and "get" type functions. An example of these functions are shown in Figure 9.1 as *check_data( ), get_data( ), change_flag( ), and put_data(obj)*. The MOV instruction is an object which begins with the following information: source, destination, copy/move (non-destructive or not). When it reaches the source, the data is appended and all that is kept is the destination information. The fact that the switch fabric can be thought of as a set of buffers will be important later in this work where various implementations of these buffers are presented.

Not shown in Figure 9.1 are objects which coordinate the interaction between the switch fabric and the SHIP. These interfaces are called "inboxes" and "outboxes". Inboxes and outboxes deal with latching data as it reaches the ship. These are used to provide a consistent interface to the switch fabric.

For this work, SystemC models of FLEET services were created. Specifically the following services were produced: Fetch SHIP, Adder SHIP, RecordStore SHIP (intermediate storage) , Literal SHIP (provides static values), and Instruction Memory. Inboxes, Outboxes, and switch fabric (instruction horn, destination horn, and source funnel) were also created but are not strictly considered services since they are not exporting their services in such a way that they can be mapped to functional descriptions. This collection of services is shown in Figure 9.2. In this figure, the interfaces for each service are shown along with what type of data they operate on.

9.2.1 FLEET Communication Structure

This section we will demonstrate another refinement technique discussed previously in Chapter 6, "depth" refinement. The purpose of the example presented here is to explore how various communication structures can be replaced in a design without changing the surrounding components. These changes are verified before the substitution using the techniques discussed in Chapter 6, Section 6.6. If the reader has not examined this portion of the work, we recommend that they do so before proceeding. This substitution then becomes "correct-by-construction". This differs from the example shown in Section 9.3 because the topologies of the architecture models in this section remain the same but the operations internally in the components are changed. This case study example is performed on a model of the FLEET architecture (described in Section 9.2). Specifically it is a manipulation of its communication structure where there exists a great deal of underspecification.

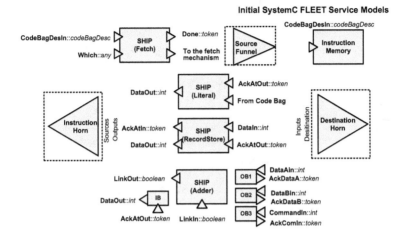

Figure 9.2: FLEET Services Created

Communication Library

The first step in developing a refinement framework is the creation of library of communication services. These services which are created first in METROPOLIS or SystemC are then transformed into LTS and have a corresponding architecture service interface. E for each LTS system S corresponds to events in the architecture service model, E_E. Events have been described in the context of the METROPOLIS environment previously in Chapter 4. The concept of event synchronization can exist as well in a SystemC model. This section will detail those architecture services in terms of their LTS representations. In all descriptions, the when an event is described as "emitted" it refers to the generation of a visible event, E_{EV}.

- *Abstract Buffer* (AB): An abstract buffer is defined to buffer data. It contains states that are representative of an $n$place buffer: $Q = \{empty, not\_empty, full\}$. $Q_o = empty$, and upon receiving a write event, it will transition to *not_empty* while emitting an event to signal a successful write. Similarly a transition occurs from *not_empty* to *full*. Read events from a consumer cause transitions to *empty* and *not_empty* states. This process also emits an event signaling a successful read. Events are not emitted when in the *full* or *empty* states for write and read request respectively. This buffer provides blocking semantics.

310

- *Copy Buffer* (CB): This architecture service structure allows consumers to copy data out of the channel without actually removing it from the channel. This requires a copy event to be emitted from the consumer. A channel transitions from *full* or *not_empty* states to *full_copy* or *not_empty_copy* and emits an event back to the consumer containing the data. In the copy states, the channel behavior is the same as the behavior at its respective *not_empty* and *full* states.

- *Random Buffer* (RB): The two previous architecture service structures assume that data organization is FIFO. This component transitions differently when there is an read event at $Q = full$. The LTS transitions to $Q = read\_choose$, where it consumes an event from an external source (i.e. random number generator), and transitions $Q = read1$ or $read2$.

Figures 9.3(a) and 9.3(b) show both the LTS for CB and RB. In each of the figures the abstract buffer (AB) is shown on the left hand side.

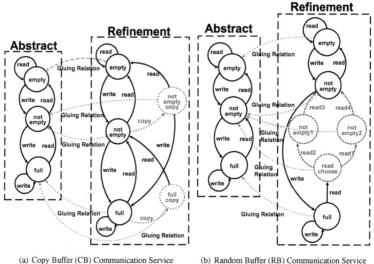

(a) Copy Buffer (CB) Communication Service   (b) Random Buffer (RB) Communication Service

Figure 9.3: LTS Communication Example #1 for FLEET

- *Delay Buffer* (DB): This service models the delay of the read and write transitions in a buffer. This is helpful in simplifying the cost model for this service as compared to the other buffers. Instead of transitioning to Q = *not_empty* when the service channel receives a write event or Q = *empty*, it will transition to Q = *writing1*, and only transition to Q = *not_empty* when it receives an *end_write* event (potentially modeled by a timer). This occurs symmetrically for read events as well.

- *Non-Blocking Buffer* (NB): AB is a blocking buffer, whereas the NB service allows a non-blocking read or write. When Q = *empty*, a read event causes a transition to Q = *nb_empty*, and it emits *read_Done* event to the consumer without data. It will take a *retr* transition back to Q = *empty*. Similarly, the service channel emits a *write_done* event when a write event is received and Q = *full* and transitions to Q = *nb_full*.

Figures 9.4(a) and 9.4(b) show both the LTS for DB and NB. In each of the figures AB is shown on the left hand side.

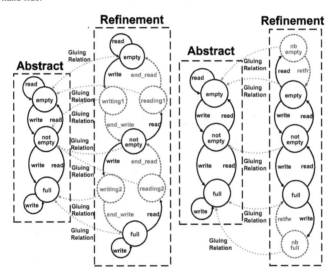

(a) Delay Buffer (DB) Communication Service (b) Non-Blocking Buffer (NB) Communication Service

Figure 9.4: LTS Communication Example #2 for FLEET

- *Larger Buffer* (LB): Currently the architectural service buffers created have an implied capacity of 2. States *not_empty* and *full* can be viewed as having 1 and 2 items respectively. It may be advantageous to have other "not-empty" states. For example, two states (or more) such as *not-empty1* and *not-empty2* can be introduced. When $Q_0 = empty$ and as write events occur, the LTS proceeds through these states to *full*. This path can not proceed back "up" the LTS since this will lead to violations of various refinement rules (see Section 6.6.1, lack of τ-divergence). Therefore there is an alternate path when a read event occurs in $Q = not\_empty2$. This has limited functionality since a return to *empty* is not possible from arbitrarily any state along this alternate path. While this is a limitation, it does allow sizing of buffers.

- *Drain Buffer* (DrB): This buffer service models the ability of a buffer to instantly fill itself or drain itself. When $Q = empty$ writes can transition as expected to $Q = not\_empty$ followed by $Q = full$. Alternately when $Q = empty$ the LTS can transition to an intermediate delay state, *d2*, which transitions immediately to $Q = full$. This is true of read events as well where $Q = full$ proceeds to *not-empty* followed by *empty* as normal. The drain operation proceeds from $Q = full$ to *d1* and finally to *empty*.

LB and DrB are shown in Figures 9.5(a) and 9.5(b). Note that Figure 9.5(b) is shown with a modified AB. DrB is not related to the native AB buffer through refinement.

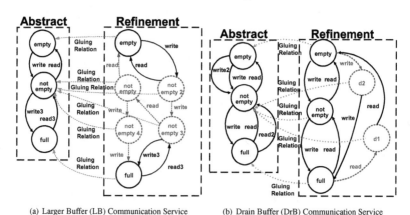

(a) Larger Buffer (LB) Communication Service   (b) Drain Buffer (DrB) Communication Service

Figure 9.5: LTS Communication Example #3 for FLEET

9.2.2 Verification Process

With the architecture communication service library complete, one must identify where there are opportunities to introduce refinement into the system. The topology of the system will not change but rather the components in the topology. This requires that the interfaces remain the same. Figure 9.6 highlights in **bold** where these opportunities exist in the FLEET architecture.

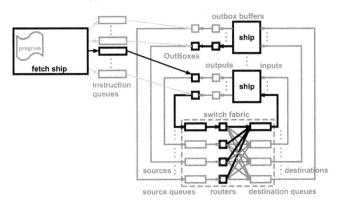

Figure 9.6: FLEET System Architecture Service Refinement Opportunities

Since the FLEET architecture is largely underspecified, there exists freedom in how to implement its communication structures. Figure 9.6 identifies the locations in FLEET where there exists the freedom to change the communication. Specifically these locations and classifications are: the Fetch SHIP , the OutBoxes , and the output of the SHIP where there are single-writer / single-reader structures. The first classification, the Fetch SHIP, acts as a producer that produces instructions to the instruction communication mechanisms and the OutBoxes consume those instructions. The second classification, the Switch Fabric, represents a multiple-writer/single-reader structure. This work's goal is to explore different ways to replace the buffers between ships, outboxes, and switch fabric. The communication service libraries developed will be used. What must be done now is to demonstrate that the modified buffers are still refinements of the original.

The refinement process for verifying architecture communication service refinement for FLEET is that which was described in Chapter 6. Using the steps outlined in Section 6.6.1, it was formally verified that each communication services in Figures 9.3(a), 9.3(b), 9.4(a), 9.4(b), and 9.5(a) are refinements of service AB given the gluing relations illustrated in the figures by the dotted arrows from each service model to AB. The service in Figure 9.5(b) is a refinement of a slightly modified AB service. This process is a verification that the communication services are a correct refinement of the abstract service in terms of control flow not data, since LTS does not infer any information on data of the system.

With the library verified, an LTS description of the entire FLEET system was created consisting of components from several smaller LTSs including the communication library just described. The FLEET system was broken into two parts for more modular and faster refinement verification (component based refinement allows for the reduction of overall transitions and states which is crucial for the running time of the algorithm). The chosen division of the entire fleet system was:

1. **The individual SHIPs with their input and output communication services** (Figure 9.7(a)). A SHIP is an LTS service component composed of $Q = \{consumer, producer\}$. It will transition from $Q = consume$ to $Q = produce$ once it receives the *read_done* event from the communication service LTS. It will then transition from *produce* to *consume* once it receives a *write_done* event from the communication service. These components are illustrated within the dotted lines of the figure. This represents the SHIP itself. Also within this space is the SHIP specific computation LTS which varies from model to model. The communication mechanisms at the producer interface and consumer interface of the SHIP each buffer the output and input data for the remaining SHIPs. These components actually represent aspects of the switch fabric. They are connected on the other interface of a producer and a consumer respectively, (integrated in the InBox and OutBox services to be described).

2. **The Fetch SHIP with all the InBoxes and OutBoxes for additional SHIPs** (Figure 9.7(b)). This part of the FLEET system consists of N InBoxes and OutBoxes (one for each SHIP input and output in the system), an instruction communication service for each InBox (a buffer service), and one Fetch SHIP. A InBox/OutBox combination (IOC) is enclosed in dotted lines in the figure and consists of two consumers and one producer LTS. Each IOC will consume an instruction from the instruction communication service (the buffer), and then according to the instruction, it consumes from its SHIP's output communication service (source of a MOV) producing data at one of the SHIP input communication services (destination of a MOV). Because LTS have no notion of data, and are only event based, it is not specified which SHIP the IOC moves data from and to. Therefore a *choice* LTS models the decision of moving from a particular SHIP output communication service to another SHIP input communication service. This choice LTS will model the information contained in a MOV instruction. The Fetch SHIP consists of N producer LTSs (one for each IOC)

and one consumer LTS. The consumer LTS will consume instructions from the memory communication service and, depending on the instruction, decide which producer should receive the data and produce data to its instruction communication service.

In order to illustrate more clearly how the SHIP, Fetch SHIP, and IOCs are connected, numbers are provided in Figure 9.7. To connect the the systems together to form the entire FLEET model, the numbered components should be viewed as corresponding between the two images. The consumer service (#1) and producer service (#2) are the services of interest for connection. Also in the figure, the buffers are labeled also with the initials of the components which can be used during the refinement process and still maintain a behavior within the specification.

Note that SynCo (described in Chapter 6) syntax defines when an LTS can take a transition in its .sync files. This is the point at which one can define that a transition is enabled when certain LTS are in certain states. Therefore, in the situations in which the system has to decide which transitions to take or which LTS's transitions to enable, a separate component was used that generates information on what choice the system should make.

LTS composition intially allowed the creation of small systems. Specifically a producer, AB, and consumer services were created as LTS objects. The producer LTS service will produce as much as the buffer allows it to, the consumer service will consume as much as the buffer allows. The buffer LTS service begins as AB and will be substituted accordingly with those library services defined earlier in the chapter. This substitution is allowed as it was verified that all refined buffer services were refinements as defined previously. Copy buffer, delay buffer, non-blocking buffer and random buffer can replace AB. However, the larger buffer (LB) was a successful refinement of a slightly modified AB. When all services were composed, this small FLEET system was shown to be a refinement (by definition the larger composed FLEET parts will also be refinements). Compositional refinement verification additionally proves that one can replace AB with any of the other communication services and maintain the refinement conditions. Table 9.1 contains a summary of the number of states and transitions in each system. The product value given is the upper bound for the total number of states existing in the system. Composition of LTS is defined in such a way that the resultant state count is considerably less than the product. The running time of SynCo is advertised as $O|SR|$ where $|SR| = |Q_R| + |T_R|$. Q_R is the number of refined states and T_R is the number of transitions in the refined system. The refined models have at most 30 states in the fetch SHIP and as few as 13 in the the delay buffer. This number is approximately double that of the abstract system but the number is still much lower than the product values. The transitions in the refined models vary between 16 and and 44. Overall these number are extremely manageable by a system with a

316

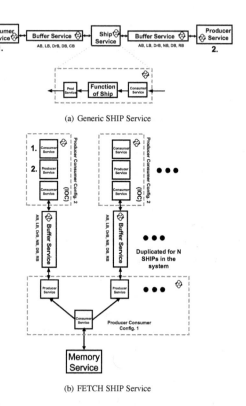

(a) Generic SHIP Service

(b) FETCH SHIP Service

Figure 9.7: LTS for Entire FLEET System Level Service Models

linear run time.

This example has illustrated that by creating very small individual LTS models (buffers, producers, consumers) of a much larger system (FLEET), an entire system can be verified. Abstract and refined models can then be interchanged freely once various state correspondences (gluing) relations have been made. The size of these systems are quite manageable and are tied to system level services by capturing the event behavior of the system level models within the individual components themselves. This process is part of a design flow which would follow the steps shown in Algorithm 7 in Chapter 6.

	Abstract			Refined		
	States		**Transitions**	**States**		**Transitions**
	Sum	**Product**		**Sum**	**Product**	
Fetch SHIP	19	648	21	30	22500	43
SHIP	16	144	18	28	8400	44
Random	8	12	11	14	108	19
Copy	8	12	11	14	120	22
Delayed	7	12	10	13	63	16
Non-Blocking	8	12	13	15	160	21

Table 9.1: FLEET LTS States and Transitions

9.3 SPI-5 Packet Processing

The previous case study illustrated aspects of compositional component based service refinement. That example demonstrated the benefits of modularity and how a communication structure's behavior could be captured efficiently. This allowed entire communication blocks to be replaced without fear of damage to overall system correctness. This section will discuss refinement verification's role in the design process as well. Specifically it examines interface based service refinement. This work is an expansion of work produced in [DRSV04]. Abstraction possibilities will be demonstrated here with an architectural design topology that goes through a number of transformations while still being shown to be a valid refinement to the initial specification.

The goal of this exercise was to analyze the architecture of an interface unit for a very high bandwidth Optical Internetworking Forum (OIF) standard, e.g., System Packet Interface Level-4 (SPI-4), Level-5 (SPI-5) [GT01] with the following requirements:

- The interface must provide the maximum bandwidth as required by the specification.

- There can be no loss of data with minimum backpressure; backpressure reduces upstream traffic flow. The architecture can generate backpressure only if the downstream system requires it.

- Determine optimally sized standard embedded memory elements. Optimal is defined as the lower bound size while functioning with no packet loss.

- The interface must support multiple input channels.

- The insertion of idle cycles (no activity) when packets are of different size must be minimized.

For this work a simple SPI-5 data generator model was designed that generates packet data every clock cycle for given number of channels. Two types of parameters are considered: architecture and application. Architecture parameters help to determine the microarchitecture parameters for various application parameters. One should choose a set of architecture parameters that match all application parameters for a given specification to ease the mapping process. Custom architecture services were created which could be parameterized to do this investigation. Additionally the architecture services were composed in a variety of ways to create a set of platforms (each at a different abstraction level) as will be described.

9.3.1 Application Parameters

Application parameters are defined as part of the functional specification. Given the specification these parameters were the aspects we felt captured the system level decisions that needed to be made.

- **Number of Channels** (N_P) - Number of PHY units. A PHY unit is a physical layer device that converts the serial optical signal to an electrical signal.

- **Data Rate/Channel** (B_P) - What configuration of PHY units can be used. The electrical signals from the PHY units are typically in a byte or multiples of bytes format.

The application parameters define what different types of PHY units the design could interface with. This is a tradeoff between flexibility and clock frequency. A smaller N_P will deliver data at a higher clock frequency, B_P, since the bandwidth must be maximized.

9.3.2 Architecture Parameters

The objective of architecture service design and development was to devise a robust architecture that will allow the application design to interface with different types of systems. To evaluate the various architectures, the following two parameters were defined:

- **Number of channels/bus** (N_B) - Number of channels that can simultaneously deliver data.

- **Bytes of data/bus** (B_B) - Number of bytes delivered from each channel.

In the simplest case $N_P = N_B$ and $B_P = B_B$, i.e., the system is configured to accept and deliver the data when all channels are equivalent. However, each channel can deliver data at a different rate. The only characteristic known is that the aggregate data from all the channels will be no more than 40 Gb/sec.

This work supports up to 16 channels. For 16 channels, each channel must be 2.5 Gb/sec, to get an aggregate of 40 Gb/sec rate ($2.5*16$). Alternatively, 4 channels can each be 10 Gb/sec. This aggregate data rate is a function of the SPI-5 specification.

The various parameters also control the internal bus width and internal clock frequency. For example:

- Bus Width (B_W) = $N_B * B_B$ - The bus width is the number of channels times the number of bytes of data for each channel.

- $N_P*B_P*C_{SYS}/B_W \rightarrow$ (Ideally small as possible); where C_{SYS}, is the system clock frequency. This indicates the backpressure needed. Values greater than 1 indicate the bus capacity has been exceeded.

An interface unit that can interact with the PHY units and deliver data to the downstream modules can now be designed. However, the effect of the decisions at this level will impact the operation and storage requirements of the design.

Example: Consider $N_B = 8$ (channels per bus) and $B_B = 4$ (bytes per channel), then $B_W = 32$ (Bytes). Then when $B_P = 4$ (data rate per channel) and $N_P = 16$ (number of channels), the data sequence is produced as shown in Table 9.2. For the first clock cycle, the 1st byte from the selected channels appears on the bus. In the second clock cycle, the 1st byte from the remaining channels is delivered.

Data Transfer Byte ($B_P = 4$)							
1st SOP Byte		2nd		3rd		4th EOP Byte	
Channels using Bus ($N_b = 8$ and $N_p = 16$)							
0-7	8-F	0-7	8-F	0-7	8-F	0-7	8-F
Clock Cycle							
C = 0	C = 1	C = 2	C = 3	C = 4	C = 5	C = 6	C = 7

Table 9.2: Example of SPI-5 Data Generation Using the Architecture and Application Parameters

For this system configuration, it will take 8 clock cycles to send 256 bytes of data over an 8 channel wide bus (16 total channels) with packets of 4 data units each. This is a simplified scenario. A more constrained implementation has been described in [RP03].

The purpose of this study was to use METROPOLIS to quickly evaluate the impact of various parameters on the entire design while minimizing the verification effort. The data generator described allows the system conditions to be quickly changed to test how modifications to the architecture topology affect the overall system performance.

9.3.3 Refinement Based Design Flow

The goal of the design flow for this case study was to (1) observe if METROPOLIS could effectively aid in the process of microarchitecture design and verification as compared to other approaches and (2) derive the architecture and application parameters described in Sections 9.3.1 and 9.3.2. The proposed design flow will simplify the microarchitecture development and help to determine which portions of the design need to be further refined with formal analysis methods.

The notion of successive platform refinement was essential in this flow. Each METROPOLIS model represented a specific platform instance. Each subsequent platform$_{i+1}$, kept a reusable abstract specification with correct behavior and equally importantly, each successive platform held the refinement relationship required with its parent platform. Theoretically any microarchitecture is a candidate for refinement. In this case, the presence of observable communication involving computation elements was required.

Platform abstraction was driven by the separation of concerns as mentioned. Beginning with the initial specification, each subsequent platform would address previous platform constraints and application and architecture parameters. At each step, refinement verification was performed. If the refinement relationship held, a set of data points concerning various metrics relevant to the design was collected. Figure 9.8 illustrates the refinement based design flow.

This methodology produced several different platforms, which exposed different aspects of the application to mapping possibilities. These platforms are referred to sequentially; they drove the microarchitecture design by revealing designs that did not meet the constraints implied by the application parameters. Simulation performance analysis drove refinement to the next platform.

9.3.4 Platform Development

The goal of platform development is to address and transform some of constraints of the previous platform and develop the optimal architecture and application parameters outlined previously. This creates a hierarchy of platforms with their corresponding successors and parents. Platforms naturally address changes to computation, communication, or coordination structure. This was natural for this ap-

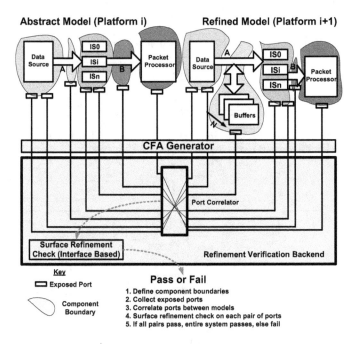

Figure 9.8: Successive Platform Refinement Methodology

plication but can be more ambiguous for other applications. METROPOLIS semantics make this relatively easy. Figure 9.9 is an illustration of all the proposed platforms.

Platform 0

Platform 0 represents the minimally constrained functionality of the initial specification. This provides the initial platform in Figure 9.9. This is a buffered producer/consumer where there is a data source (producer), some internal storage (buffer) and a packet processor (consumer). There is communication (A, B) but no notion of what architectural form they take (i.e. bus, shared memory, etc). There is only the notion of direction (read or write) and that A and B can only be accessed by one element per unit time. The initial system presents what is we term "constraint 0":

Constraint 0 - Only complete packets can be delivered to the packet processors. Partial packets have to remain in the internal storage or dropped based on other system requirements.

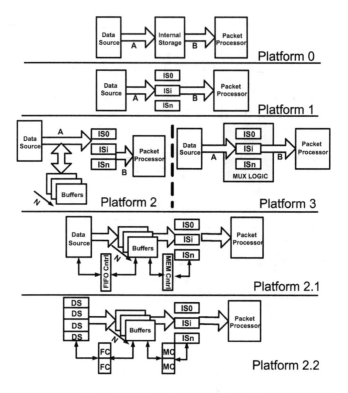

Figure 9.9: Platform Development for SPI-5

Inherent constraints (9.1, 9.2, and 9.3) are reflected by the application topology (where DS = Data Source; IS = Internal Storage; PP = Packet Processor):

$$MaxRateProduction(DS) \leq MinRateConsumption(IS) \tag{9.1}$$

$$MaxCapacity(IS) \geq MaxProduction(DS) - MinConsumption(PP) at any instant t \tag{9.2}$$

$$DataFormat(DS) = DataFormat(IS) = DataFormat(PP) \tag{9.3}$$

Equations 9.1 and 9.2 ensure that this is a lossless communication mechanism while 9.3 captures the fact that these are primitive communication mechanisms in which data is merely transferred not transformed. The next platform should look to transform some of these constraints. This transformation needs to occur to move the platform to a level which not only is closer to a real implementation but also one in which simulation performance results will be meaningful.

Platform 1

The internal storage for each channel depends on the data rate of the channel. A simple implementation due to this constraint can be stated as a set of refined constraints on the internal storage.

Constraint 1 - B_P is an application parameter; hence the internal memory must allow storage space for each channel to be dynamically adjusted. Aggregate data rate of 40Gb/sec must be preserved. The number of divisions (N_M) must equal the number of PHY units, i.e., $N_M = N_P$.

With the aggregate data rate and different data rate per PHY units, application parameters were combined as in Table 9.3.

Data Rate/Phy, B_P	Number of Channels, N_P
40 GB/Sec	1
10 GB/Sec	4
2.5 GB/Sec	16
1.25 GB/Sec	64
625 MB/Sec	256

Table 9.3: SPI-5 Application Parameter Interaction

METROPOLIS simulations indicated that for a large number of channels, the current bus architecture would not be sufficient. Therefore it was decided to restrict N_P to 1, 4 and 16.

As with the previous platform (platform 1) there are still constraints but now they generate a relationship between platforms. These constraints can be derived from the topology as before as shown in number 9.5 or from Metropolis semantics as shown in number 9.4.

$$Coordination(Platform1) > Coordination(Platform0) \qquad (9.4)$$

$$Services(Platform1) = Services(Platform0) \qquad (9.5)$$

The fact that constraint 9.5 requires that the platform have the same number of services coupled with constraint 9.4's observation of increased coordination, manifests itself as a change to the IS service. Initially it was a SCSI service. It will now become a MCSI service with each component now becoming a segmented aspect of the internal storage. This relation indicates that platform 1 will require more explicit coordination with equal processes. This will restrict behaviors, which hold a refinement relationship.

Platform 2 and Platform 3

Analysis using the above set of constraints imposes strict timing based on the clock frequency. For a large memory this will be a difficult constraint to meet. The constraint of platform 1 needs to be further refined or implemented differently. As the constraint based, successive refinement process proceeds, implementation related considerations dominate. The refined constraint can now be stated as:

Constraint 2 -The data rate and number of channel based internal storage should have pipelined writes.

The implementation with this constraint leads to:

- Using a mux-based logic organization as shown in platform 3. This scheme was not implemented due to lack of a formal refinement relationship (as discovered by the refinement verification process).

- Using an external buffer to intermediately store incoming packets (read transaction) and then pass them to the internal storage (write transaction), as shown in platform 2.

The coordination introduced in platform 1 manifests itself as control logic as shown in platform 3. This makes the coordination explicit but does not ensure refinement due to the addition of a component whose behavior is outside of the specification. Communication refinement was needed as well as reverting to a previous communication refinement of the internal storage (IS) as in platform 2.

As Figure 9.9 shows, if communication (A) is actually refined into buffers as in platform 2 then there is no need for platform 3. As hoped, this will prevent the continued growth of the coordination overhead introduced in platform 1 and the refinement of the IS into internal memory does not change the platform properties in platform 0. The design will now proceed from platform 2.

Platform 2.1

Subsequent METROPOLIS simulation analysis indicated that during peak times the read transactions dominated the system. Therefore in progressing to the next platform a constraint should be developed which will improve on this situation:

Constraint 3 - The pipelined write transaction should be independent of the read transaction.

Platform 2.1 recognizes that coordination must be added in order to manage buffers and for constraint 3 to be realized. This coordination will require two units of control introducing added coordination. This coordination will further constrain the behavior into the refinement relationship. Figure 9.9 shows this refinement became the two additional component objects added in order to provide buffer management. These are new components which are added to the buffer service. This transforms the buffer service from a MCSI service to a MCMI service.

At this point, few architecture parameters are changing, but the refinement is proceeding more closely to a final implementation.

Platform 2.2

The "final" constraint on the system was added to have independently operating PHY units. This is important because it was desirable to ensure that there were no assumptions built into the data generation and internal bus organization. The final constraint can be stated as:

Constraint 4 - Packet generation from various channels should be independent activities.

This refinement is performed on the data source and implements the application parameters, that is:

$$Number of DS = N_P \qquad (9.6)$$

$$Size of DS = B_P \qquad (9.7)$$

Ultimately this is simply an addition of components to an already MCMI service. Platform 2.2 shows a final refinement of the microarchitecture for this investigation. This computation refinement requires a coordination refinement in order to process this data properly. Therefore additional METROPOLIS quantity managers will be needed as well.

Notice that the DS block now is made up of multiple blocks. This requires a similar transformation for the FIFO Control (FC) and the memory control (MC). This final refinement will be by design a refinement of all previous platforms before it and was verified as such by the refinement verification process used throughout this section.

9.3.5 METROPOLIS Models

For the purposes of simplicity, a one-to-one mapping between functional processes and architectural services was carried out. This mapping required the construction of architectural models for
each of the platforms presented. METROPOLIS architecture service models were derived to represent
platforms 2, 2.1 and 2.2. Figure 9.10 shows a diagram of the "final" model, platform 2.2. METROPOLIS
mapping processes are provided for the DS, FC, MC, and the FIFO scheduler (FS) processes in the functional model. Parameterized, custom made architecture services were created to provide computation
services (DS, FC, MC, FS) in platform 2.2. Also METROPOLIS media reflect memory elements (buffers).
Also provided in the figure are the quantity managers for each of the services. This illustrates how the
scheduled and scheduling netlists are partitioned.

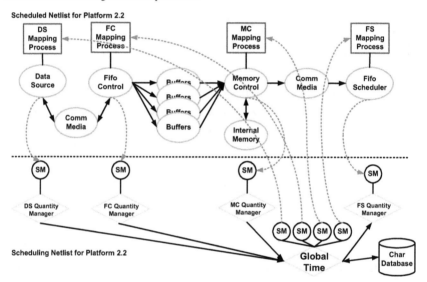

Figure 9.10: METROPOLIS Architecture Model for Platform 2.2

9.3.6 Results

After the creation of a subsequent platform$_{i+1}$, the second step to continue the development
process was refinement verification. The procedure, in keeping with the successive platform refinement

methodology, was concurrent with each subsequent platform development. The platform$_{i+1}$ is considered only if the answer to the refinement question, (Platform$_{i+1}$, Platform$_i$) was YES.

Refinement verification required the creation of a control flow automata (CFA) for both the abstract and the refined model to capture the behaviors, B, of each model. The CFA creation can be done via a backend service in METROPOLIS that extracts this information automatically (Figure 9.8). This process was covered in detail in Chapter 6, Section 6.5.

A trace, a, is determined by the traversal the CFA. This represents a potential execution of the model. Once the set of traces, B, for each model is determined, the refinement verification stage simply ensures that the behavior of the refined model is a subset of the abstract behavior.

Refinement verification via the CFA creation process is, for each process, P, in the model, M :

1. Create a CFA with the Metropolis Backend for Platform$_i$ (Ab) and Platform$_{i+1}$ (Ref).

2. Identify a cycle in the CFA, this is a trace a.

3. Add, a, to the set of behaviors, B.

4. Continue until all cycles are identified. Do this for each CFA in the abstract and refined models.

5. Compare the behaviors B_{ref} to the abstract behavior B_{ab} for the corresponding CFAs.

6. If $B_{ref} \subseteq B_{ab}$ return **YES**; Else return **NO**.

Figure 9.11 shows the control flow automata for two particular architecture services in platform 2.1 (abstract) and 2.2 (refined). FIFO scheduler is just one example of the 4 architecture service models shown in Figure 9.10. The circles are the control locations, Q. Control location 1 is the initial location, q_0. The operations, O_p, on each transition, $\rightarrow$, are specific function calls used in the model (denoted by "()") or Boolean predicates. The cycles in these CFAs represent possible execution traces of the model and are show in Table 9.4.

Naturally since these are cyclic graphs there must be some notion that each cycle may be subsequently followed by any other cycle in the set infinitely often. This work uses ω to denote this. Therefore, the abstract FIFO scheduler behavior, B_{Ab}, is $\{$**T1, T2, T3, T4**$\}^{\omega}$ and the refinement behavior, B_{ref}, is $\{$**T1, T3, T4**$\}^{\omega}$. Notice that the FIFO scheduler trace has a function call, $qData()$, which also is denoted with a ω. This is due to the loop shown in the graph containing finitely many calls to this function. This demonstrates the nested use of ω. The creation of the CFA is automatic and the evaluation of the traces via graph traversal is automated as well as discussed in Chapter 6, Section 6.5. This demonstrates refinement verification in the design flow prior to creating another platform and gathering data.

328

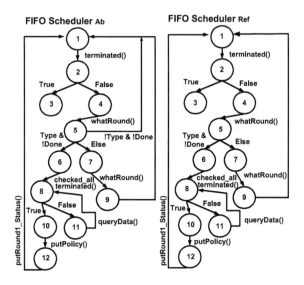

FIFO Scheduler Ab   FIFO Scheduler Ref

Figure 9.11: Sample Control Flow Automata for Abstract and Refined FIFO Scheduler

Figure 9.12 provides a sample of the data analysis possible in the design. This figure shows FIFO occupancy between subsequent platforms (2.1 on the left side and 2.2 on the right side) in combination with changes in both architecture (N_B) and application (B_P) parameters. The number of channels (x-axis) varies in increments of 1, 4, or 8. Each channel size count was coupled with 4 data rate/channel values (1, 2, 3, 4). Notice that the FIFO occupancy (y-axis) in the refined model (2.2) is bounded by the worst case (highest occupancy) in the abstract model. The data in the refined model actually indicates that FIFO occupancy is unaffected by B_P and that for all 4 settings the occupancy is the same as $B_P = 1$ in the abstract model. This type of data analysis will drive platform development in the future and demonstrates the usefulness of design exploration.

9.4   Distributed Supervisory Control System

Distributed systems in the automotive domain consist of distributed controls applications deployed onto heterogeneous distributed architectures. These architectures consist of electronic control units (ECUs) connected with standardized buses. The mapping problem for these systems involves allo-

Trace	FIFO Scheduler Traces (Function Calls)			
T1	Terminated()			
T2	Terminated()	wRnd()$^\omega$		
T3	Terminated()	wRnd()$^\omega$	wRnd()$^\omega$	
T4	Terminated()	wRnd()$^\omega$	Terminated()$^\omega$	qData ()$^\omega$
T4 cont.	putPolicy()	PR1S()$^\omega$		
$B_{ref} = \{T1, T3, T4\} \subseteq B_{ab} = \{T1, T2, T3, T4\} \rightarrow$ Refinement!				

Table 9.4: Traces from FIFO Scheduler CFAs

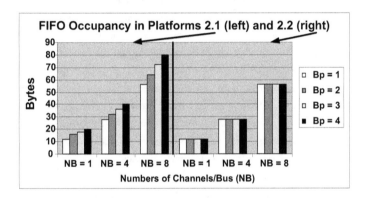

Figure 9.12: FIFO Occupancy Data for Platform 2.1 and 2.2

cating the processing tasks onto ECUs, the messages to buses, and configuring the execution of both. A variety of technical and business factors have contributed to making this problem especially challenging.

First, automobiles are increasingly becoming differentiated based on their electronic content, especially the type of active safety functionality they feature. It is estimated that a majority of the innovation and cost of a new vehicle now resides in the electronics [LH02]. Due to the large volumes, automotive electronics are price-conscious, meaning that up-front optimization effort is valuable if system costs can ultimately be reduced.

However, system-level optimization has not always been feasible in the automotive domain. Typically, hardware and software is bundled together in modules that are sold to the system integrators - the carmakers. Recently, there has been an effort from the carmakers to de-link and standardize the hardware and software portions of the system. This is being carried out under the auspices of the AU-TOSAR [HSF$^+$04] initiative. Not only does AUTOSAR enable the carmakers to carry out system-level optimization, it also enables the suppliers to compete more effectively [SV07]. The de-linking of software from hardware facilitates the modeling and optimization that will be described here.

In this section, we describe a case study where a automotive safety application is deployed onto a distributed automotive platform.

9.4.1 Stability Control Application

The embedded automotive applications considered in this work are active safety applications. These applications collect data from 360° sensors around the vehicle to understand the positioning of surrounding objects and detect hazardous conditions. On hazard detection, the active safety functions attempt to inform the driver or provide control overlays to reduce the risk to the occupants of the vehicle. Most of these functions are high-level controls which drive low-level actuation loops.

Application descriptions can be viewed as directed graphs, with nodes representing function blocks and edges representing data dependencies. Data dependencies are messages that are sent between blocks. The application description is further characterized by end-to-end latency constraints along certain paths from sources to sinks. These constraints bound the execution time of certain chains of tasks to occur within the predefined time. The active-safety application considered here has been obtained through collaboration with General Motors.

The application is a limited-by-wire system that implements a supervisory control layer over the steering, braking and suspension systems. The objective is to integrate active vehicle control subsystems to provide stability and comfort to the occupants. The high-level view of the functionality is shown in Figure 9.13. The supervisor plays a command augmentation role for braking, suspension and steering by using sensors to collect data from the environment. This supervisory two-tier control architecture enables a flexible and scalable design where new chassis control features can be easily added into the system by only changing the supervisory logic.

9.4.2 Architectural Platform

Architectural platforms for the systems we consider are highly distributed. Distributed archi-tectures supporting the execution of hard real-time applications are common not only for automotive, but also for avionics and industrial control systems. To provide design-time guarantees on timing constraints, different design and scheduling methodologies are used. For instance, avionics systems are often built

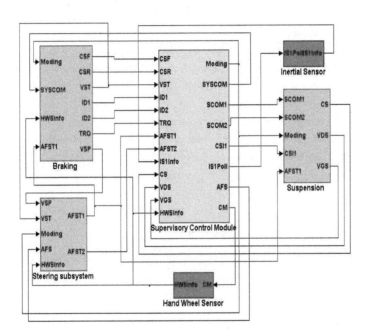

Figure 9.13: Functional Model for Distributed Supervisory Control System

based on static, time-driven schedules. Due to resource efficiency and ultimately price concerns, many automotive systems are designed based on run-time priority-based scheduling of tasks and messages. In this work, we consider architectures which feature two standards supporting this model: the OSEK operating system standard [OSE06] and the CAN bus arbitration model [Bos91]. OSEK OSs provide preemptive priority-based run-time task execution while CAN buses allow non-preemptive priority-based run-time message transmission.

ECUs feature CAN controllers which provide send and receive functionality. The number of buffers available in the CAN controller is strongly correlated with price, and typical CAN controllers have a limited number of send/receive buffers. This necessitates different tasks on the ECU to share CAN controller buffer space. Due to this limited buffer space and because ECUs are not synchronized with each other, message loss and duplication may occur.

Different interaction models may be implemented at the interface between any two resource domains (such as an ECU and a bus). The simplest interaction model consists of periodic activation with asynchronous communication, where all interacting tasks are activated periodically and communicate by means of asynchronous buffers based on non-blocking read/write semantics. Similarly, message transmission is triggered periodically and each message contains the latest values of the signals that are mapped into it.

9.4.3 Choosing the Model of Computation

To choose an appropriate MoC for systems in this domain, we will first consider the current design methodology. In current industrial practice, the application is typically described in Matlab/Simulink. Simulation is typically carried out to validate the application model. The architectural model is not captured explicitly during the design flow. Code generation from the application model to the architectural model is carried out using RealTime Workshop (RTW). However, RTW assumes that the generated code is deployed on uniprocessor architectures. In practice, since the architectural platform can lose and duplicate messages, the semantics of the architecture and the functionality are not compatible. Therefore, a correct-by-construction deployment is not guaranteed. The system needs to be re-validated, even if the functionality has already been tested.

In current industrial practice, system validation is aided by two techniques: overdesign and in-vehicle testing. Overdesign involves sending additional messages between tasks to compensate for expected message loss. Regardless of overdesign, in-vehicle testing is required to assure the designer that the system works as expected. Unfortunately, in-vehicle testing is expensive and occurs late in the design cycle.

Finding a suitable MoC to capture automotive systems involves a tradeoff between expressiveness and analyzability, just as in the multimedia domain. One choice is to keep the more analyzable synchronous reactive model of computation for the application, and add wrappers to the architecture to ensure compatibility. The second choice is to expose architectural non-idealities in the application model.

The first choice involves making the architecture compatible with the synchronous reactive model. This may involve adding synchronization capabilities between different ECUs in the architecture, expanding buffer sizes for the transmission, or adding retransmission capabilities to the architectural platform. These options directly increase either architectural cost or utilization, but allow better analysis.

The second choice involves making the lack of synchronization, message loss, and message duplication visible in the functional model. This "lossy" MoC does not require any wrappers to be defined for the architectural model. However, since the functional model becomes more expressive under this MoC, the analysis that can be carried out is much weaker. We explore the second MoC further in this work, since it more easily captures the current industrial situation, and leads to interesting automated mapping problems.

More specifically, the execution model considered in this work is the following. Input data (generated by a sensor, for instance) is available at one of the system's ECUs. A periodic activation signal from a local clock triggers the computation of an application task on this ECU. Local clocks on different ECUs are not synchronized. The task reads the input data, computes intermediate results, and writes them to the output buffer from where they can be read by another task or used for assembling the data content of a message. Messages - also periodically activated - transfer the data from the output buffer on the current ECU over the bus to an input buffer on another ECU. Tasks may have multiple fan-ins and messages can be multicast. Eventually, task outputs are sent to a system output (an actuator, for instance).

In the lossy MoC, concurrent processes communicate through FIFOs. FIFOs are fixed length and may nondeterministically delete or duplicate messages. FIFO access is non-blocking. The MoC is timed, so processes are triggered periodically, each with a specific rate and a relative phase. If all FIFOs are one-place buffers and processes are triggered at the same rate and with the same phase, then the MoC is equivalent to the synchronous reactive model. Otherwise, depending on the configuration used for the rates, phase offsets, FIFO sizes, and message duplication and deletion policies, the functional model can behave as it would when mapped to an actual architectural platform.

When implementing feedback control applications in this fashion, the (quasi) periodic stream of actuator commands may be based on sensor data taken a variable number of samples in the past, depending on how the various clocks align. For this reason, the control algorithms are typically designed

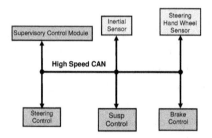

Figure 9.14: Architectural Model for Distributed Supervisory Control System

favoring robustness over performance. Techniques like time-stamping and sequence counters are sometimes used at the application level to compensate for variations and to improve robustness. Nonetheless, hard bounds on latency and periodicity are provided as implementation requirements.

9.4.4 Manual Design Space Exploration

The lossy MoC has been used to model a distributed supervisory control system from General Motors. Implementation of the MoC and detailed architectural modeling is carried out in METROPOLIS. A more complete description of this case study is given in [ZDSV+06a].

The architecture consists of 6 ECUs and includes 2 smart sensors connected over a high-speed CAN communication bus. The two sensors are the hand wheel sensor which obtains steering position and the inertial sensor for yaw rate lateral acceleration measurements. The interfaces to the body and powertrain vehicle subsystems are not modeled. The architectural model is shown in Figure 9.14.

Details of the ECU modeling are shown in Figure 9.15. Multiple software tasks can execute on the ECU. Initially, they interact with the middleware, which provides both location and access transparency. Location transparency means that the data sources and sinks remain hidden from the software tasks. For instance, communication may be mapped to the CAN bus for remote tasks or a local buffer in the case of communication between multiple tasks on the same ECU. Access transparency means that the internal representation as well as access policies for shared communication resources are also hidden from the software tasks. The RTOS model implements an OSEK-compliant priority-based preemptive scheduler. The CAN driver transfers messages between the middleware buffers and the CAN controller. The CAN controller has bus sender and receiver processes that execute concurrently with the software tasks.

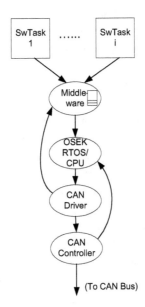

Figure 9.15: Details of ECU Modeling

A variety of options can be explored during design space exploration including:

- Allocation of tasks to ECUs

- Priority assignments for tasks and messages

- Packing of signals to messages

- Configuration for CAN controllers and drivers

For this system, when carrying out mapping between the functionality and the architecture, there is a one-to-one mapping between the tasks in the application and the ECUs in the architecture. For this case study, only the CAN controller and driver configuration are varied.

The objective of design space exploration is to mitigate the effects of priority inversion within the system. Priority inversion is a well known but often overlooked problem in embedded real-time systems. The basic scenario involves three or more tasks and a shared resource.

When the shared resource needs mutually exclusive access to a shared resource, it is generally accepted that a higher priority task cannot execute until the resource is "released", even if the task "occupying" the resource has lower priority: no preemption can take place. This phenomenon is not referred to as priority inversion, even though the higher priority task may need to wait for the completion of the lower priority task. The actual priority inversion comes from the presence of one or more intermediate priority tasks that do not need the shared resource and that may preempt or delay the completion of the lower priority task, which in turn delays the completion of the higher priority task. This latter phenomenon occurs between lower priority tasks and the higher priority task(s) without any mutual exclusion constraints that would otherwise justify it.

In the models used for this case study, this inversion may occur between messages of different priority originating from the same ECU, when transmit buffers in the CAN controller are shared between them. It is fairly common to have a single transmit buffer shared by all transmitted messages from a single ECU, making this problem especially acute.

If a high priority message is queued in the middleware while a low priority message resides in the transmit buffer of the CAN controller, the high priority message will be blocked until the low priority message is successfully transmitted. Again, this is not the inversion per se, but when intermediate priority messages are transmitted on the CAN bus by other nodes, they effectively delay the transmission of the low priority message and, consequently, of the high priority message.

In the case study, described further in [ZDSV$^+$06a], differently sized elements are used for the CAN controller transmit buffers. The system model accurately captures the effect on varying the transmit buffer sizes on the transfer times for messages of low, medium, and high priorities.

9.5 Experimental Vehicle System Synthesis

The correct deployment of active-safety applications on distributed automotive architectures requires end-to-end latency deadlines to be met. This is challenging since deadlines must be enforced across a set of ECUs and buses, each of which supports multiple functionality. The need for accommodating legacy tasks and messages further complicates the scenario.

In this section, we *automatically* assign task and message periods for distributed automotive systems. This is accomplished by leveraging schedulability analysis within a convex optimization framework to simultaneously assign periods and satisfy end-to-end latency constraints. Our approach is applied to an industrial case study as well as an example taken from the literature and is shown to be both effective and efficient.

9.5.1 Experimental Vehicle

This application supports advanced distributed functions with end-to-end computations collecting data from 360° sensors. The actuators consist of the throttle, brake and steering subsystems and of advanced HMI (Human-Machine Interface) devices. A total of 92 tasks exchange 196 messages in the block diagram for the application.

End-to-end deadlines are placed over paths between 12 pairs of source-sink tasks in the application. A change of data at the input of the path must lead to a corresponding change at the path output within the specified end-to-end latency. Most of the paths follow a six-stage structure: sensor preprocessing & sensory fusion, object detection, selection of target objects in the environment, core functions, vehicle longitudinal & lateral controls with actuator arbitration & planning, and, finally, low-level loops of the actuators themselves. Most of the intermediate stages are shared among the tasks. Therefore, the blocks in the application graph are quite densely connected. Despite the small number of source-sink pairs, there are 222 unique paths among them.

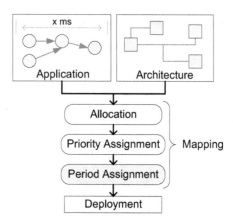

Figure 9.16: Period assignment within the overall design flow

9.5.2 Design Flow

The period assignment problem addressed in this work tackles a part of the larger design flow shown in Figure 5.7. Mapping deploys functional blocks to tasks and tasks to ECUs. Correspondingly, signals are mapped into local communication or messages that are exchanged over the buses. We further divide the mapping step into three stages: allocation, priority assignment, and period assignment. Allocation is the first stage and assigns tasks to ECUs and messages to buses. Each task is allocated to a single ECU while each message is allocated to a single bus. The second stage assigns priorities to both tasks and messages. The last stage assigns periods to task and messages.

In this work, we restrict the focus to the period assignment stage. Given an allocation and priority assignment for both tasks and messages, our approach automatically assigns periods for all tasks and messages in order to satisfy the end-to-end latency requirements. The results of period assignment may trigger design iterations over the allocation and/or priority assignment stages when a feasible solution cannot be found or when the design can be further improved by changing the allocation or reassigning the priorities (for example, following the rate monotonic rule).

9.5.3 Prior Work

Both static and dynamic priority, distributed as well as centralized scheduling methods have been proposed in the past for distributed systems. Static and centralized scheduling is typical of time triggered design methodologies, like the Time-Triggered Architecture (TTA) [KDK$^+$89] and its network protocol TTP and of implementations of synchronous reactive models, including Esterel and Lustre [BCE$^+$03]. Also, the recent FlexRay standard [Fle06] for high speed communication in automotive systems provides two transmission windows, one dedicated to time-driven periodic streams with static design-time assignment of transmission slots, and the other for asynchronous event-driven communication.

Priority-based scheduling is also very popular in control applications and is supported by the native CAN network arbitration protocol. The worst case transmission latencies of CAN messages (with timing constraints) have been analyzed and discussed in past research work [TBW95]. Also, the OSEK operating system standard for automotive applications supports not only priority scheduling, but also resource sharing with predictable blocking times. Priority-based scheduling of single processor systems has been thoroughly analyzed with respect to worst case response time and feasibility conditions [HKL94].

End-to-end deadlines have been discussed in research work in the context of both single-processor as well as distributed architectures. The synthesis of task parameters (activation rates and offsets) and (partly) of task configuration itself in order to guarantee end-to-end deadlines in single processor applications is discussed in [GHS95]. Later, the work has been tentatively extended to distributed systems [SH96] where a set of design patterns are applied to meet the deadlines using offset-based scheduling.

The periodic activation model with asynchronous communication can be analyzed quite easily in the worst case, because it allows the decomposition of the end-to-end schedulability problem into local problem instances, one for each resource (ECU or bus). This is not true in the case of data-driven activation models, where local schedulers have cross dependencies due to the propagation of activation signals. In this case, the problem of distributed hard real-time analysis has been first addressed by the holistic model [PEP02] based on the propagation of the release jitter along the computation path.

While the prior work provides analysis procedures with reduced pessimism, the synthesis problem is today largely open, except for [RJE05], where the authors discuss the use of genetic algorithms for optimizing allocation and priority assignments with respect to a number of constraints, including end-to-end deadlines and jitter.

9.5.4 Representation

The systems we consider can be represented with a weighted directed graph (O, L) and a set $\mathcal{R}$. O is the set of vertices denoting the schedulable objects (tasks and messages), L is the set of edges representing the flow of information (data dependencies), and $\mathcal{R}$ is a set of shared resources supporting the execution of the tasks (ECUs) and the transmission of messages (buses).

- $O = \{o_1, \ldots, o_n\}$ is the set of schedulable *objects* implementing the computation and communication functions of the system. An object o_i represents either a task or a message and is characterized by two parameters: a maximum time requirement c_i and a resource R_j to which it is allocated $(o_i \rightarrow R_j)$. All objects are scheduled according to their priority and a total order exists between the priorities of all objects on each resource. The object is periodically activated with a period t_i. r_i is the worst case response time of o_i, representing the largest time interval from the activation of the object to its completion in case it is a task, or its arrival at the destination in case it is a message. The response time of an object includes its own time requirement as well as the time spent waiting to gain access to the resource.

- $L = \{l_1, \ldots, l_m\}$ is the set of *links*. A link $l_i = (o_h, o_k)$ connects an object o_h (the source) to object o_k (the sink). One object can be the source or sink of many links. At the end of its execution or transmission, an object delivers results (task) or its data content (message) on all outgoing links. For any link, the sink object is activated by a periodic timer and, when it executes, reads the latest signal value that was transmitted over the link.

- $\mathcal{R} = \{R_1, \ldots, R_z\}$ is the set of logical resources that can be used by the objects to carry out their computations. Resources are either ECUs or buses and are scheduled with a priority-based scheduler.

A *path* p is a finite sequence of objects $(p \in O^*)$ that, starting from $o_i = src(p)$, reaches $o_j = snk(p)$ with a link between every pair of adjacent objects. o_i is the path's source and o_j is the sink. Sources are activated by external events, while sinks activate actuators. Multiple paths may exist between each source-sink pair. The worst case end-to-end latency incurred when traversing a path p is denoted as ℓ_p. The *path deadline* for p, denoted by d_p, is an application requirement that may be imposed on selected paths.

The graph in Figure 9.17 can be used to explain the representation. It consists of 8 tasks and 5 messages allocated to 3 ECUs and 1 bus respectively. For each object, the time requirement is given by the value inside the object, while the priority is given by the value beside it. Three paths are present in the example; two of which have deadlines associated with them. Shaded nodes denote external events or actuators.

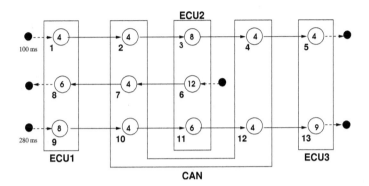

Figure 9.17: An example system representation for vehicle system synthesis

Object Schedulability

Object schedulability requirements in the system ensure that each task and message is processed every activation. This requirement enforces the assumption that objects are not queued for later processing. This assumption is compatible with all modes of the OSEK standard. The constraint that must be met is:

$$r_i \le t_i \qquad \forall i \in O$$

Resource Utilization

Resource utilization constraints place an upper bound on the fraction of time a resource may spend processing its objects. The utilization must always be less than 100%, and may be constrained further due to the designer-imposed restrictions. Resource utilization is calculated as:

$$\sum_{i:o_i \to R_j} \frac{c_i}{t_i} \le u_j \qquad \forall R_j \in \mathcal{R}$$

To calculate the utilization on a resource R_j, we take the sum of the processing time divided by the period for all objects which are allocated to that resource ($i : o_i \to R_j$). u_j is the utilization bound for the resource and may be set to less than 1 for reasons such as future extensibility, where the ability to add additional tasks or messages to the resource late in the design cycle is important.

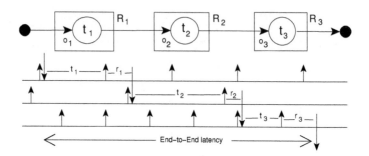

Figure 9.18: End-to-End Latency Calculation

End-to-End Latency

The worst case end-to-end latency can be computed for each path by adding the worst case response times and the periods of all the objects in the path:

$$\ell_p = \sum_{k:o_k \in p} t_k + r_k$$

In the worst case, as shown in Figure 9.18, an external event arrives immediately after the completion of the first instance of task o_1. The event data will be read by the task on its next instance and the result will be produced after its worst case response time, that is, $t_1 + r_1$ time units after the arrival of the external event. Since there is no coordination between tasks on separate resources, the situation repeats in the worst case for each link in the path. To get more precise results, the best case response time v_i of any predecessor object o_i should be subtracted from the period t_i in the previous formula. However, in most cases, including the case studies in Section 9.5.6, $v_i \ll t_i$ and v_i can be ignored.

For multiple communicating tasks with harmonic periods on the same ECU, the analysis can be less pessimistic if we assume that the designer can select the relative activation phase of all tasks. In case the sink task is activated with a relative phase with respect to the source equal to its worst case response time, then the contribution of the pair to the end-to-end latency can possibly be reduced. Let o_1 and o_2 be two tasks on the same ECU that appear (in that order) in a path with an end-to-end deadline. If $t_1 = kt_2$ is satisfied, where $k \in \mathbb{N}^+$, then t_2 is *oversampled-harmonic* with respect to t_1. Similarly, if $kt_1 = t_2$, where $k \geq 2$, then t_2 is *undersampled-harmonic* with respect to t_1. Latency analysis for these situations is developed in [NGK$^+$07] and summarized in Table 9.5.

Condition	Path Fragment Latency
Non-local or non-harmonic	$r_1 + t_1 + r_2 + t_2$
Local oversampled-harmonic	$r_1 + t_1 + r_2$
Local undersampled-harmonic	$r_1 + r_2 + t_2$

Table 9.5: Latency over local harmonic path fragments

Response Time Analysis

The key in adjusting object periods to meet end-to-end latency constraints is determining the relationship between object periods and response times. Response time analysis is also important in the calculation of object schedulability. The response time relationships are similar, but not identical, for tasks and messages. The analysis in this section summarizes work from [HKL94] and [TBW95].

Task Response Times   In a system with preemption and priority-based scheduling, the worst case response time r_i for a task $o_i \in T$ depends on the computation time requirement c_i for the task itself as well as the interference from higher priority tasks on the same resource. r_i can be calculated using the following recurrence:

$$
\begin{aligned}
w_i(q) &= (q+1)c_i + \sum_{j \in hp(i)} \left\lceil \frac{w_i(q)}{t_j} \right\rceil c_j \\
r_i &= \max_q \{ w_i(q) - qt_i \} \\
\forall q &= 0 \dots q^* \text{ until } r_i(q^*) \le t_i
\end{aligned}
\tag{9.8}
$$

Where $j \in hp(i)$ refers to the set of higher priority tasks on the same resource. The need of evaluating the first q instances inside the busy period is caused by the uncertainty about the instance which causes the worst case response time. A lower bound on the worst case response time can be obtained by restricting the calculation to the first instance ($q = 0$). This bound is tight in case tasks complete their work within a single period, i.e. $r_i \le t_i$ for all o_i. In this case, the formula can be simplified as:

$$
r_i = c_i + \sum_{j \in hp(i)} \left\lceil \frac{r_i}{t_j} \right\rceil c_j \qquad \forall o_i \in T
\tag{9.9}
$$

Note that the term $\left\lceil \frac{r_i}{t_j} \right\rceil$ indicates the maximum number of preemptions from a higher priority task j. The numerator indicates the amount of time that the task is vulnerable to preemption, whereas the denominator indicates how often the higher priority task j is activated. The ceiling function is used since this is a worst-case analysis.

344

Message Response Times   Worst case message response times are calculated similarly to worst case task response times. The main difference is that message transmission on the CAN bus is not preemptable. Therefore, a message o_i may have to wait for blocking time b_i, which is $\max_{j \in lp(i)} c_j$ where $lp(i)$ is the set of lower priority messages that are allocated to the same bus as o_i. Likewise, the message itself is not subject to preemption from higher priority messages during its own transmission time c_i. The response time relationship is:

$$
\begin{aligned}
& w_i(q) = b_i + qc_i + \sum_{j \in hp(i)} \left\lceil \frac{w_i(q)}{t_j} \right\rceil c_j \ (w_i > 0) \\
& r_i = \max_q \{ c_i + w_i(q) - qt_i \} \\
& \forall q = 0 \ldots q^* \text{ until } r_i(q^*) \leq t_i
\end{aligned}
\tag{9.10}
$$

Again, a lower bound on r_i can be computed by only considering the first instance ($q = 0$) and the formula is simplified as:

$$
r_i = c_i + b_i + \sum_{j \in hp(i)} \left\lceil \frac{r_i - c_i}{t_j} \right\rceil c_j \qquad \forall o_i \in \mathcal{M}
\tag{9.11}
$$

In calculating the number of preemptions from a higher priority message j, the difference from Equation 9.9 is that a message is not vulnerable to preemption while it is being transmitted.

9.5.5   Period Optimization Approach

From the relationships given in Equations 9.9 and 9.11, it is apparent that the response time of each object is related to the periods of higher priority objects on the same resource. Intuitively, reducing the period of an object will increase the response times of other objects with lower priorities on the same resource. The end-to-end latencies of multiple paths may be affected as a result. Also, modifying object periods also affects the utilization of the resource. Lowering the period for an object increases the utilization of the resource. Finally, lower periods also makes object schedulability more difficult.

If object periods are modified individually, then achieving convergence is difficult, since any change to one period affects many others. Instead, we concentrate on mathematical programming (MP) techniques, which simultaneously consider modifications to the periods of all objects.

The benefits of a MP optimization approach are particularly relevant to the period synthesis problem. First, in assigning periods, there are a large number of interdependencies between the objects on different paths. Considering one path at a time is not guaranteed to find a feasible, let alone optimal, solution. MP approaches consider all constraints simultaneously. Next, and more importantly,

MP approaches can be customized with system-specific issues by simply adding additional constraints. Whereas other solution mechanisms are brittle to changes in the problem assumptions, MP approaches can adapt to different problem assumptions or partial solutions. For example, the existence of legacy tasks and messages whose periods are fixed or otherwise restricted can be handled quite easily with additional constraints.

This section is organized as follows. First, the problem is captured with a generic mathematical programming formulation. Next, two specialized forms of mathematical programming - geometric programming (GP) and mixed-integer geometric programming (MIGP) - are described. The period optimization problem is defined as an MIGP and a GP approximation is developed. Approximation error is reduced by an iterative procedure.

Mathematical Programming Formulation

The period assignment problem is defined over the following sets: the objects O, which are partitioned into messages $\mathcal{M}$ and tasks $\mathcal{T}$, the set of resources $\mathcal{R}$, and the paths with end-to-end constraints $\mathcal{P}$. All objects $o_i \in O$ have associated computation time parameters c_i, lower bounds on periods n_i, and upper bounds on periods x_i. Additionally, messages $o_i \in \mathcal{M}$ have associated blocking times b_i. Path deadlines d_p are specified for all $p \in \mathcal{P}$. u_j are the maximum permitted utilization values for all resources $R_j \in \mathcal{R}$. The main decision variables for all $o_i \in O$ are the periods t_i while the response times r_i are used as helper variables.

The problem to be solved can be formulated as follows:

$$min. \qquad \sum_{o_i \in O} r_i \qquad\qquad\qquad (9.12)$$

$$s.t. \qquad \sum_{k:o_k \in p} t_k + r_k \le d_p \qquad \forall p \in \mathcal{P} \qquad (9.13)$$

$$r_i = c_i + \sum_{j \in hp(i)} \left\lceil \frac{r_i}{t_j} \right\rceil c_j \qquad \forall o_i \in \mathcal{T} \qquad (9.14)$$

$$r_i = c_i + b_i + \sum_{j \in hp(i)} \left\lceil \frac{r_i - c_i}{t_j} \right\rceil c_j \quad \forall o_i \in \mathcal{M} \qquad (9.15)$$

$$r_i \le t_i \qquad \forall o_i \in O \qquad (9.16)$$

$$\sum_{i:o_i \to R_j} \frac{c_i}{t_i} \le u_j \qquad \forall R_j \in \mathcal{R} \qquad (9.17)$$

$$n_i \le t_i \qquad \forall o_i \in O \qquad (9.18)$$

$$t_i \le x_i \qquad \forall o_i \in O \qquad (9.19)$$

The objective function can be selected according to the optimization goals. 9.12 corresponds to the minimization of average response time over all objects in the system. However, a different choice related to the extensibility of the solution can also be used. For instance, minimizing the maximum resource utilization.

Constraint 9.13 ensures that the path deadlines are met. Note that the less pessimistic path latencies from Section 9.5.4 can be substituted here when possible. Constraints 9.14 and 9.15 relate the node response times to the computation times and periods, according to Equations 9.9 and 9.11. Constraint 9.16 adheres to the assumption that response times are lower than object periods and enforces object schedulability. Resource utilization is bounded by Constraint 9.17.

Finally, even when there are no explicit end-to-end deadlines imposing a constraint on the maximum execution periods of tasks and messages, such bounds may be specified separately – especially for feedback control applications – as in Constraints 9.18 and 9.19.

Depending on system-specific situations, additional constraints may be added that relate the periods of different objects. For instance, periods for two objects o_i and o_j may be constrained to be equal, i.e. $t_i = t_j$, or with a given oversampling ($t_i = nt_j$) or undersampling ($mt_i = t_j$) ratio (where n and m are positive integer constants). A more generic requirement might be to ensure that the objects are undersampling or oversampling with some unknown integer proportionality k between the periods. For example, $t_i = kt_j$ where $k \in \mathbb{Z}^+$. If such constraints are defined over adjacent tasks on the same resource, the less conservative analysis from Section 9.5.4 can be used.

Geometric Programming

Geometric programming (GP) is a special form of convex programming [BV04]. GPs have polynomial time computational complexity and can be solved very efficiently by a variety of off-the-shelf solvers. After [BKVH06], a GP in standard form is:

$$
\begin{aligned}
\text{minimize} \quad & f_0(x) \\
\text{subject to} \quad & f_i(x) \le 1 \quad i = 1,\ldots,m \\
& g_i(x) = 1 \quad i = 1,\ldots,p
\end{aligned}
$$

where $x = (x_1,\ldots,x_n)$ is a vector of positive real-valued decision variables. f is a set of *posynomial* functions, while g is a set of *monomial* functions. A posynomial is the sum of monomials, where a monomial function m has the following form:

$$m(x) = cx_1^{a_1} x_2^{a_2} \ldots x_n^{a_n} \qquad c > 0, a_i \in \mathbb{R}$$

If x contains both integral and real-valued decision variables, the resulting problem is a mixed-integer geometric program (MIGP). Unlike GPs, MIGPs are not convex and cannot be efficiently solved.

In this work, we make use of the gpposy [KKMB06] solver to solve GPs. Solver interfacing is handled by the Yalmip [Lof04] framework, which can overlay a branch-and-bound approach to solve MIGP problems as well.

Mixed Integer Geometric Programming Formulation

Based on the original mathematical programming formulation, we can transform it into a mixed integer geometric program with some slight changes.

$$\begin{aligned}
& \text{min.} && \sum_{o_i \in O} r_i && && (9.20) \\
& \text{s.t.} && \frac{\ell_p}{d_p} \leq 1 && \forall p \in \mathcal{P} && (9.21) \\
& && \frac{c_i + \sum_{j \in hp(i)} z_{ij} c_j}{r_i} \leq 1 && \forall o_i \in \mathcal{T} && (9.22) \\
& && \frac{c_i + b_i + \sum_{j \in hp(i)} z_{ij} c_j}{r_i} \leq 1 && \forall o_i \in \mathcal{M} && (9.23) \\
& && \frac{r_i}{t_i} \leq 1 && \forall o_i \in O && (9.24) \\
& && \sum_{i:o_i \to R_j} \frac{c_i}{t_i \times u_j} \leq 1 && \forall R_j \in \mathcal{R} && (9.25) \\
& && \frac{n_i}{t_i} \leq 1 \qquad \frac{l_i}{x_i} \leq 1 && \forall o_i \in O && (9.26) \\
& && \frac{r_i}{t_j \times z_{ij}} \leq 1 && \forall o_i \in \mathcal{T} && (9.27) \\
& && \frac{r_i}{t_j \times z_{ij} + c_i} \leq 1 && \forall o_i \in \mathcal{M} && (9.28)
\end{aligned}$$

Constraints 9.21, 9.24, 9.25, and 9.26 are simple reformulations of their counterparts from the original formulation. z_{ij} is a new set of integer variables which captures the number of preemptions from a higher priority object j on a lower priority object i on the same resource. Note that the integrality of these variables forces the formulation to be a MIGP. Constraints 9.27 and 9.28 determine the values of these variables.

To enable this formulation to be compatible with the standard form of MIGP, we need to carry out a simple change of variables. This change of variables replaces the term $r_i - c_i$ with a new variable r_i' for all messages ($\forall o_i \in \mathcal{M}$).

Approximation

Since MIGP problems are very difficult to solve, we approximate the MIGP period optimization problem with a GP formulation. In order to cast the problem into a GP form, the interference variables z_{ij} are relaxed to real-valued variables and parameters $0 \leq \alpha_{ij} \leq 1$ are added to them. For clarity, let the approximated response time variables be s_i; then, Constraints 9.27 and 9.28 from the MIGP become:

$$\frac{s_i}{t_j(z_{ij}+\alpha_{ij})} \leq 1 \quad \forall o_i \in \mathcal{T} \tag{9.29}$$

$$\frac{s_i}{t_j(z_{ij}+\alpha_{ij})+c_i} \leq 1 \quad \forall o_i \in \mathcal{M} \tag{9.30}$$

Thus, the GP approximation consists of the objective function 9.20 with s_i in place of r_i, Constraints 9.21) – 9.26 (also with s_i in place of r_i) and Constraints 9.29 and 9.30.

If the values of all α_{ij} are 1, then the approximation is always conservative, i.e. $s_i \geq r_i$. If some $\alpha_{ij} < 1$, no such guarantees can be made. Clearly, the accuracy of the approximation depends upon the α parameters that are used.

Fixed integer harmonicity constraints can be handled directly within the GP formulation, whereas variable integer harmonicity constraints require a branch-and-bound approach with much higher complexity. However, if the number of such constraints is small, the impact on overall runtime is not prohibitive. The Yalmip framework used to solve the MPs can handle such Mixed Integer Geometric Programs without additional modifications.

Reducing Approximation Error

The α parameters in the GP formulation represent the degree of conservatism used for the approximation of the response times. Setting all $\alpha_{ij} = 1$ is a safe, but pessimistic approximation that may produce an infeasible problem instance. In this section, an iterative procedure is presented to find α parameters that preserve feasibility with reduced conservatism.

Given some set of α parameters, if the GP is feasible, optimal t_i values from the GP solution can be obtained. We can obtain the r_i values by substituting these t_i values into Constraints 9.9 and 9.11. For all $o_i \in O$, let e_i represent the relative error between the estimated and actual response times, i.e. $e_i = \frac{s_i - r_i}{r_i}$. If all $e_i \geq 0$, then the optimal GP solution results in a feasible solution to the exact problem, while if all $e_i = 0$, then the GP solution is not only feasible, but optimal. If some $e_i < 0$, then the GP has underestimated some response times and Constraints 9.21 or 9.24 in the exact problem may have been violated.

An iterative procedure can be used to assign the α parameters. A new GP problem is solved during each iteration, and the e_i values are used to recalculate the α parameters for the subsequent iteration. The procedure is summarized in Algorithm 9.

The input parameter to the procedure is f, which represents the maximum permissible estimation error. At initialization, all α_{ij} are conservatively assigned to 1. Inside the loop, the GP problem is solved and the estimated response times and assigned periods are obtained. If the problem is infeasible, then all α values are scaled, and a new GP problem is solved during the next iteration. If the GP problem in the current iteration is feasible, then the exact response times are calculated with Constraints 9.9 and 9.11. The relative error e_i and possible violations to Constraint 9.24 can then be calculated. Next, α_{ij} values are adjusted based on e_i, and are saturated either at 0 or 1 if necessary. After all exact response times have been calculated, violations to path constraints 9.21 can be checked. If none of the constraints have been violated, and if the maximum absolute estimation error is lower than the limit for all objects, the procedure terminates, otherwise the next iteration is executed with the modified α values. An iteration limit may also be specified.

9.5.6 Case Studies

The period optimization approach is validated in this section with two case studies. The first is an experimental vehicle system that incorporates advanced active safety features, as described in Section 9.5.1. The second case study is a fault tolerant distributed system taken from [TBW92].

9.5.7 Active Safety Vehicle

The architecture consists of 29 ECUs connected with 4 CAN buses, with speeds ranging from 25kb/s to 500kb/s. Worst case execution time estimates have been obtained for all tasks. Message length and bus speed is used to calculate the maximum transmission time for all CAN messages. Each ECU is allocated from 1 to 22 tasks and each CAN bus is allocated from 14 to 105 messages. The system graph contains a total of 604 links.

The deadlines are set at 300 ms for 9 of the 12 source-sink pairs, at 200 ms for two pairs, and at 100 ms for one pair. For 9 pairs of local tasks over 2 ECUs, harmonicity constraints with fixed integer constants are present. Some task and message rates are bounded explicitly, due to controller requirements and maximum sampling rates from sensors. To provide for future extensibility and a safety margin, maximum utilization parameters u_i from (9.25) are set at 70% for all ECUs and buses.

The system configuration used is a snapshot from an early study of the possible architecture

---

Algorithm 9: Iterative Period Assignment Procedure

---

1 Input Parameter = f /*acceptable error bound   */

2 **forall** $o_i \in O$ **do**

3      $\alpha_{ij} = 1$;
    /*conservative initialization   */

4 **while** *true* **do**

5   (s, t) = GP(α) /*solve the GP   */

6   **if** *GP is infeasible* **then**

7   **forall** $o_i \in O$ **do**

8            $\alpha_{ij} = \frac{1}{2}\alpha_{ij}$ /*reduce α values   */

9   **else**

10        $vior = 0$;

11        $viol = 0$;

12   **forall** $o_i \in O$ **do**

13   calculate r_i using fixpoint ;

14           $e_i = \frac{s_i - r_i}{r_i}$ /*relative approximation error   */

15   **if** $r_i > t_i$ **then**

16              $vior = vior + 1$ /*schedulability constraint violation   */

17           $\alpha_{ij} = \alpha_i - e_i$ /*α values for next iteration   */

18   ensure $0 \le \alpha_{ij} \le 1$;

19   **forall** $p \in \mathcal{P}$ **do**

20   **if** $\ell_p > d_p$ **then**

21              $viol = viol + 1$ /*path constraint violation   */

22   **if** $viol = 0 \wedge vior = 0 \wedge (\forall o_i \in O, \max(|e_i|) < f)$ **then**

23   exit ;

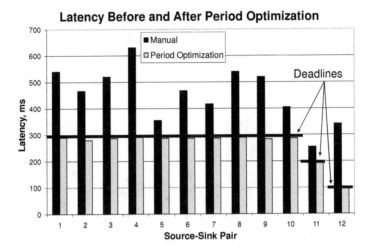

Figure 9.19: Period optimization meets all deadlines

configurations, in which the periods of task and messages had not been finalized. The preliminary manual estimates are based on designer intuition. These initial period assignments, in the worst case, do not meet any of the deadlines as shown in Figure 9.19.

Starting with all the α parameters equal to 1, we perform a GP optimization. The results of this optimization are also shown in Figure 9.19. All 222 paths between the 12 source-sink pairs meet their deadlines. The GP problem takes 24 seconds to solve on a 1.6 GHz Pentium M processor with 768 MB of RAM. The GP period assignments are quite different from the manual ones; the average period increases by 90%.

To determine the effectiveness of the iterative procedure, we can track the reduction in $\max(|e_i|)$, $\forall o_i \in O$ across several iterations. The results are shown in Figure 9.20. 15 iterations of Algorithm 9 are shown on the x-axis. The y-axis (with a logarithmic scale) shows the *maximum* absolute estimation error for the response time estimate used within the GP formulation. The *average* estimation error, not shown, drops from 6.98% to 0.009% during these same 15 iterations. Overall, the maximum estimation error is reduced by a factor of 102, while the average estimation error decreases by a factor of 780. The discrepancy between the approximated ($\sum_{o_i \in O} s_i$) and actual ($\sum_{o_i \in O} r_i$) objective values drops from 27.1% during the first iteration to 0.0045% during the final iteration.

Since the runtime per iteration is independent of the α values, the total solver time for 15

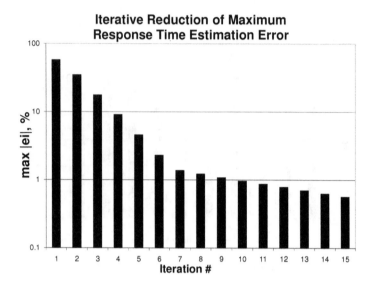

Figure 9.20: Iterative reduction in maximum estimation error

iterations is 6 minutes. Even though the α values are reduced below 1, (9.21) and (9.24) from the exact problem are not violated during any of the 15 iterations.

Finally, we can relax the 9 harmonicity constraints from fixed integer constants to integer variables. This changes the problem from a GP to a Mixed Integer GP. The *bnb* solver within Yalmip applies a branch-and-bound procedure to find the solution, and the solution time increases to 227 seconds per iteration.

Fault-tolerant Distributed System

This system is based on the example given in [TBW92] and contains task replicas allocated to different ECUs for fault tolerance. The system consists of 43 tasks and 36 messages deployed onto an architecture with 8 ECUs and a single bus. The bus is assumed to run at 250kb/s. Initial period assignments for tasks are taken from the example, while initial message periods are assumed to be equal to the source task periods. Task and message priorities are assigned using the rate monotonic rule. The initial end-to-end latencies for six paths in the system are noted.

The experiments for this system are concerned not just with meeting end-to-end delay constraints, but with reducing the path latencies as much as possible while meeting resource utilization bounds. Utilization bounds are set at 70% for each of the 9 resources, and deadlines for the six paths are set to their initial latencies.

First, we attempt to minimize average path latency on the six paths by modifying the objective function. After 15 iterations, each of which takes 1.25 seconds, the average path latency is reduced by 45%. The average utilization for the 8 ECUs is increased from 56% to 61% while the bus utilization is reduced from 74% to 52%. Next, we carry out six more experiments where we minimize each of the individual path latencies separately. The latencies for each of the six paths can be decreased an additional 17% to 63%, for a total reduction ranging between 55% and 70% from the initial latencies.

These experiments demonstrate that it is possible to customize the approach for a modified flow where the designer is interested in minimizing specific path latencies. Even without modifying allocations or priority assignments, period assignment alone is capable of significantly affecting end-to-end latencies in the system.

9.5.8 Conclusions

The continuing proliferation of distributed automotive functionality and architectures complicates the mapping process for these systems. This work provides an optimization procedure that automates the period assignment stage within mapping. First, by leveraging schedulability analysis, we develop an MIGP formulation that is applicable for systems with run-time priority-based scheduling. Next, the MIGP formulation is approximated by a GP formulation and the approximation error between the two formulations is reduced with an iterative procedure. The approach has been applied to two case studies and shown to be efficient, accurate, and extensible. In the future, this work will be integrated with the earlier mapping stages of the design flow shown in Figure 9.16 in order to carry out joint allocation, priority assignment and period assignment. We are also considering synthesizing hybrid data-driven and periodic activation models [ZNP+07] for such systems.

9.6 Conclusions

The distributed systems domain features concurrent components communicating via messages. The first two case studies considered in this chapter are concerned with distributed systems-on-chip. The latter two case studies are concerned with automotive systems, where the electronic control units are distributed throughout the vehicle. The conclusion from the first case study is that by replicating

IP for a highly distributed system and relying on refinement verification to validate that IP, the task of microprocessor verification can be simplified tremendously. The second case study demonstrates how to leverage the formal semantics of METROPOLIS to carry out system validation when real-time constraints are present. The two case studies in the automotive domain feature active safety applications deployed on distributed CAN-based architectures. In the third case study, a model of computation has been chosen for this class of systems and verified with respect to its accuracy at capturing the design space. The model of computation exposes the architectural non-idealities, allowing for more efficient implementations, at the cost of reduced functional verification capabilities. The fourth case study automates a part of the mapping problem for such systems by developing an approach based on geometric programming.

Chapter 10

Conclusions and Future Directions

"A conclusion is the place where you got tired thinking." – Martin Henry Fischer

Human history has been shaped by our constant drive to improve. Whether it be in athletic competition, business growth, or the development of new technologies, we are constantly looking for new opportunities. It comes as no surprise then to see the tremendous advancement in electronic systems over the past 20 years. From the early beginnings of personal computing to the nearly ubiquitous presence of embedded electronics today, those involved in the development of such systems have a right to be proud of their achievements. However, there is much work still to be done. Just as we have seen RTL flows and tools begin to give way to ESL methodologies, so will we see ESL give way at some point to new technologies. Recently a new area termed *Cyber Physical Systems* or CPS is emerging. CPS looks at integrating physical processes with computing systems. Applications of CPS are seen in biological, aviation, automotive, and industrial systems. With this new direction also come a slew of new challenges and opportunities. The authors of this book look forward to the future and all that it will bring with it. We would never dare to predict the future other than to say it is going to be "exciting".

This book has begun to lay the path to creating a mature ESL environment. Frameworks, design flows, and case studies have all been presented which give potential solutions to some of the problems facing the community. We hope that this work will be built upon and made more robust. We have discussed how ESL can be classified and created several ways in which offerings in this area can be organized and described. We have presented the theory of Platform-based design and shown in detail two frameworks which support it. Design flows have been shown using these frameworks while still maintaining the required separation between design aspects required of next generation design environments. Supporting techniques of design refinement and characterization have illustrated how to effectively "bridge the gap" posed by increasing abstraction levels in today's designs. Finally, we have provided a number of detailed explorations into case studies in the multimedia and distributed systems domains.

We hope that the reader now has a better sense of the state of the art in this area and is fully armed to go forth and tackle some of the problems posed in this book. We now conclude the book with future directions in the areas of theoretical foundations, design flows and needed case studies.

10.1 Future Directions

"Well, sure, the Frinkiac-7 looks impressive, don't touch it, but I predict that within 100 years, computers will be twice as powerful, 10,000 times larger, and so expensive that only the five richest kings of Europe will own them." – Prof. John I.Q. Nerdelbaum Frink, Jr., Beloved Simpsons character

10.1.1 Theoretical Directions

This section will discuss a number of potential areas for investigation involving the more theoretical aspects of the work presented throughout the book.

An interesting refinement problem when using the Micro and MacroProperties described in Chapter 6, Section 6.4 was how to select a minimal set of MacroProperties to cover the required set of Micro and MacroProperties. A potential solution is to frame the problem similarly to a covering problem in logic synthesis. Each MicroProperty could be viewed as a minterm. MacroProperties could be viewed as cubes. Cubes would be constructed in such a way that the MacroProperties at a higher level absorb cubes of lower level MacroProperties. The problem then of course becomes selecting a minimal set of cubes that cover all minterms. This can be accomplished with any number of heuristic and exact algorithms such as a heuristic PLA minimizer such as Espresso [RSV87].

Another refinement extension that we are interested in is the expression of MicroProperties and MacroProperties as assertions in a language such as SystemVerilog [Acc07]. The assertions would be created in such a way that if an assertion is generated it reveals the fact that a property has not been held. Assertion based verification could be a powerful way to introduce a more efficient event based verification scheme into the design flow.

In METRO II, constraints are defined over events generated in the first phase of execution. These constraints are then resolved in the third phase of execution. Currently only very basic constraints have been used. These include rendezvous constraints and constraints to enforce blocking read and write operations to shared FIFO elements. What needs to be done is an investigation into the expressiveness of the current constraint mechanisms in METRO II, a creation of a large library of constraints, and finally design guidelines on how to use these constraints to both correctly design a system as well as create an efficient design environment.

Architecture service models currently are not created in a very standardized way. This is a positive aspect in that the design frameworks remain flexible and offer a wide range of choices concerning the model of computation used to specify the system. However if one wants to reason about one service vs. another service in a formal way it can be difficult. Ideally the state of an architecture service could be extracted. This state could serve as a point of comparison for a wide variety of services. If the state is captured in a rigorous, formal manner, a variety of model checking techniques could be applied. Adding state to an architectural model would also help in the process of architectural validation.

In this work we illustrated that not only are embedded architectural platforms becoming more parallel, but the general-purpose platforms used to carry out the automated mapping are becoming more parallel as well. This means that the algorithms used to carry out the automated mapping need to scale with the increasing number of processors as well. Inherently sequential algorithms that enjoy favorable runtimes today will see their relative advantage diminish in the future. Simulation and optimization algorithms that can be partitioned into fairly independent units of work will fare better.

For simulation, the key is to simulate concurrent portions of the system using separate processes. This is, for instance, supported by the execution model of SystemC. However, the execution semantics of SystemC requires that the kernel execute in an interleaving manner with the processes in the model [MRH$^+$01], reducing the possible speedup on multiple processors. Unfortunately, the multi-phase nature of METRO II only exacerbates these issues.

For optimization, the challenge is to concurrently explore different part of the design space. Any approach can be modified to take advantage of concurrent execution resources, but randomized algorithms such as simulated annealing and mathematical programming approaches such as integer programming are particularly well-suited to handle this challenge. Even though these may not necessarily provide the best uniprocessor performance, they may exhibit better scaling.

10.1.2 Improved Design Flows

This section will discuss a benefits and disadvantages of the presented design flows as well as future work to improve some aspects of these design flows to make them both more robust and powerful.

The first unexpected benefit was the power and usefulness of METROPOLIS events. For example, events could be captured easily to produce structures used in verifying the refinement of architecture services. Since METROPOLIS uses events to signify both the start and end of an action, it is very convenient to observe both the termination of an action as well as the nesting of actions. For example *begin_func1, begin_func2, end_func2, begin_func1* is a trace demonstrating a nested function call. Communication both between services and within the service itself is explicitly scheduled using events and therefore, it became very easy to extract the both CFAs and LTS structures from event sequences in the models. Event scheduling can be enforced as well to add determinism to the CFAs and LTSs. Additionally, events were a very efficient mechanism for the annotation of simulation performance and made the the characterization process described not only easy but almost "free" from a simulation overhead standpoint.

The second unexpected benefit was the scalability of the characterization process. The characterization process presented was not only able to be almost fully parallelized in its creation but also it was agnostic to the system that the tools were a part of (Unix or Windows for example). This occurs since each permutation of a design instance is independent from the last. Secondly, the way in which the Xilinx tools are created, the design template has no notion of operating system or hardware platform. The description also allows itself to be updated to new IP instances and device targets with a few simple changes to instance version declarations which can be accomplished with a simple SED unix script command. In this way, the thousands of permutation instances created in this thesis can be updated for future tool releases or device revisions with a simple script run only once.

The third benefit was how easily the composition of architecture instances from collections of METROPOLIS media was. Initially, it might have been assumed that METROPOLIS processes were the natural object of choice for services. However, the proposed method of only using processes as mapping tasks for the functional model, and composing services (SCSI, MCSI, MCMI) from media worked extremely well. This was due to the fact that (1) media can communicate directly to each other (processes can not) and (2) media implement interfaces (which then are extended directly by ports).

The disadvantages these design flows grew from some of the issues related to the tools used to implement the design flow (e.g. METROPOLIS) more than being inherent in the actual flow. Many of those unique to METROPOLIS will be addressed in METRO II. For example, the confusing design process that resulted from METROPOLIS quantity managers being overburdened with the tasks of both scheduling and annotation has been resolved in METRO II. Additionally, the mapping effort was extremely high in METROPOLIS due to the fact that specific event relations between functional and architectural models had to be specified manually. This process will be improved in the future. However there are two sets of disadvantages which will continue across design tools.

The first set of disadvantages is the lack of possible automation in the refinement verification flow. There are two very obvious places in which designer expertise is needed and automation is not easy (if at all possible). The first example is in the creation of the witness module required during "surface" refinement. This module is required by the interface based tools and requires that the designer be aware of the operation of both the abstract and refined models. The issue arises since the designer must "convert" all the private variables of the abstract reactive module to interface variables. This conversion can be non-trivial (it is much more than a syntactic change) and may require a great deal of designer effort and thought. For large designs the effort may quickly outweigh the benefits. This is one of the reasons a separate ELIF based KISS flow was provided. The second refinement automation difficulty is in the specification of refinement properties for "vertical" refinement (event based flow). Each property described will work for a family of refined architectures but will need to be recreated to reflect the components and interfaces in the event that other objects are used in other designs. It is also not clear how to rank the MacroProperties in terms of which require less effort to prove *a-priori* in relation to each other. This ranking will be required by any heuristic algorithm wishing the prove them in an efficient manner. It also needs to be clearly shown that a generated MacroProperty requires less effort than the sum of its implied MicroProperties to prove.

Another refinement area particularly ripe for future work is in the integration of the techniques in the design flow for compositional component based refinement. As it currently stands there are large portions which are automated but this process is not complete. The compositional component refinement flow is not even tied together with a set of rudimentary scripts much less a presented as an automatic solution. In order to do this, LTS (.fts file), gluing relations (.inv file), and synchronization (.sync file) generation would have to be automated. The first of these should be tied more closely with the model directly to ensure that it is correct-by-construction. This transformation could be done by transversing the models to collect system variables to represent system states and events as labels for transitions. The other two files could minimally be generated by reading from a system specification. This information could generate the syntax used for the tool being employed (in this thesis it is SynCo).

The second set of disadvantages is in the characterization flow. Recall that the database is composed of three portions. Two of these, "execution time for processing sequential code" and "physical timing", can be obtained by automation. For example instruction set simulators can obtain the former and the flow described in this work the latter. However, the third category, "transaction timing" is typically obtained by understanding the bus protocol of the architecture being created. In the case of this thesis, the CoreConnect bus numbers were added manually after a careful examination of the protocol. In the event that another bus or switch mechanism was used, as similar manual analysis would have to be performed.

Cost model specification for the services currently is static. This information comes primarily from the characterization process. It would be ideal to provide a more formal declarative specification mechanism on top of this. For example, bus transactions execution time is currently a function of (1/bus clock speed) * bus cycles. A declarative constraint such as execution time $< 50ns$ would imply a number of bus cycles (given a clock speed) or a clock speed (given a cycle count). This constraint could be used to enforce a specific performance given the fact that the designer wishes to build that enforcement in the scheduling mechanism as opposed to the model itself. This may be of use when the component being characterized is very abstract (early in the design process perhaps) or if the component is part of a testbench which is only being used to simulate the environment and not actually targeted for synthesis.

Another undesirable aspect of the characterization design flows which can continue to be improved upon is the characterization of general purpose processing architecture services. In particular the automation, of the flows presented in Chapter 7, Figure 7.4. The flow described for generating data from instruction set simulators for processing elements like ARM9 currently is very manual. Large portions of this flow could be improved upon with the development of several scripts to not only tie tools together better but also to extract information from the architecture service models. A way to specify a collection of core routines automatically is key. Each of these routines should have the ability to be set to N different characterization flows each which generate a different cost. Subsequently each routine should then have N different cost models which can be selected dynamically at runtime.

It would be ideal as well to populated the characterizer database used to increase accuracy with computation timing data directly taken from an instruction set simulator (ISS) after running a set of benchmark applications. Currently only a few applications have been profiled (H.264 and MJPEG). This set should be significantly expanded if this work is to be of future use. Additionally, the transaction timing information only includes a small set of bus transactions for the PLB and OPB. Again this should be expanded in the face of additional applications.

The third set of issues relates to the robustness of the design flows. A design component which has been discussed in METRO II but is still being developed is an adaptor. This should allow a component using one model of computation to interact with a component in another model of computation. Such components are key to realizing the advantages of METRO II for heterogeneous systems.

Currently, this flow is also very heavily targeting FPGAs. It would be nice to extend this to FPAAs as well as ASIPs. This extension would require more library services to be built and augmenting the characterization flow to work with the tools used to program those devices.

Synthesis of the architectural services to traditional VHDL or Verilog IP would also be of interest. This transformation would involve beginning with small synthesizable constructs and building services from these. Aspects of this work have been started with researchers at UCLA as part of the Xpilot work [CCF+05].

Finally, as with any design flow of this size, the most important extension that can be done is more and more testing. As more designs are created with this flow and compared to their implementations, the better the entire process will become. It is our hope that eventually modeling takes the place of rapid prototyping and EDA reaches the point at which modeling data is the primary contributor to the design exploration process. It is in the push to realize this goal that this book's contribution can most clearly be seen.

10.1.3 Future Case Studies

This section will discuss two case studies currently under development. These each are in the METRO II design environment and should be completed by the end of 2008.

A case study currently under development is a h.264 decoder in the METRO II design environment. This begins with an application model. This model began as pure C code and has been "wrapped" as a collection of METRO II components. The architecture model for the exploration is a set of tasks, an RTOS model, and a collection of processing elements which have been preprofiled regarding their performance costs and processing elements which run core routines at runtime to determine the cost of the service. These two models will be mapped together and simulated. The goal of this case study will not only be to determine the best architecture and mapping scheme for the given application but also to investigate the performance of METRO II simulation. Of particular interest is the relationship between events generated as service requests, events related to synchronization mechanisms, and the overall efficiency of simulation. Issues such as simulation runtime and expressiveness of the current METRO II constraint mechanisms will be investigated.

Another case study in METRO II will examine a model of a heating and cooling system in a building. A model of the heating and cooling system will be described in a Modelica [FB02] model. This includes several components. One component is a sensor. This will read the current temperature of the room. The other component is an actuator. This will turn on the heating and cooling mechanisms for the room. Finally there is a controller. The controller will query the sensor for the temperature and interact with the actuator based on the designer's goal regarding the overall climate in the room. Figure 10.1 illustrates the system. The model for the environment is non-trivial. Not only do the dynamics of the room by itself have to be consided but also the interaction with the other rooms that it is connected to. The temperature in adjoining rooms, the loss or increase of heat from windows in the room, and the pressure difference in adjoining rooms is all captured in the Modelica model. The controller, actuator, and sensor all are modeled in METRO II. Using the an interface called CORBA (Common Object Re-

362

quest Broker Architecture) the Modelica model can be interfaced with during runtime with METRO II simulation. Modelica allows for continuous time to be modeled while the METRO II functional modeling elements will be connected to architecture services. These services will provide overall costs and performance information from the simulation (e.g. bandwidth, latency). This information will be fed to COSI [PCSV08] which will perform communication synthesis. The results of COSI will be used to generate a better mapping of functional components to architectural services as well as provide insight into a better overall device topology to improve performance. This case study will be very valuable as we will show METRO II interacting with a number of tools.

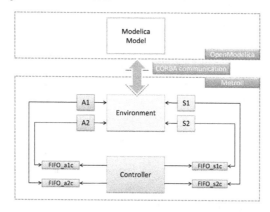

Figure 10.1: Proposed Heating and Cooling System in METRO II

Bibliography

[Acc07]   Accellera. *System Verilog*. World Wide Web, http://www.systemverilog.org, 2007.

[AG06]   Samar Abdi and Daniel Gajski. Verification of system level model transformations. *International Journal of Parallel Programming*, 34(1):29–59, 2006.

[AH99]   Rajeev Alur and Thomas A. Henzinger. Reactive modules. *Formal Methods in System Design: An International Journal*, 15(1):7–48, July 1999.

[AH03]   Rajeev Alur and Thomas A. Henzinger. *Hierarchical Verification*, chapter 8. Draft, March 2003.

[AHM$^+$98] Rajeev Alur, Thomas A. Henzinger, Freddy Y. C. Mang, Shaz Qadeer, Sriram K. Rajamani, and Serdar Tasiran. Mocha: Modularity in model checking. In *CAV '98: Proceedings of the 10th International Conference on Computer Aided Verification*, pages 521–525, 1998.

[Alt04]   Altera. *Altera FGPAs*. World Wide Web, http://www.altera.com, 2004.

[Ana04]   Anadigm. *Anadigm FPAAs*. World Wide Web, http://www.anadigm.com, 2004.

[ARM06]   ARM. *ARM Processor*. World Wide Web, http://www.arm.com, 2006.

[BA97]   Doug Burger and Todd M. Austin. The simplescalar tool set, version 2.0. *SIGARCH Comput. Archit. News*, 25(3):13–25, 1997.

[Bal02]   Laurie Balch. Ever resilient, eda is growing. *EE Times*, July 2002.

[BBL$^+$01] Felice Balarin, Jerry Burch, Luciano Lavagno, Yosinori Watanabe, Roberto Passerone, and Alberto Sangiovanni-Vincentelli. Constraints specification at higher levels of abstraction. In *Proceedings of HLDVT'01*, page 129. IEEE Computer Society, 2001.

[BBS91]   Randal E. Bryant, Derek L. Beatty, and Carl-Johan H. Seger. Formal hardware verifi-
            cation by symbolic ternary trajectory evaluation. In *DAC '91: Proceedings of the 28th
            conference on ACM/IEEE Design Automation*, pages 397–402, New York, NY, USA,
            1991. ACM Press.

[BCE$^+$03]   Albert Benveniste, Paul Caspi, Stephen A. Edwards, Nicolas Halbwachs, Paul Le Guer-
            nic, and Robert de Simone. The synchronous languages 12 years later. *Proceedings of
            the IEEE*, 91, January 2003.

[BCG$^+$97]   Felice Balarin, Massimiliano Chiodo, Paolo Giusto, Harry Hsieh, Attila Jurecska, Lu-
            ciano Lavagno, Claudio Passerone, Alberto Sangiovanni-Vincentelli, Ellen Sentovich,
            Kei Suzuki, and Bassam Tabbara. *Hardware-software Co-design of Embedded Systems:
            The POLIS Approach*. Kluwer Academic Publishers, 1997.

[BELP95]   Greet Bilsen, Marc Engels, Rudy Lauwereins, and J.A. Peperstraete. Cyclo-Static Data
            Flow. In *Proc. ICASSP'95*, volume 5, page 3255, Detroit, USA, 1995.

[Ben96]   Armin Bender. MILP based task mapping for heterogeneous multiprocessor system. In
            Proceedings of EURO-DAC, september 1996.

[BHJ$^+$96]   Felice Balarin, Harry Hsieh, Attila Jurecska, Luciano Lavagno, and Alberto Sangiovanni-
            Vincentelli. Formal verification of embedded systems based on CFSM networks. In
            Proceedings of the Design Automation Conference, June 1996.

[BHKR05]   Pavle Belanovic, Martin Holzer, Bastian Knerr, and Markus Rupp. Automated verifica-
            tion pattern refinement for virtual prototypes. In *Conference of Design of Circuits and
            Integrated Systems*, Lisbon, Portugal, November 2005.

[BHL$^+$03]   Felice Balarin, Harry Hsieh, Luciano Lavagno, Claudio Passerone, Alberto Sangiovanni-
            Vincentelli, and Yoshinori Watanabe. Metropolis: An integrated environment for elec-
            tronic system design. *IEEE Computer*, April 2003.

[BHLM02]   Joseph Buck, Soonhoi Ha, Edward A. Lee, and David G. Messerschmitt. Ptolemy: A
            framework for simulating and prototyping heterogeneous systems. *Readings in Hard-
            ware/Software Co-Design*, pages 527–543, 2002.

[BK85]   Jan A. Bergstra and Jan Willem Klop. Algebra of communicating processes with abstrac-
            tion. *Theory of Computer Science*, 37:77–121, 1985.

[BKVH06]   Stephen P. Boyd, Seung Jean Kim, Lieven Vandenberghe, and Arash Hassibi. A tutorial
             on geometric programming. *Optimization and Engineering*, 2006.

[BLea02]   Felice Balarin, Luciano Lavagno, and et al. Concurrent Execution Semantics and Se-
             quential Simulation Algorithms for the Metropolis Metamodel. *Proc. 10th Int'l Symp.
             Hardware/Software Codesign*, pages 13–18, 2002.

[BLP+02a]   Felice Balarin, Luciano Lavagno, Claudio Passerone, Alberto Sangiovanni-Vincentelli,
             Yosinori Watanabe, and Guang Yang. Concurrent execution semantics and sequential
             simulation algorithms for the metropolis meta-model. In *Proceedings of the Tenth Inter-
             national Symposium on Hardware/Software Codesign*, May 2002.

[BLP+02b]   Felice Balarin, Luciano Lavagno, Claudio Passerone, Alberto L. Sangiovanni-Vincentelli,
             Marco Sgroi, and Yosinori Watanabe. Modeling and designing heterogeneous systems.
             In *Concurrency and Hardware Design, Advances in Petri Nets*, pages 228–273, London,
             UK, 2002. Springer-Verlag.

[BMA05]   Brian Bailey, Grant Martin, and Thomas Anderson. *Taxonomies for the Development and
             Verification of Digital Systems*. Springer, 2005.

[BMP07]   Brian Bailey, Grant Martin, and Andrew Piziali. *ESL Design and Verification*. Morgan-
             Kaufmann, 2007.

[Bos91]   Robert Bosch. CAN specification, version 2.0. Stuttgart, 1991.

[BPPSV05]   Felice Balarin, Roberto Passerone, Alessandro Pinto, and Alberto L. Sangiovanni-
             Vincentelli. A formal approach to system level design: Metamodels and unified design
             environments. In *3rd ACM and IEEE International Conference on Formal Methods and
             Models for Co-Design (MEMOCODE 2005)*, pages 155–163, July 2005.

[Buc93]   Joseph T. Buck. *Scheduling Dynamic Dataflow Graphs with Bounded Memory Using the
             Token Flow Model*. PhD thesis, EECS Department, University of California, Berkeley,
             1993.

[Buc94]   Joseph T. Buck. Static scheduling and code generation from dynamic dataflow graphs
             with integer- valued control streams. In *Proceedings of the 28th Asilomar Conference on
             Signals, Systems, and Computers*, november 1994.

[BV04]   Stephen P. Boyd and Lieven Vandenberghe. *Convex optimization*. Cambridge University Press, 2004.

[CBP+05]   L.P. Carloni, F. De Bernardinis, C. Pinello, A. Sangiovanni-Vincentelli, and M. Sgroi. Platform-based design for embedded systems. In *The Embedded Systems Handbook*. CRC Press, 2005.

[CCF+05]   Deming Chen, Jason Cong, Yiping Fan, Guoling Han, Wei Jiang, and Zhiru Zhang. xPilot: A platform-based behavioral synthesis system. In *SRC TechCon'05*, November 2005.

[CCH+99]   H. Chang, L. Cooke, M. Hunt, G. Martin, A. McNelly, and L. Todd. *Surviving the SOC Revolution: A Guide to Platform-Based Design*. Kluwer Academic Publishers, 1999.

[CDH+05]   Xi Chen, Abhijit Davare, Harry Hsieh, Alberto Sangiovanni-Vincentelli, and Yosinori Watanabe. Simulation based deadlock analysis for system level designs. In *Design Automation Conference*, June 2005.

[CFH+06]   Jason Cong, Yiping Fan, Guoling Han, Wei Jiang, and Zhiru Zhang. Platform-based behavior-level and system-level synthesis. In *International SOC Conference*, pages 199–202, September 2006.

[CGJ+02]   Massimiliano Chiodo, Paolo Guisto, Attila Jurecska, Harry C. Hsieh, Alberto Sangiovanni-Vincentelli, and Luciano Lavagno. Hardware-software codesign of embedded systems. pages 313–323, 2002.

[CGL93]   Edmund M. Clarke, Orna Grumberg, and David E. Long. Verification tools for finite state concurrent systems. In J.W. de Bakker, W.-P. de Roever, and G. Rozenberg, editors, *A Decade of Concurrency-Reflections and Perspectives*, volume 803, pages 124–175, Noordwijkerhout, Netherlands, 1993. Springer-Verlag.

[CKH+00]   Pierluigi Crescenzi, Viggo Kann, Magnús Halldórsson, Marek Karpinski, and Gerhard Woeginger. A compendium of NP optimization problems, 20 March 2000.

[CLJ+01]   William S. Coates, Jon K. Lexau, Ian W. Jones, Scott M. Fairbanks, and Ivan E. Sutherland. Fleetzero: An asynchronous switching experiment. In *ASYNC '01: Proceedings of the 7th International Symposium on Asynchronous Circuits and Systems*, pages 173–182, Washington, DC, USA, 2001. IEEE Computer Society.

[Cora]   Xilinx Corporation. http://www.xilinx.com.

[Corb]   Xilinx Corporation. Microblaze processor reference guide.

[Cor05]   Xilinx Corporation. Fast simplex link (fsl) bus (v2.00a), December 2005.

[CPT03]   Andrew S. Cassidy, JoAnn M. Paul, and Donald E. Thomas. Layered, multi-threaded, high-level performance design. In *DATE '03: Proceedings of the Conference on Design, Automation and Test in Europe*, page 10954, Washington, DC, USA, 2003. IEEE Computer Society.

[CRS03]   Pablo E. Coll, Celso E. Ribeiro, and Cid C. De Souza. Multiprocessor scheduling under precedence constraints: Polyhedral results. Technical Report 752, Opt. Online, October 12, 2003.

[Dat05]   Gartner Dataquest. User wants and needs survey. *Gartner Dataquest*, 1992-2005.

[Dat08]   Gartner DataQuest. *Market Trends: ASIC and FPGA, Worldwide*, 1q05 update edition, 2002-2008.

[Dav07]   Abhijit Davare. *Automated Mapping for Heterogeneous Multiprocessor Embedded Systems*. PhD thesis, EECS Department, University of California, Berkeley, 2007.

[DCZ+06]   Abhijit Davare, Jike Chong, Qi Zhu, Douglas Densmore, and Alberto Sangiovanni-Vincentelli. Classification, customization, and characterization: Using milp for task allocation and scheduling. Technical Report UCB/EECS-2006-166, EECS Department, UC Berkeley, Dec. 11 2006.

[DDM+07]   Abhijit Davare, Douglas Densmore, Trevor Meyerowitz, Alessandro Pinto, Alberto Sangiovanni-Vincentelli, Guang Yang, and Qi Zhu. A next-generation design framework for platform-based design. In *Design and Verification Conference (DV-CON'07)*, February 2007.

[DDSV06a]   Doug Densmore, Adam Donlin, and Alberto Sangiovanni-Vincentelli. Fpga architecture characterization for system level performance analysis. In *Design Automation and Test Europe (DATE)*, pages 734–739, March 2006.

[DDSV06b]   Douglas Densmore, Adam Donlin, and Alberto L. Sangiovanni-Vincentelli. FPGA architecture characterization for system level performance analysis. In *DATE06*, Munich, Germany, March 6–10, 2006.

[DDSZ04]   Abhijit Davare, Douglas Densmore, Vishal Shah, and Haibo Zeng. Simple case study in metropolis. Technical Report UCB.ERL 04/37, University of California, Berkeley, September 2004.

[DeH00]   Andre DeHon. The density advantage of configurable computing. In *IEEE Computer*, April 2000.

[Den04]   Douglas Densmore. Metropolis Architecture Refinement Styles and Methodology. Technical Report UCB/ERL M04/36, University of California, Berkeley, CA 94720, September 14, 2004.

[Den07]   Douglas Densmore. *A Design Flow for the Development, Characterization, and Refinement of System Level Architectural Services*. PhD thesis, EECS Department, University of California, Berkeley, 2007.

[Des07a]   CoFluent Design. *CoFluent Studio*. World Wide Web, http://www.cofluentdesign.com, 2007.

[Des07b]   Mirabilis Design. *Visual Sim*. World Wide Web, http://www.mirabilisdesign.com, 2007.

[Des07c]   Summit Design. *System Architect*. World Wide Web, http://www.sd.com, 2007.

[DGK94]   Srinivas Devadas, Abhijit Ghosh, and Kurt Keutzer. *Logic synthesis*. McGraw-Hill, Inc., New York, NY, USA, 1994.

[Dic06]   Merriam-Webster Online Dictionary. *Heterogeneous*. World Wide Web, http://www.merriam-webster.com (1 Aug. 2003), 2006.

[DLMM04]   Tatjana Davidovic, Leo Liberti, Nelson Maculan, and Nena Mladenovic. Mathematical programming-based approach to scheduling of communicating tasks. Technical report, GERAD, December 15, 2004.

[Don04]   Adam Donlin. Transaction level modeling: Flows and use models. In *CODES+ISSS '04: Proceedings of the 2nd IEEE/ACM/IFIP International Conference on Hardware/Software Codesign and System Synthesis*, pages 75–80, New York, NY, USA, 2004. ACM Press.

[DP02]   B. A. Davey and H. A. Priestley. *Introduction to Lattices and Order.* Cambridge University Press, 2002.

[DPSV06]   Douglas Densmore, Roberto Passerone, and Alberto Sangiovanni-Vincentelli. A platform-based taxonomy for esl design. *IEEE Design and Test of Computers*, 23(5):359–374, 2006.

[DRSV04]   Douglas Densmore, Sanjay Rekhi, and Alberto Sangiovanni-Vincentelli. Microarchitecture development via metropolis successive platform refinement. In *Design Automation and Test Europe (DATE)*, pages 346–351, February 2004.

[DRW98]   Robert P. Dick, David L. Rhodes, and Wayne Wolf. TGFF: task graphs for free. In *CODES*, pages 97–101, 1998.

[DSP08]   Douglas Densmore, Alena Simalatsar, and Roberto Passerone. A methodology for architecture exploration and performance analysis using system level design languages and rapid architecture profiling. In *Third International IEEE Symposium on Industrial Embedded Systems (SIES)*, La Grande Motte, France, June 11–13, 2008.

[DZMSV05]   Abhijit Davare, Qi Zhu, John Moondanos, and Alberto L. Sangiovanni-Vincentelli. JPEG encoding on the Intel MXP5800: A platform-based design case study. In *Proceedings of the 3rd Workshop on Embedded Systems for Real-time Multimedia*, pages 89–94, 2005.

[DZN+07]   Abhijit Davare, Qi Zhu, Marco Di Natale, Claudio Pinello, Sri Kanajan, and Alberto L. Sangiovanni-Vincentelli. Period optimization for hard real-time distributed automotive systems. In *DAC*, pages 278–283. IEEE, 2007.

[DZSV06]   Abhijit Davare, Qi Zhu, and Alberto L. Sangiovanni-Vincentelli. A platform-based design flow for kahn process networks. Technical Report UCB/EECS-2006-30, EECS Department, University of California, Berkeley, Mar 2006.

[EKRZ04]   Don Edenfeld, Andrew B. Kahng, Mike Rodgers, and Yervant Zorian. 2003 technology roadmap for semiconductors. *IEEE Computer*, 37(1):47–56, 2004.

[ES03]   Niklas Een and Niklas Sorensson. An extensible SAT-solver. In *International Conference on Theory and Applications of Satisfiability Testing (SAT), LNCS*, volume 6, 2003.

[ES06]   IBM Engineering and Technology Services. *Time to Market.* World Wide Web, http://www-03.ibm.com/technology/businessvalue/timetomarket.shtml, 2006.

[FB02]   Peter Fritzson and Peter Bunus. Modelica-a general object-oriented language for con-
           tinuous and discrete-event system modeling and simulation. In *SS '02: Proceedings of
           the 35th Annual Simulation Symposium*, page 365, Washington, DC, USA, 2002. IEEE
           Computer Society.

[FGK93]   R. Fourer, D. M. Gay, and B. W. Kernighan. *AMPL – A Modeling Language for Mathe-
           matical Programming*. The Scientific Press, South San Francisco, 1993.

[Fle06]   Flexray. Protocol specification v2.1 rev. a. available at http://www.flexray.com, 2006.

[For93]   MPI Forum. MPI: A message passing interface. In *Proceedings of Supercomputing '93*,
           pages 878–883, Portland, OR, Nov 1993. IEEE CS Press.

[fS99]   International Technology Roadmap for Semiconductors. *1999 Update ITRS*.
           http://www.itrs.net, 1999.

[fS04]   International Technology Roadmap for Semiconductors. *2004 Update ITRS*.
           http://www.itrs.net, 2004.

[Gar96]   David Garlan. Style-based refinement for software architecture. In *Joint Proceedings
           of the Second International Software Architecture Workshop (ISAW-2) and International
           Workshop on Multiple Perspectives in Software Development (Viewpoints '96) on SIG-
           SOFT '96 workshops*, pages 72–75, New York, NY, USA, 1996. ACM Press.

[GB03]   M. Geilen and T. Basten. Requirements on the Execution of Kahn Process Networks. In
           P. Degano, editor, *Proc. of the 12th European Symposium on Programming*, 2003.

[GG02]   Martin Grajcar and Werner Grass. Improved constraints for multiprocessor system
           scheduling. In *DATE*, page 1096. IEEE Computer Society, 2002.

[GGB97]   Jie Gong, Daniel D. Gajski, and Smita Bakshi. Model refinement for hardware-software
           codesign. *ACM Transactions on Design Automation of Electronic Systems*, 2(1):22–41,
           1997.

[GHS95]   Richard Gerber, Seongsoo Hong, and Manas Saksena. Guaranteeing real-time require-
           ments with resource-based calibration of periodic processes. *IEEE Trans. on Software
           Engineering*, 21(7):579–592, July 1995.

[GK83]   D. D. Gajski and R. H. Kuhn. Guest editor's introduction: New VLSI tools. *IEEE Computer*, December 1983.

[GK05]   Matthias Gries and Kurt Keutzer. *Building ASIPs: The Mescal Methodology*. Springer-Verlag New York, Inc., Secaucus, NJ, USA, 2005.

[GLL99]   A. Girault, B. Lee, and E.A. Lee. Hierarchical finite state machines with multiple concurrency models. *IEEE Trans. on Computer-Aided Design of Integrated Circuits and Systems*, 18(6):742–760, June 1999. Research report UCB/ERL M97/57.

[GLLK79]   R. L. Graham, E. L. Lawler, J. K. Lenstra, and A. H. G. Rinnooy Kan. Optimization and approximation in deterministic sequencing and scheduling: A survey. *Ann. Discrete Mathematics*, 5:287–326, 1979.

[Goe05]   Richard Goering. Esl may rescue eda, analysts say. *EE Times*, June 2005.

[Gri04]   Matthias Gries. Methods for Evaluating and Covering the Design Space during Early Design Development. *Integration, the VLSI Journal, Elsevier*, 38(2):131–183, December 2004.

[GT01]   K. Gass and R. Tuck. System packet interface level 5. *Optical Internetworking Forum Contribution*, OIF(2001.134), November 2001.

[GWH05]   Nolan Goodnight, Rui Wang, and Greg Humphreys. Computation on programmable graphics hardware. *IEEE Computer Graphics and Applications*, 25(5):12–15, 2005.

[Har87]   David Harel. Statecharts: A visual formalism for complex systems. *Science of Computer Programming*, 8(3):231–274, June 1987.

[Hen03]   Jörg Henkel. Closing the soC design gap. *IEEE Computer*, 36(9):119–121, 2003.

[HJKH03]   Michael Horowitz, Anthony Joch, Faouzi Kossentini, and Antti Hallapuro. H.264/avc baseline profile decoder complexity analysis. *IEEE Transactions on Circuits and Systems for Video Technology*, 13(7):704–716, 2003.

[HJM+02]   Thomas A. Henzinger, Ranjit Jhala, Rupak Majumdar, George C. Necula, Gregoire Sutre, and Westley Weimer. Temporal safety proofs for systems code. In *Proceedings of the 14th International Conference on Computer-Aided Verification (CAV)*, pages 526–538. Lecture Notes in Computer Science 2404, Springer-Verlag, 2002.

372

[HKK04]   Bernd Hardung, Thorsten Kölzow, and Andreas Krüger. Reuse of software in distributed
            embedded automotive systems. In *EMSOFT '04: Proceedings of the 4th ACM interna-
            tional conference on Embedded software*, pages 203–210, New York, NY, USA, 2004.
            ACM.

[HKL94]   M. Gonzalez Harbour, M. Klein, and J. Lehoczky. Timing analysis for fixed-priority
            scheduling of hard real-time systems. *IEEE Transactions on Software Engineering*, 20(1),
            January 1994.

[HLY⁺06]   Soonhoi Ha, Choonseung Lee, Youngmin Yi, Seongnam Kwon, and Young-Pyo Joo.
            Hardware-software codesign of multimedia embedded systems: The peace. *12th IEEE
            International Conference on Embedded and Real-Time Computing Systems and Applica-
            tions (RTCSA'06)*, 0:207–214, 2006.

[HNO97]   Lance Hammond, Basem A. Nayfeh, and Kunle Olukotun. A single-chip multiprocessor.
            IEEE Computer, 30(9):79–85, 1997.

[Hoa78]   Charles A. R. Hoare. Communicating sequential processes. *Communications of the ACM*,
            21(8):666–677, 1978.

[Hol97]   Gerard J. Holzmann. The model checker SPIN. *IEEE Trans. on Software Engineering*,
            23(5):279–295, May 1997.

[HS00]   Gary D. Hachtel and Fabio Somenzi. *Logic Synthesis and Verification Algorithms*.
            Kluwer Academic Publishers, Norwell, MA, USA, 2000.

[HS02]   Soha Hassoun and Tsutomu Sasao, editors. *Logic Synthesis and Verification*, chapter 12
            SAT and ATPG: Algorithms for Boolean Decision Problems. Kluwer Academic Publish-
            ers, 2002.

[HSF⁺04]   H. Heinecke, K.-P. Schnelle, H. Fennel, J. Bortolazzi, L. Lundh, J. Leflour, J.-L. Mate,
            K. Nishikawa, and T. Scharnhorst. Automotive open system architecture an industry-
            wide initiative to manage the complexity of emerging automotive e/e-architectures. In
            Proceedings of Convergence 2004, October 2004.

[IBM99]   IBM. *CoreConnect Bus Architecture*, white paper edition, 1999.

[IBM03]   IBM. *OPB Bus Functional Model Toolkit*, 6th edition, version 3.5 edition, June 2003.

[IHK⁺01]   John Davis II, Christopher Hylands, Bart Kienhuis, Edward A. Lee, Jie Liu, Xiaojun Liu, Lukito Muliadi, Steve Neuendorffer, Jeff Tsay, Brian Vogel, and Yuhong Xiong. Ptolemy ii : Heterogeneous concurrent modeling and design in java. Technical Report UCB/ERL M01/12, EECS Department, University of California, Berkeley, 2001.

[ijg]   Independent JPEG group, http://www.ijg.org.

[Ini07]   Open SystemC Initiative. *SystemC*. World Wide Web, http://www.systemc.org, 2007.

[int]   Intel MXP5800 Digital Media Processor Product Brief, Intel Corporation, 2004.

[Int04]   Intel. *Intel Flash Memory*. World Wide Web, http://www.intel.com/design/flash, 2004.

[Int06a]   Intel. *Intel Pentium 4 Processor*. World Wide Web, http://www.intel.com/products/processor/pentium4, 2006.

[Int06b]   Intel. *Intel PXA270 Processor for Embedded Computing*. World Wide Web, http://www.intel.com/design/embeddedpca/applicationsprocessors/302302.htm, 2006.

[JM9]   MPEG4 AVC Reference Software JM92. http://www.m4if.org/index.php, mpeg industry forum.

[JO03]   Charles P. Poole Jr. and Frank J. Owens. *Introduction to Nanotechnology*. Wiley, 2003.

[JS05]   A. Jantsch and I. Sander. Models of computation and languages for embedded system design. *IEE Proceedings - Computers and Digital Techniques*, 152(2):114–129, 2005.

[JSRK05]   Yujia Jin, Nadathur Rajagopalan Satish, Kaushik Ravindran, and Kurt Keutzer. An automated exploration framework for fpga-based soft multiprocessor systems. In *Proceedings of the 2005 International Conference on Hardware/Software Codesign and System Synthesis (CODES-05)*, pages 273–278, September 2005.

[KA99]   Yu-Kwong Kwok and Ishfaq Ahmad. Static Scheduling Algorithms for Allocating Directed Task Graphs to Multiprocessors. *ACM Comput. Surv.*, 31(4):406–471, 1999.

[Kah74]   G. Kahn. The Semantics of a Simple language for Parallel Programming. In *Proceedings of IFIP Congress*, pages 471–475. North Holland Publishing Company, 1974.

[Kah05]   Jim Kahle. The cell processor architecture. In *38th Annual IEEE/ACM International Symposium on Microarchitecture (MICRO-38 2005) Keynote Address*, 2005.

[KDK+89]   Hermann Kopetz, Andreas Damm, Christian Koza, Marco Mulazzani, Wolfgang Schw-
             abl, Christoph Senft, and Ralph Zainlinger. Distributed fault-tolerant real-time systems:
             The MARS approach. *IEEE Micro*, 9(1):25–40, February 1989.

[KDvdWV02]   Bart Kienhuis, Ed F. Deprettere, Pieter van der Wolf, and Kees A. Vissers. A methodology
             to design programmable embedded systems - the Y-chart approach. volume 2268 of
             Lecture Notes in Computer Science, pages 18–37. Springer, 2002.

[kei]   Keil. http://www.keil.com.

[KES+00]   E.A. de Kock, G. Essink, W.J.M. Smits, P. van der Wolf, J.Y. Brunel, W.M. Kruijtzer,
             P. Lieverse, and K.A. Vissers. YAPI: Application Modeling for Signal Processing Sys-
             tems. *Proceedings of the 37th Design Automation Conference*, 2000.

[KKMB06]   Kwangmoo Koh, Seungjean Kim, Almir Mutapcic, and Stephen Boyd. gpposy: A matlab
             solver for geometric programs in posynomial form. Technical report, Stanford University,
             May 2006.

[KL03a]   Olga Kouchnarenko and Arnaud Lanoix. Refinement and verification of synchronized
             component-based systems. In *FME 2003: Formal Methods, Lecture Notes in Computer
             Science*, volume 2805/2003, pages 341–358. Springer Berlin / Heidelberg, 2003.

[KL03b]   Olga Kouchnarenko and Arnaud Lanoix. Synco: a refinement analysis tool for synchro-
             nized component-based systems. In Margaria T., editor, *FM'03 Tool Exhibition Notes*,
             pages 47–51, Pisa, Italy, September 2003.

[KLM+97]   Gregor Kiczales, John Lamping, Anurag Mendhekar, Chris Maeda andCristina Lopes,
             Jean-Marc Loingtier, and John Irwin. Aspect-oriented programming. In Mehmet Akşit
             and Satoshi Matsuoka, editors, *Proceedings European Conference on Object-Oriented
             Programming*, volume 1241, pages 220–242. Springer-Verlag, Berlin, Heidelberg, and
             New York, 1997.

[KM77]   G. Kahn and D.B. MacQueen. Coroutines and networks of parallel processes. In *Pro-
             ceedings of IFIP Congress*, pages 993–998. North Holland Publishing Company, 1977.

[KMN+00]   K. Keutzer, S. Malik, A. R. Newton, J. Rabaey, and A. Sangiovanni-Vincentelli. System
             level design: Orthogonolization of concerns and platform-based design. *IEEE Transac-*

tions on Computer-Aided Design of Integrated Circuits and Systems, 19(12), December 2000.

[KMPS05]   V. S. Anil Kumar, Madhav V. Marathe, Srinivasan Parthasarathy, and Aravind Srinivasan. Scheduling on unrelated machines under tree-like precedence constraints. In *Proceedings of APPROX-RANDOM 2005*, volume 3624 of *Lecture Notes in Computer Science*, pages 146–157. Springer, 2005.

[Kri05]   Ravi Krishnan. Future of embedded systems technology. *BCC Research*, June 2005.

[KSLB03]   G. Karsai, J. Sztipanovits, A. Ledeczi, and T. Bapty. Model-integrated development of embedded software. *Proceedings of the IEEE*, 91(1):145–164, January 2003.

[Kue02]   Kurt Kuetzer. Programmable platforms will rule. *EETimes*, September 2002.

[KWD+06]   Shinjiro Kakita, Yosinori Watanabe, Douglas Densmore, Abhijit Davare, and Alberto Sangiovanni-Vincentelli. Functional model exploration for multimedia applications via algebraic operators. In *Sixth International Conference on Application of Concurrency to System Design (ACSD)*, June 2006.

[Kwo07]   Oh-Hyun Kwon. Perspective of the future semiconductor industry: Challenges and solutions. In *Keynote Address at the 44th Design Automation Conference*, June 2007.

[Lam05]   David Lammers. Shift to 65 nm has its costs. *EE Times*, July 11 2005.

[Lan07]   Unified Modeling Language. *UML*. World Wide Web, http://www.uml.org, 2007.

[Lee06]   Edward A. Lee. The problem with threads. *Computer: IEEE Computer*, 39, 2006.

[LH02]   Gabriel Leen and Donal Heffernan. Expanding automotive electronic systems. *IEEE Computer*, 35(1):88–93, 2002.

[Lib05]   Leo Liberti. Compact linearization for bilinear mixed-integer problems. Technical Report 1124, Opt. Online, May 6, 2005.

[LKA+95]   David C. Luckham, John L. Kenney, Larry M. Augustin, James Vera, Doug Bryan, and Walter Mann. Specification and analysis of system architecture using rapide. *IEEE Transactions on Software Engineering*, 21(4):336–355, apr 1995.

[LM87]   Edward A. Lee and David G. Messerschmitt. Static Scheduling of Synchronous Data Flow Programs for Digital Signal Processing. *IEEE Trans. Comput.*, 36(1):24–35, 1987.

[LMB⁺01] Akos Ledeczi, Miklos Maroti, Arpad Bakay, Gabor Karsai, Jason Garrett, Charles Thomason, Greg Nordstrom, Jonathan Sprinkle, and Peter Volgyesi. The generic modeling environment. In *IEEE Workshop on Intelligent Signal Processing*, May 2001.

[Lof04]   J. Lofberg. Yalmip : A toolbox for modeling and optimization in MATLAB. In *Proc. of the CACSD Conference*, Taipei, 2004.

[LP95]   E.A. Lee and T.M. Parks. Dataflow Process Networks. In *Proceedings of the IEEE, vol.83, no.5*, pages 773 – 801, May 1995.

[LST87]   Jan Karel Lenstra, David B. Shmoys, and Éva Tardos. Approximation algorithms for scheduling unrelated parallel machines. In *28th Annual Symposium on Foundations of CS*, pages 217–224, Los Angeles, California, 12–14 October 1987. IEEE.

[LSV98]   Edward A. Lee and Alberto Sangiovanni-Vincentelli. A framework for comparing models of computation. *IEEE Transactions on Computer Aided Design*, 17(12), June 1998.

[LSvdWD01] Paul Lieverse, Todor Stefanov, Pieter van der Wolf, and Ed Deprettere. System Level Design With Spade: An m-jpeg Case Study. In *Proceedings of the 2001 IEEE/ACM international conference on Computer-aided design*, pages 31–38. IEEE Press, 2001.

[LV95]   David C. Luckham and James Vera. An event-based architecture definition language. *IEEE Transactions on Software Engineering*, 21(9):717–734, 1995.

[LvdWD01] Paul Lieverse, Pieter van der Wolf, and Ed Deprettere. A trace transformation technique for communication refinement. In *CODES '01: Proceedings of the Ninth International Symposium on Hardware/Software Codesign*, pages 134–139, New York, NY, USA, 2001. ACM Press.

[LvdWDV01] Paul Lieverse, Pieter van der Wolf, Ed Deprettere, and Kees Vissers. A methodology for architecture exploration of heterogeneous signal processing systems. *Journal of VLSI Signal Processing for Signal, Image and Video Technology*, 29(3):197–207, November 2001. Special issue on SiPS'99.

[LXL01]   Xiaojun Liu, Yuhong Xiong, and Edward A. Lee. The Ptolemy II Framework for Visual Languages. In *Proceedings of the IEEE 2001 Symposia on Human Centric Computing Languages and Environments (HCC'01)*, page 50. IEEE Computer Society, 2001.

[Mak00]   Tsugio Makimoto. The rising wave of field programmability. In *International Conference on Field Programmable Logic and Applications (FPL)*, pages 1–6, 2000.

[Man04]   Tets Maniwa. Focus report: Electronic system-level (ESL) tools. *Chip Design*, April/May 2004.

[Mas01]   Paul Master. Worldphone challenges designers. *EE Times*, September 25 2001.

[MB06]   Anthony Massa and Michael Barr. *Programming Embedded Systems*, chapter 1. O'Reilly Publishers, October 2006.

[MBR02]   K. Masselos, S. Blionas, and T. Rautio. Reconfigurability requirements of wireless communication systems. In *IEEE Workshop on Heterogeneous Reconfigurable Systems on Chip*, 2002.

[McM93]   Kenneth L. McMillan. *Symbolic Model Checking*. Kluwer Academic Publishers, Norwell, MA, USA, 1993.

[Mic04]   Cypress Microsystems. *Cypress Microsystems Home Page*. World Wide Web, http://www.cypressmicro.com/corporate/corporate.htm, 2004.

[MKM+02]   Andrew Mihal, Chidamber Kulkarni, Matthew Moskewicz, Mel Tsai, Niraj Shah, Scott Weber, Yujia Jin, Kurt Keutzer, Christian Sauer, Kees Vissers, and Sharad Malik. Developing architectural platforms: A disciplined approach. *IEEE Design and Test*, 19(6):6–16, 2002.

[MORT96]   Nenad Medvidovic, Peyman Oreizy, Jason E. Robbins, and Richard N. Taylor. Using object-oriented typing to support architectural design in the c2 style. In *SIGSOFT '96: Proceedings of the 4th ACM SIGSOFT symposium on Foundations of Software Engineering*, pages 24–32, New York, NY, USA, 1996. ACM Press.

[MP03]   Philippe Magarshack and Pierre G. Paulin. System-on-chip beyond the nanometer wall. In *Proceedings of the Design Automation Conference*, pages 419–424, June 2003.

378

[MPC⁺97]   Nelson Maculan, Stella C. S. Porto, Celso Carneiro, Ribeiro Cid, and Carvalho Souza. A new formulation for scheduling unrelated processors under precedence constraints, April 28 1997.

[MQR95]   Mark Moriconi, Xiaolei Qian, and R. A. Riemenschneider. Correct architecture refinement. *IEEE Transactions on Software Engineering*, 21(4):356–3, 1995.

[MRH⁺01]   W. Mueller, J. Ruf, D. Hofmann, J. Gerlach, T. Kropf, and W. Rosenstiehl. The simulation semantics of systemc, 2001.

[MS05]   Maher N. Mneimneh and Karem A. Sakallah. Principles of sequential-equivalence verification. *IEEE Design and Test*, 22(3):248–257, 2005.

[MSVSL08] Trevor Meyerowitz, Alberto Sangiovanni-Vincentelli, Mirko Sauermann, and Dominik Langen. Source level timing annotation and simulation for a heterogeneous multiprocessor. In *DATE08*, Munich, Germany, March 10–14 2008.

[Mur]   Praveen K. Murthy. Multiprocessor DSP code synthesis in Ptolemy. Technical Report ERL-93-66, University of California, Berkeley.

[Mut08]   Ann Steffora Mutschler. Eda revenue inches up in q4. *EDN.com*, April 2008.

[NGK⁺07]   Marco Di Natale, Paolo Giusto, Sri Kanajan, Claudio Pinello, and Patrick Popp. Architecture exploration for time-critical and cost-sensitive distributed systems. In *Proceedings of the SAE Conference*, 2007.

[NGMG98]   Ratan Nalumasu, Rajnish Ghughal, Abdelillah Mokkedem, and Ganesh Gopalakrishnan. The test model-checking approach to the verification of formal memory models of multiprocessors. In *Computer Aided Verification*, pages 464–476, 1998.

[NLK03]   Nir Naor, Yoav Lerman, and Melanie Kessler. Forte/fl user guide. Technical report, Intel Corporation, January 2003.

[NS96]   Stephen G. Nash and Ariela Sofer. *Linear and Nonlinear Programming*. McGraw-Hill, January 1996.

[NTL04]   Anders Nilsson, Eric Tell, and Dake Liu. An accelerator architecture for programmable multi-standard baseband processors. In *Proceedings of Wireless Networks and Emerging Technologies (WNET)*, 2004.

[NW88]   G. L. Nemhauser and L. A. Wolsey. *Integer and Combinatorial Optimization*. John Wiley and Sons, New York, 1988.

[OH05]   Kunle Olukotun and Lance Hammond. The future of microprocessors. *ACM Queue*, September 2005.

[O'R07]   O'Reilly. *Perl.Com: The Source for PERL*. World Wide Web, http://www.perl.com, 2007.

[OSE06]   OSEK. OS version 2.2.3 specification. Available at http://www.osek-vdx.org, 2006.

[Par95]   T. Parks. *Bounded Scheduling of Process Networks*. PhD thesis, University of California, Berkeley, 1995.

[PBL99]   Jose L Pino, Shuvra S. Bhattacharyya, and Edward A. Lee. A hierarchical multiprocessor scheduling framework for. Technical report, Berkeley, CA, USA, 1999.

[PCSV08]   Alessandro Pinto, Luca Carloni, and Alberto L. Sangiovanni-Vincentelli. Cosi: A public-domain design framework for the design of interconnection networks. Technical Report UCB/EECS-2008-22, EECS Department, University of California, Berkeley, Mar 2008.

[PEP02]   Traian Pop, Petru Eles, and Zebo Peng. Holistic scheduling and analysis of mixed time/event-triggered distributed embedded systems. In *10th International Symposium on Hardware/Software Codesign (CODES 2002)*, pages 187–192, Estes Park, Colorado, USA, May 6-8 2002.

[PEP06]   Andy D. Pimentel, Cagkan Erbas, and Simon Polstra. A systematic approach to exploring embedded system architectures at multiple abstraction levels. *IEEE Transactions on Computers*, 55(2):99–112, 2006.

[PHL+01]   Andy D. Pimentel, Louis O. Hertzberger, Paul Lieverse, Pieter van der Wolf, and Ed F. Deprettere. Exploring Embedded-Systems Architectures With Artemis. *Computer*, 34(11):57–63, 2001.

[Pin04]   Alessandro Pinto. Metropolis Design Guidelines. Technical Report UCB/ERL M04/40, University of California, Berkeley, CA 94720, September 14, 2004.

[PPer]   Shiv Prakash and Alice C. Parker. SOS: Synthesis of application-specific heterogeneous multiprocessor systems. *J. Parallel Distrib. Comput.*, 16(4):338–351, 1992, December.

[Pro04]   Third Generation Partnership Project. General universal mobile telecommunications system (umts) architecture. Technical Specification TS 23.101, 3GPP, December 2004.

[Pro07]   Prosilog. *Nepsys*. World Wide Web, http://www.prosilog.org, 2007.

[PT02]   J. Paul and D. Thomas. A Layered, Codesign Virtual Machine Approach to Modeling Computer Systems. In *Proceedings of the conference on Design, automation and test in Europe*, page 522. IEEE Computer Society, 2002.

[RB03]   Francois Remond and Pierre Bricaud. Set top box soc design methodology at stmicro-electronics. In *DATE '03: Proceedings of the conference on Design, Automation and Test in Europe*, page 20220, Washington, DC, USA, 2003. IEEE Computer Society.

[RJE05]   Razvan Racu, Marek Jersak, and Rolf Ernst. Applying sensitivity analysis in real-time distributed systems. In *Proceedings of the 11th Real Time and Embedded Technology and Applications Symposium*, pages 160–169, San Francisco (CA), U.S.A., March 2005.

[RL04]   Chris Rowen and Steve Leibson. *Engineering the Complex SOC: Fast, Flexible Design with Configurable Processors*. Prentice Hall PTR, June 2004.

[Roh03]   Ronald A. Rohrer. DAC, moore's law still drive EDA. *IEEE Design & Test of Computers*, 20(3):99–100, 2003.

[RP03]   Sanjay Rekhi and Rangarajan Sri Purasai. The next level of abstraction: Evolution in the life of an asic design engineer. *Synopsys Users Group (SNUG), San Jose*, 2003.

[RSSJ03]   Tarvo Raudvere, Ingo Sander, Ashish Kumar Singh, and Axel Jantsch. Verification of Design Decisions in ForSyDe. In *Proceedings of the 1st IEEE/ACM/IFIP international conference on Hardware/software codesign and system synthesis*, pages 176–181. ACM Press, 2003.

[RSV87]   Richard L. Rudell and Alberto L. Sangiovanni-Vincentelli. Multiple-valued minimization for pla optimization. *IEEE Transactions on CAD of Integrated Circuits and Systems*, 6(5):727–750, 1987.

[SBC00]   Nagaraj Shenoy, Prithviraj Banerjee, and Alok N. Choudhary. A system-level synthesis algorithm with guaranteed solution quality. In *DATE*, page 417. IEEE Computer Society, 2000.

[SC82]   A. P. Sistla and E. M. Clarke. The complexity of propositional linear temporal logics. In *Proceedings of the fourteenth annual ACM symposium on Theory of computing*, pages 159–168. ACM Press, 1982.

[SH96]   M. Saksena and S. Hong. Resource conscious design of distributed real-time systems – an end-to-end approach. In *Proc. IEEE Int'l Conf on Engineering of Complex Computer Systems*, 1996.

[SJ04]   Ingo Sander and Axel Jantsch. System modeling and transformational design refinement in forsyde. In *IEEE Transactions on Computer-Aided Design*, volume 23, January 2004.

[SM02]   Jack Shandle and Grant Martin. Making embedded software reusable for SoCs. *EE Times*, March 2002.

[SN05]   G. Smith and D. Nadamuni. *ESL Landscape 2005*. Gartner Dataquest, 2005.

[SNBW05] Gary Smith, Daya Nadamuni, Laurie Balch, and Nancy Wu. Report on worldwide eda market trends. *Gartner Dataquest*, December 2005.

[Sol07]   Beach Solution. *EASI-Studio*. World Wide Web, http://www.beachsolutions.com, 2007.

[Son07]   Sonics. *Sonics Studio*. World Wide Web, http://www.sonicsinc.com, 2007.

[SSL$^+$92] Ellen M. Sentovich, Kanwar J. Singh, Luciano Lavagno, Cho Moon, Rajeev Murgai, Alexander Saldanha, Hamid Savoj, Paul R. Stephan, Robert K. Brayton, and Alberto Sangiovanni-Vincentelli. Sis: A system for sequential circuit synthesis. Technical report, University of California, Berkeley, 1992.

[ST93]   David B. Shmoys and Éva Tardos. Scheduling unrelated machines with costs. In *SODA*, pages 448–454, 1993.

[ST98]   D. Skillicorn and D. Talia. Models and languages for parallel computation. *ACM Computing Surveys*, 30(2):123–169, 1998.

[Sut06]   Ivan E. Sutherland. Fleet - a one-instruction computer. *University of California, Berkeley*, 2006.

[SV02]   Alberto Sangiovanni-Vincentelli. Defining platform-based design. *EEDesign*, February 2002.

[SV07]   Alberto L. Sangiovanni-Vincentelli. Quo vadis sld: Reasoning about trends and chal-
           lenges of system-level design. In *Proceedings of the IEEE*, volume 95, pages 467–506,
           march 2007.

[SVM01]   Alberto Sangiovanni-Vincentelli and Grant Martin. Platform-based design and software
           design methodology for embedded systems. *IEEE Design and Test of Computers*, pages
           23–33, November-December 2001.

[SVSK01]   Patrick Schaumont, Ingrid Verbauwhede, Majid Sarrafzadeh, and Kurt Keutzer. A quick
           safari through the reconfigurable jungle. In *Design Automation Conference (DAC)*, June
           2001.

[Sys07]   VaST Systems. *Comet/Meteor*. World Wide Web, http://www.vastsystems.com, 2007.

[SZT⁺04]   Todor Stefanov, Claudiu Zissulescu, Alexandru Turjan, Bart, and Ed Deprettere. System
           Design Using Kahn Process Networks: The Compaan/Laura Approach. In *Proceedings
           of the conference on Design, automation and test in Europe*, page 10340. IEEE Computer
           Society, 2004.

[TBW92]   Ken Tindell, Alan Burns, and Andy J. Wellings. Allocating hard real-time tasks: An
           NP-hard problem made easy. *Real-Time Systems*, 4(2):145–165, 1992.

[TBW95]   Ken Tindell, Alan Burns, and A. J. Wellings. Calculating controller area network (can)
           message response times. *Control Eng. Practice*, 3(8):1163–1169, 1995.

[Tea04]   The Metropolis Project Team. The metropolis meta model version 0.4. Technical Report
           UCB/ERL M04/38, University of California, Berkeley, September 2004.

[Tec07]   MLDesign Technologies. *MLDesigner*. World Wide Web, http://www.mldesigner.com,
           2007.

[TGS⁺03]   Jean-Pierre Talpin, Paul Le Guernic, Sandeep Kumar Shukla, Rajesh Gupta, and Frederic
           Doucet. Polychrony for formal refinement-checking in a system-level design methodol-
           ogy. In *ACSD '03: Proceedings of the Third International Conference on Application of
           Concurrency to System Design*, pages 9–19, Washington, DC, USA, 2003. IEEE Com-
           puter Society.

[Tom03]   Mark F. Tompkins. Optimization techniques for task allocation and scheduling in distrib-
           uted multi-agent operations. Master's thesis, MIT, June 2003.

[Tur02]   Jim Turley. *The Essential Guide to Semiconductors*, chapter 5. Prentice Hall Publishers, December 2002.

[VH05]   Jay Vleeschhouwer and Woojin Ho. The state of eda; just slightly up for the year to date. *Technical and Design Software, The State of the Industry*, December 2005.

[vHPPH01]   A. van Halderen, S. Polstra, A. Pimentel, and L. Hertzberger. Sesame: Simulation of embedded system architectures for multi-level exploration. In *In Proc. of the conference of the Advanced School for Computing and Imaging (ASCI)*, pages 99–106, May 2001.

[Wal91]   Gregory K. Wallace. The JPEG Still Picture Compression Standard. *j-CACM*, 34(4):30–44, April 1991.

[Wan84]   Z. Wang. Fast algorithms for the discrete w transform and for the discrete fourier transform. In *IEEE Transactions on Acoustics, Speech, & Signal Processing*, volume ASSP-32, pages 803 – 816, August 1984.

[web08]   Dictionary.com   website.   *Taxonomy.*   World   Wide   Web, http://dictionary.reference.com/browse/taxonomy (28 May 2008), 2008.

[Wol05]   Wayne Wolf. *Computers as Components: Principles of Embedded Computing System Design*. Morgan Kaufmann, 2005.

[WSBL03]   T. Wiegand, G.J. Sullivan, G. Bjntegaard, and A. Luthra. Overview of the H.264/AVC Video Coding Standard. *IEEE Transactions on Circuits and Systems for Video Technology*, 13(7):560–576, 2003.

[Xil02]   Xilinx. *Virtex II Pro Platform FPGA Handbook*, ug120 (v2.0) edition, October 2002.

[Xil03a]   Xilinx. *FIFOs using Virtex II Block RAM*, xapp258 (v1.3) edition, January 2003.

[Xil03b]   Xilinx. *PowerPC Processor Reference Guide*, edk 6.1 edition, September 2003.

[Yos02]   Junko Yoshida. Philips semi see payoff in platform-based design. *EETimes*, October 2002.

[ZDSV⁺06a] Haibo Zeng, Abhijit Davare, Alberto Sangiovanni-Vincentelli, Sampada Sonalkar, Sri Kanajan, and Claudio Pinello. Design space exploration of automotive platforms in metropolis. In *Proceedings of the Society of Automotive Engineers Congress*, April 2006.

[ZDSV06b]   Qi Zhu, Abhijit Davare, and Alberto Sangiovanni-Vincentelli. A semantic-driven synthesis flow for platform-based design. In *Fourth ACM-IEEE International Conference on Formal Methods and Models for Codesign (MEMOCODE'06)*, July 2006.

[ZNP+07]   Wei Zheng, Marco Di Natale, Claudio Pinello, Paolo Giusto, and Alberto Sangiovanni-Vincentelli. Synthesis of task and message activation models in real-time distributed automotive systems. In *Proc. of Design Automation and Test, Europe*, 2007.

www.ingramcontent.com/pod-product-compliance
Lightning Source LLC
LaVergne TN
LVHW022259060326
832902LV00020B/3170